Elements

II		IB	IIB	IIIA	IVA	VA	VIA	VIIA	0
$)$	$(d^8s^2)^†$	s^1d^{10}	s^2	s^2p^1	s^2p^2	s^2p^3	s^2p^4	s^2p^5	s^2p^6
									2 He 4.003
				5 B 10.81	6 C 12.01	7 N 14.01	8 O 15.999	9 F 18.99	10 Ne 20.18
				13 Al 26.98	14 Si 28.09	15 P 30.97	16 S 32.06	17 Cl 35.45	18 Ar 39.95
7 3	28 Ni 58.71	29 Cu 63.54	30 Zn 65.37	31 Ga 69.72	32 Ge 72.59	33 As 74.92	34 Se 78.96	35 Br 79.91	36 Kr 83.80
 .91	46 Pd 106.4	47 Ag 107.87	48 Cd 112.40	49 In 114.82	50 Sn 118.69	51 Sb 121.75	52 Te 127.60	53 I 126.90	54 Xe 131.30
7 r 2.2	78 Pt 195.09	79 Au 196.97	80 Hg 200.59	81 Tl 204.37	82 Pb 207.19	83 Bi 208.98	84 Po	85 At	86 Rn

ner Transition Elements-f

n	63 Eu 151.96	64 Gd 157.25	65 Tb 158.92	66 Dy 162.50	67 Ho 164.93	68 Er 167.26	69 Tm 168.93	70 Yb 173.04	71 Lu 174.97
u	95 Am	96 Cm	97 Bk	98 Cf	99 Es	100 Fm	101 Md	102 No	103 Lw

PERGAMON INTERNATIONAL LIBRARY
of Science, Technology, Engineering and Social Studies

The 1000-volume original paperback library in aid of education industrial training and the enjoyment of leisure

Publisher: Robert Maxwell, M.C.

ORGANO-TRANSITION METAL COMPOUNDS AND RELATED ASPECTS OF HOMOGENEOUS CATALYSIS

Comprehensive Inorganic Chemistry

ORGANO-TRANSITION METAL COMPOUNDS AND RELATED ASPECTS OF HOMOGENEOUS CATALYSIS

B. L. Shaw and N. I. Tucker

Chapter 53 of
Comprehensive Inorganic Chemistry

PERGAMON PRESS

OXFORD   .   NEW YORK   .   TORONTO
SYDNEY   .   PARIS   .   BRAUNSCHWEIG

Pergamon Press Offices:

U.K.	Pergamon Press Ltd., Headington Hill Hall, Oxford, OX3 0BW, England
U.S.A.	Pergamon Press Inc., Maxwell House, Fairview Park, Elmsford, New York 10523, U.S.A.
CANADA	Pergamon of Canada Ltd., 207 Queen's Quay West, Toronto 1, Canada
AUSTRALIA	Pergamon Press (Aust.) Pty. Ltd., 19a Boundary Street, Rushcutters Bay, N.S.W. 2011, Australia
FRANCE	Pergamon Press SARL, 24 rue des Ecoles, 75240 Paris, Cedex 05, France
WEST GERMANY	Pergamon Press GmbH, D-3300 Braunschweig, Postfach 2923, Burgplatz 1, West Germany

First edition 1973

Reprinted, with corrections, from Comprehensive Inorganic Chemistry, 1975

Library of Congress Catalog Card No. 77-189736

Printed in Great Britain by A. Wheaton & Co, Exeter
ISBN 0 08 018872 9 (Hard cover)
ISBN 0 08 018871 0 (Flexicover)

CONTENTS

v

Contents of Comprehensive Inorganic Chemistry

Independent Opinion

PREFACE

The excellent reception that has been accorded to *Comprehensive Inorganic Chemistry* since the simultaneous publication of the five volumes of the complete work has been accompanied by the plea that sections should be made available in a form that would enable specialists to purchase copies for their own use. To meet this demand the publishers have decided to issue selected chapters and groups of chapters as separate editions. These chapters will, apart from the corrections of misprints and the addition of prefatory material and individual indices, appear just as they did in the main work. Extensive revision would delay publication and greatly raise the cost, so limiting the circulation of these definitive reviews.

A. F. TROTMAN-DICKENSON
Executive Editor

53. ORGANO-TRANSITION METAL COMPOUNDS AND RELATED ASPECTS OF HOMOGENEOUS CATALYSIS

B. L. SHAW and N. I. TUCKER

The University of Leeds

LIST OF ABBREVIATIONS

acac	acetylacetonato	Et	ethyl
Bu	n-butyl	Me	methyl
Bu^t	t-butyl	NBD	norbornadiene
COD	1,5-cyclo-octa-1,5-diene	Ph	phenyl
COT	cyclo-octatetraene	Pr	n-propyl
Cp	cyclopentadienyl	Pr^i	isopropyl
diphos	Ph$_2$PCH$_2$CH$_2$PPh$_2$	py	pyridine
dipy	dipyridyl		

INTRODUCTION

The field of organo-transition metal chemistry is one of the largest and most active areas of modern chemistry; yet before 1951 it was relatively unimportant. The field is an old one since Zeise's salt, $K[PtCl_3(C_2H_4)]H_2O$, was known in 1830 and a few alkyl derivatives of gold and platinum have been known since the early 1900s. The discovery of ferrocene (dicyclopentadienyliron) in 1951 and Ziegler–Natta catalysis about the same time, prompted much more interest in the field. Soon after the synthesis of ferrocene, many other metals were shown to form stable cyclopentadienyl- and later arene-derivatives. The discovery of ways of stabilizing transition metal–carbon σ-bonds followed a little later.

Although the existence of π-allylic complexes was not established until as recently as 1960, this is now an important area of transition metal chemistry. The field of olefin, polyolefin and acetylene complexes, relatively small in 1960, is now very large indeed. More recent still has been the discovery of carbene, benzyne, carbide and other unusual complexes.

In this account of the chemistry of organo-transition metal complexes the compounds are classified mainly from the number of carbon atoms which are actually bonded to the metal and to each other. For example, section 1 deals with alkyl, aryl, ethynyl, acyl and carbene complexes (one carbon atom); section 2 with olefin and chelating diolefin (e.g. cyclo-octa-1,5-diene) complexes and complexes formed from acetylene; section 3 with π-allylic complexes; section 4 with conjugated diolefinic complexes, e.g. containing buta-diene or cyclobutadiene; section 5 with cyclopentadienyls; section 6 with arene complexes;

and section 7 with tropylium complexes. In section 8 are discussed complexes formed from cyclo-octatetraene or azulene even though fewer than eight carbon atoms are bonded to a metal atom; carbollide complexes are also discussed in this section.

The uses of organo-transition metal complexes as homogeneous catalysts and as intermediates in organic syntheses are discussed in the appropriate section. Such aspects of organo-metallic chemistry are currently of great interest and importance.

For an excellent and full account of organo-transition metal complexes the reader is recommended to refer to the book by Green[1]. This gives many hundreds of references up to 1967. The book by Pauson[2] serves as an introductory text. There are also many articles on organo-transition metal compounds and related topics in the review journals.

1. ALKYLS, ARYLS, ACETYLIDES, FLUOROCARBON COMPLEXES, CARBIDES

1.1. ALKYLS AND ARYLS[3, 4]

1.1.1. Preparations and General

1.1.1a. *Using Alkylating or Arylating Reagents*

This is the most commonly used method of synthesis and generally consists of replacing a halide ligand on the transition metal by an alkyl or aryl group, leaving the other ligands unaffected. Organolithium reagents are more reactive than Grignard reagents in this replacement, but frequently either may be used[5].

Examples:

$$trans\text{-}NiBr_2(PEt_2Ph)_2 \xrightarrow{\text{mesityl MgBr}} trans\text{-}Ni(mesityl)_2(PEt_2Ph)_2$$

$$CrCl_3 \cdot 3THF \xrightarrow{\text{PhMgBr}} CrPh_3 \cdot 3THF$$

$$\left.\begin{array}{l} cis\text{-}PtCl_2(PEt_3)_2 \\ trans\text{-}PtCl_2(PEt_3)_2 \end{array}\right\} \xrightarrow{\text{PhLi}} \left\{\begin{array}{l} cis\text{-}PtPh_2(PEt_3)_2 \\ trans\text{-}PtPh_2(PEt_3)_2 \end{array}\right.$$

$$PtCl_2 \text{ (chelating diolefin)} \xrightarrow{\text{MeMgCl}} PtMe_2 \text{ (chelating diolefin)}$$

$$TiCl_2Cp_2 \xrightarrow{\text{MeLi}} TiMe_2Cp_2$$

$$2TiCl_4 + Al_2Cl_2Me_4 \longrightarrow 2TiCl_3Me + 2AlCl_3$$

$$FeI(CO)_2Cp + HgPh_2 \longrightarrow FePh(CO)_2Cp$$

Sometimes the alkyl or aryl group may be generated from a transition metal complex[6], e.g.

NaMn(CO)₅ reacts similarly.

[1] M. L. H. Green, *Organometallic Compounds*, Vol. 2, *The Transition Elements*, Methuen (1968).
[2] P. L. Pauson, *Organometallic Chemistry*, Edward Arnold, London (1967).
[3] I. I. Kritskaya, *Russian Chem. Rev.* **35** (1966) 167.
[4] G. W. Parshall and J. J. Mrowca, *Advances Organomet. Chem.* **7** (1968) 157.
[5] J. Chatt and B. L. Shaw, *J. Chem. Soc.* A (1960) 1718.
[6] R. F. Heck, *J. Am. Chem. Soc.* **90** (1968) 313.

It is possible to effect only partial replacement of the halogens

$$PtCl_4 + HgMe_2 \rightarrow [PtClMe_3]_4$$

$$NbCl_5 + ZnMe_2 \rightarrow NbCl_2Me_3$$

$$trans\text{-}RuCl_2(diphos)_2 + Al_2R_6 \rightarrow cis\text{-}RuClR(diphos)_2$$

$$(R = Me, Et \text{ or } Ph)$$

However, for alkyl– or aryl–platinum(II) or alkyl–iridium(III) complexes it is usually better to replace all the halogens and then to cleave them selectively using halogen acid or halogen[7], e.g.

$$IrCl_3L_3 \xrightarrow{LiMe_1} IrMe_3L_3 \xrightarrow{I_2} IrIMe_2L_3 \xrightarrow{I_2} IrI_2MeL_3 \quad (L = PMe_2Ph)$$

Triphenylphosphine has also been used as an arylating agent,.e.g. $Pd(PPh_3)_4$ and $PdCl_2$ when heated together in dimethyl sulphoxide at $130°$ give a 90% yield of *trans*-$PdClPh(PPh_3)_2$, together with a diphenylphosphide-bridged species.

1.1.1b. *From Organic Halides and Complex Metal Anions*

Metal carbonyls can frequently be reduced to anions which will react with alkyl halides to give alkyl–metal carbonyls[7a, 7b]. Usually the anion is prepared *in situ*, e.g.

$$Mn_2(CO)_{10} \xrightarrow{Na/Hg} Na[Mn(CO)_5] \xrightarrow{MeI} MnMe(CO)_5$$

$$[Mo(CO)_3Cp]_2 \xrightarrow{Na/Hg} Na[Mo(CO)_3Cp] \xrightarrow{RX} MoR(CO)_3Cp \quad (R = alkyl)$$

$$[Fe(CO)_2Cp]_2 \rightarrow Na[Fe(CO)_2Cp] \xrightarrow{BrCH_2C:CH} Fe(CH_2C:CH)(CO)_2Cp \text{ [8]}$$

A semiquantitative study of the reaction of complex metal anions with alkyl halides gives the order of nucleophilicity to be $[Fe(CO)_2Cp]^- > [Ru(CO)_2Cp]^- > [Ni(CO)Cp]^- > [Re(CO)_5]^- > [W(CO)_3Cp]^- > [Mn(CO)_5]^- > [Mo(CO)_3Cp]^- > [Cr(CO_3Cp]^- > [Co(CO)_4]^- > [Cr(CN)(CO)_5]^- > [MoCN(CO)_5]^- > [WCN(CO)_5]^-$.

The method may be used to synthesize aryl– or vinyl–metal complexes, but the reactions may be sluggish and the yields low.

Onium compounds may be used in place of halides, e.g.

$$[Fe(CO)_2Cp]^- + Ph_3S^+BF_4^- \rightarrow FePh(CO)_2Cp$$

1.1.1c. *Additions to Coordinated Olefins and Other Insertion Reactions*

In its most commonly occurring form this reaction involves the addition of a metal–ligand bond across a coordinated olefin or acetylene, i.e. either an olefin (acetylene) insertion reaction or a nucleophilic attack on a coordinated olefin, i.e.

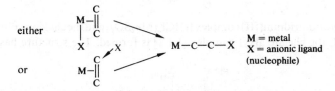

M = metal
X = anionic ligand
(nucleophile)

---

[7] B. L. Shaw and A. C. Smithies, *J. Chem. Soc.* (1967) 1047.

[7a] R. B. King, *Advances Organomet. Chem.* **2** (1964) 157.

[7b] F. Calderazzo, R. Ercoli and G. Natta, in *Organic Syntheses Via Metal Carbonyls*, Vol. 1 (eds. I. Wender and P. Pino), Interscience, New York (1968), p. 4.

[8] P. W. Jolly and R. Pettit, *J. Organomet. Chem.* **12** (1968) 491.

It is likely that insertion gives a *cis*-addition across the double (or triple) bond and that nucleophilic attack gives a *trans*-addition. Also in this class of reaction is the insertion of methylene formed from diazomethane:

$$MnH(CO)_5 + CH_2N_2 \rightarrow MnMe(CO)_5$$
$$IrCl(CO)(PPh_3)_2 + CH_2N_2 \rightarrow IrCH_2Cl(CO)(PPh_3)_2$$

Diazoacetic ester reacts similarly to give compounds containing the grouping MCH_2COOEt (M = metal).

Examples of insertions of olefins or acetylenes into metal–hydrogen bonds to give alkyls include the following:

$$PtHCl(PEt_3)_2 \underset{\text{heat}}{\overset{C_2H_4}{\rightleftharpoons}} PtClEt(PEt_3)_2$$

This reaction is reversible and catalysed by stannous chloride in the presence of which equilibrium is reached in 30 min at 1 atm/25°. On heating the dideuterio compound $PtCl(CD_2CH_3)(PEt_3)_2$, however, a mixture of deuterioethylenes, $PtHCl(PEt_3)_2$ and $PtDCl(PEt_3)_2$ is produced[9]. This suggests that the reversible insertion step occurs several times before coordinated ethylene leaves the complex. An unusual synthesis of alkyl–platinum complexes occurs when lithium tetrachloroplatinate(II) is heated with olefins such as oct-1-ene in a formic acid/dimethylformamide mixture. The n-octyl complex $PtCl(C_8H_{17})CO \cdot DMF$ is formed, presumably by addition of a platinum hydride species across the double bond. Other olefins react similarly.

Examples of hydrides of other metals adding to olefins or acetylenes to give alkyls[10] are summarized below:

$$RhCl(PPh_3)_3 \xrightarrow{HCl} RhHCl_2(PPh_3)_2$$

$$CH_2:CH_2 \swarrow \qquad \searrow HC:CH$$

$$RhCl_2Et(PPh_3)_2 \qquad RhCl_2(CH:CH_2)(PPh_3)_2$$
$$MnH(CO)_5 + CH_2:CHCH:CH_2 \rightarrow CH_3CH:CHCH_2Mn(CO)_5$$

Addition of pyridine and acrylonitrile to rhodium trichloride in ethanol gives $RhCl_2\{CH(CN)Me\}py_3$, formed presumably by addition of a rhodium hydride species to the double bond of acrylonitrile. Ethylene oxide gives a β-hydroxyethyl complex with $CoH(CO)_4$:

$$CoH(CO)_4 + CH_2\overset{\diagdown \diagup}{\underset{O}{}}CH_2 \rightarrow Co(CH_2CH_2OH)(CO)_4$$

From the hydridoiridium(III) complex $IrHCl_2(Me_2SO)_3$ and the chalcone $PhCOCH:CHPh$ in isopropanol, the chelate alkyl complex [1.1.1] is formed. The structure has been proved

$$Ph-C\overset{CH_2}{\underset{\|}{\diagup}}\overset{}{\diagdown}CHPh$$
$$O-----IrCl_2(Me_2SO)_2$$

[1.1.1]

[9] J. Chatt, R. S. Coffey, A. Gough and D. T. Thompson, *J. Chem. Soc.* A (1968) 190.
[10] M. C. Baird, J. T. Mague, J. A. Osborn and G. Wilkinson, *J. Chem. Soc.* A (1967) 1347.

by X-ray diffraction. This alkyl complex is an intermediate in the catalysed reduction of the chalcone by isopropanol in the presence of the iridium hydride.

Attack by bromine on an *o*-allylphenyldiphenylphosphine derivative of gold(I) bromide [1.1.2] gives a chelated alkyl derivative of gold(III) [1.1.3].

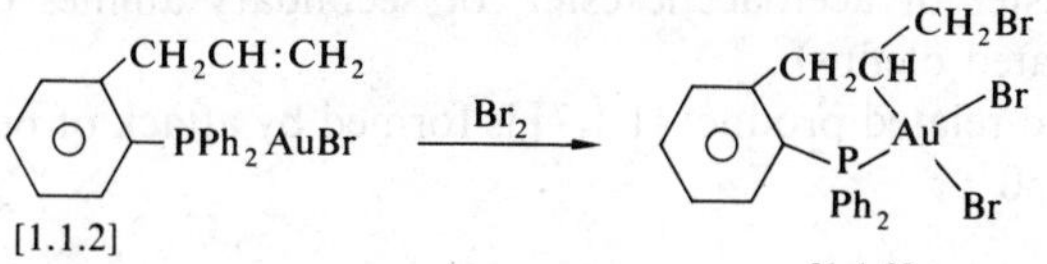

[1.1.2]

[1.1.3]

Treatment of the dimethyl(*o*-allylphenyl)arsine complex of platinum(II) [1.1.4] with bromine followed by ethanol gives the alkyl–platinum(IV) complex [1.1.5], the structure of which has been determined by X-ray diffraction[11]. A remarkable skeletal rearrangement from allyl to isoalkyl occurs, probably during the ethanol treatment.

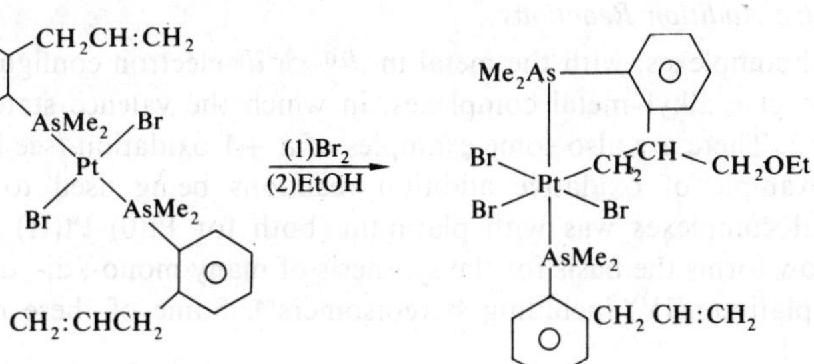

[1.1.4]

[1.1.5]

An example of addition across an acetylenic bond to give a vinylic complex [1.1.6] occurs when acetylenic amines of the type $Me_2NCR_1R_2C:CR_3$ are treated with palladium chloride in methanol[12]:

$$PdCl_2 + Me_2NCR_1R_2C:CR_3 \rightarrow$$

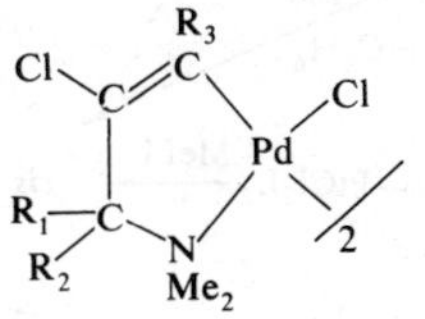

[1.1.6]

Chelating dienes coordinated to palladium(II) or platinum(II) are susceptible to attack by nucleophiles, forming a metal–carbon σ-bond, i.e.

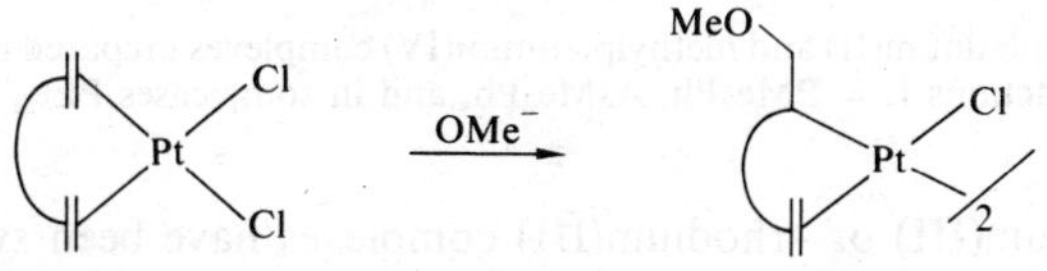

[11] M. A. Bennett, G. J. Erskine, J. Lewis, R. Mason, R. S. Nyholm, G. B. Robertson and A. D. C. Towl, *Chem. Commun.* (1966) 395.
[12] T. Yukawa and S. Tsutsumi, *Inorg. Chem.* **7** (1968) 1458.

Examples of chelating dienes include dicyclopentadiene (the first example), norbornadiene, COD and dipentene. The attack of the methoxy group is *exo* as shown by X-ray diffraction and n.m.r. studies, i.e. effectively *trans*-addition across the double bond occurs. Many other nucleophiles will attack coordinated dienes[13]; these include the acetate anion, the anions of malonic ester or acetoacetic ester, or secondary amines (see section 2 on the reactions of coordinated olefins).

An interesting and related product [1.1.7] is formed by attack of perdeuteriopyridine on $PtCl_2(C_2H_4)py$ at $-50°$:

$$\underset{[1.1.7]}{D\!\!\bigcirc\!\!N\!-\!CH_2CH_2\!-\!PtCl_2py}$$

1.1.1d. *Oxidative Addition Reactions*

Many metal complexes, with the metal in d^{10}- or d^8-electron configuration, will add on alkyl halides to give alkyl–metal complexes, in which the valence state of the metal has increased by $+2$. There are also some examples of a $+1$ oxidation (see below).

The first example of oxidative addition reactions being used to synthesize alkyl–transition metal complexes was with platinum, both for Pt(0)–Pt(II) and Pt(II)–Pt(IV). The reaction now forms the basis for the synthesis of many mono-, di-, tri- and tetra-methyl derivatives of platinum(IV) including stereoisomers[14]. Some of these reactions are summarized in Fig. 1.

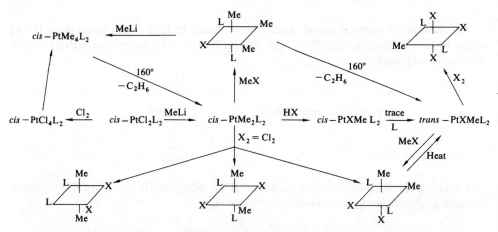

FIG. 1. Some methylplatinum(II) and methylplatinum(IV) complexes prepared by oxidative addition and methylation reactions $L = PMe_2Ph$, $AsMe_2Ph$, and in some cases PEt_3, PPr_3^n. $X =$ halogen.

Many alkyl–iridium(III) or –rhodium(III) complexes have been synthesized by adding alkyl halides to planar iridium(I) or rhodium(I) complexes. Additions to iridium(I) complexes of the type *trans*-$IrX(CO)L_2$ have been extensively studied ($X =$ halogen, $L =$ tertiary

[13] D. White, *Organomet. Chem. Rev.* **3** (1968) 497.
[14] J. D. Ruddick and B. L. Shaw, *J. Chem. Soc.* A (1969) 2964.

phosphine or tertiary arsine). With L = PPh$_3$ the additions appear to be *cis*[15] but with L = PMePh$_2$, PMe$_2$Ph or AsMe$_2$Ph, the additions have definitely been shown to be *trans*[16, 17], i.e.

where X, X' = halogen.

There is also a very large entropy of activation associated with the reaction. Allylic halides have been shown to add initially *cis* to the complex IrCl(CO)(PMe$_2$Ph)$_2$, but the resultant σ-allyl complex isomerizes via a π-allyl cation to an isomeric σ-allyl complex in which the overall addition is *trans* (Fig. 2). The initial attack on the iridium possibly occurs via the allylic double bond and not on the carbon atom to which the bromide is attached[18].

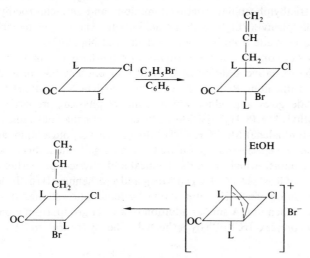

FIG. 2. The stereochemistry of the addition of allyl bromide to *trans*-IrCl(CO)(PMe$_2$Ph$_2$)$_2$.
L = PMe$_2$Ph.

Rhodium complexes of similar type, i.e. *trans*-RhX(CO)L$_2$, will also add methyl halides. The complex RhCl(PPh$_3$)$_3$ undergoes addition with methyl iodide, with loss of PPh$_3$. The product is RhI$_2$Me(PPh$_3$)$_2$.

In the above-mentioned oxidative additions of d^8 systems aliphatic phosphine ligands are better at promoting the addition than triphenylphosphine; this is probably due to both electronic and steric factors. Tertiary arsines seem to be better than tertiary phosphines.

There are also several syntheses of alkyl–transition metal complexes based on addition of alkyl halides to metal atoms with d^{10} electron configurations, particularly with platinum(0) and palladium(0). Methyl iodide readily reacts with Pt(PPh$_3$)$_3$ to give *trans*-PtIMe(PPh$_3$)$_2$.

15 M. A. Bennett, R. J. H. Clark and D. L. Milner, *Inorg. Chem.* **6** (1967) 1647.
16 J. P. Collmann, *Accounts of Chem. Research* **1** (1968) 136.
17 A. J. Deeming and B. L. Shaw, *J. Chem. Soc.* A (1969) 1128.
18 A. J. Deeming and B. L. Shaw, *J. Chem. Soc.* A (1969) 1562.

The reaction has been extended to the addition of other halides, particularly unsaturated halides such as β-bromostyrene, di-iodoacetylene, phenylethynyl bromide and some chloroethylenes (di-, tri- and tetra-), and also to cyclohexyl or trityl halides[19].

Addition of tetrachloroethylene to $Pt(PPh_3)_3$ gives initially a tetrachloroethylene complex, which then rearranges to a chloro (trichlorovinyl) complex of unknown stereochemistry[20]:

$$Pt(PPh_3)_3 + \xrightarrow{Cl_2C:CCl_2} Pt(Cl_2C:CCl_2)(PPh_3)_2 \rightarrow PtCl(ClC:CCl_2)(PPh_3)_2$$

The kinetics of the second (isomerization) step have been studied and the mechanism found to depend on the solvent. In hydroxylic solvents the rate-determining step is close to S_N1 and thought to involve solvolysis of the tetrachloroethylene ligand to give a carbonium ion intermediate. In benzene as solvent, migration of chlorine from carbon to platinum occurs directly without formation of a carbonium ion intermediate.

What is probably an oxidative addition of tetrachloroethylene to nickel(0) is involved in the synthesis of the very stable $trans$-$NiCl(CCl:CCl_2)(PEt_3)_2$ from (1) a mixture of nickel acetylacetonate, triethylphosphine triethylaluminium and tetrachloroethylene or (2) by boiling $NiCl(2$-allylphenyl$)(PEt_3)_2$ with tetrachloroethylene[21]. The remarkable di(trityl)-nickel(II) can also be made from hexaphenylethane and $Ni(COD)_2$.

An important development in the formation or synthesis of alkyl– or aryl–transition metal bonds is the oxidative addition of C–H to a metal. The first observation of such a reaction was with ruthenium. Reduction of cis- or $trans$-$RuCl_2\{Me_2PCH_2CH_2PMe_2\}_2$ with sodium naphthylide gives a product which from its physical properties appears to be cis-$Ru^{II}H(2$-naphthyl$)\{Me_2PCH_2CH_2PMe_2\}_2$. Its chemical behaviour indicates a ruthenium (0) complex $Ru(naphthalene)(Me_2PCH_2CH_2PMe_2)_2$, however, and a rapid tautomeric equilibrium between the two species was postulated[22]. On being heated, this product loses naphthalene to give a complex which from its chemical and physical properties appears to be a tautomeric mixture of $Ru^0(Me_2PCH_2CH_2PMe_2)_2$ and a ruthenium(II) hydride species formed by the oxidative addition of a C–H group (shown by deuteration studies to come from one of the PCH_3s) to the ruthenium. A similar addition of a C–H group, from a phosphine ligand, occurs when the complex $IrCl(PPh_3)_3$ is heated. The hydro-iridium(III) product [1.1.8]

[1.1.8]

has been characterized.

The hydro-cobalt–N_2 species, $CoH(N_2)(PPh_3)_3$, when treated with deuterium at 25° has up to nineteen of its hydrogens replaced by deuterium to give $CoD(N_2)L_3$ (where L is triphenylphosphine with its $ortho$-hydrogens replaced by deuterium). The sequence of oxidative addition and elimination steps outlined in Fig. 3 is thought to be involved in the replacement of one hydrogen by deuterium; eventually all the $ortho$-hydrogens will be replaced.

[19] C. D. Cook and G. S. Jauhal, *Can. J. Chem.* **45** (1967) 301.
[20] W. J. Bland, J. Burgess and R. D. W. Kemmit, *J. Organomet. Chem.* **15** (1968) 217.
[21] R. G. Miller, D. R. Fahey and D. P. Kuhlman, *J. Am. Chem. Soc.* **90** (1968) 6248.
[22] J. Chatt and J. M. Davidson, *J. Chem. Soc.* (1965) 843.

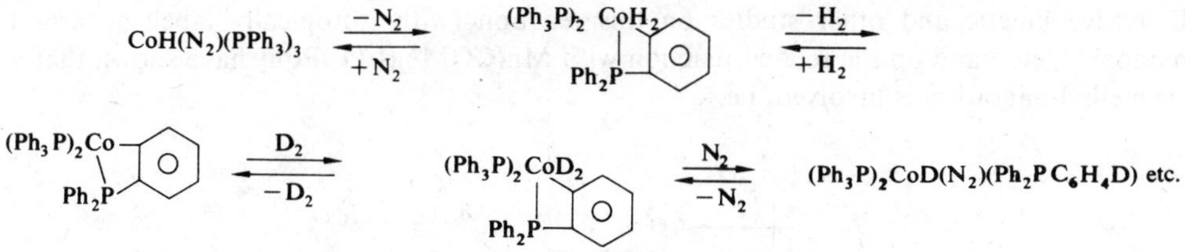

FIG. 3. Showing how the *ortho*-hydrogens of triphenylphosphine in $CoH(N_2)(PPh_3)_3$, when treated with D_2, are replaced by deuterium[23].

It has also been suggested that in the ruthenium-chloride-catalysed dimerization of acrylonitrile a key step is the formation of a hydro(2-cyanovinyl)ruthenium species by fission of a C–H bond of acrylonitrile.

The reaction of *N,N*-dimethylbenzylamine with the chloropalladate ion to give the chelated arylpalladium species [1.1.9] may involve intermediate oxidative addition of the *ortho*-C–H or electrophilic attack by the palladium.

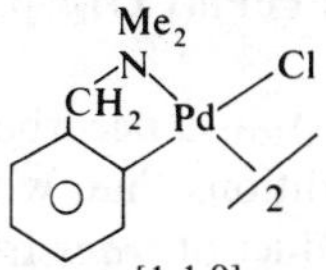

[1.1.9]

One-electron oxidative addition reactions giving alkyl–metal complexes have been reported for vanadium(II), chromium(II), and cobalt(II). Alternatively, the reactions might be regarded as nucleophilic attack on the alkyl halide by the metal complex (which is in a low valence state and carries a high electron density), e.g.

$$VCp_2 + MeBr \rightarrow VMeCp_2 + VBrCp_2$$
$$Cr_{aq}^{2+} + PhCH_2I \rightarrow [CrCH_2Ph(H_2O)_5]^+ + [CrI(H_2O)_5]^+$$

There are many examples of such reactions with the pentacyanocobalt(II) anion[24], e.g.

$$[Co(CN)_5]^{3-} \xrightarrow{RX} [CoR(CN)_5]^{3-} + [CoX(CN)_5]^{3-}$$

HC:CH   X = Cl or Br;   R = Me, $PhCH_2$, $CH_2CH:CH_2$, $CH:CH_2$, etc.

$$[(CN)_5CoCH:CHCo(CN)_5]^{6-}$$

The reactions with alkyl halides are first order in both reactants.

1.1.1e. *From Acyl, Sulphonyl or Arylazo Complexes*

Complexes containing the arrangement M–U–R, where M = metal; U = unsaturated grouping CO, SO_2, N:N; and R = alkyl or aryl, can eliminate the unsaturated grouping U on heating, giving an alkyl– or aryl–metal complex. There are several examples of acyl-metal complexes decomposing in this way and more recently some sulphonyl and diazo complexes have been shown to behave similarly.

The first example was

$$Mn(COMe)(CO)_5 \xrightarrow{heat} MnMe(CO)_5$$

[23] G. W. Parshall, *J. Am. Chem. Soc.* **90** (1968) 1669.
[24] J. Kwiatek and J. K. Seyler, *J. Organomet. Chem.* **3** (1965) 433.

Extensive kinetic and other studies on this reaction (with isotopically labelled carbon monoxide, etc.) and on the related reaction with $Mn(COMe)(CO)_4PPh_3$ have shown that a *cis*-methyl migration is involved, i.e.

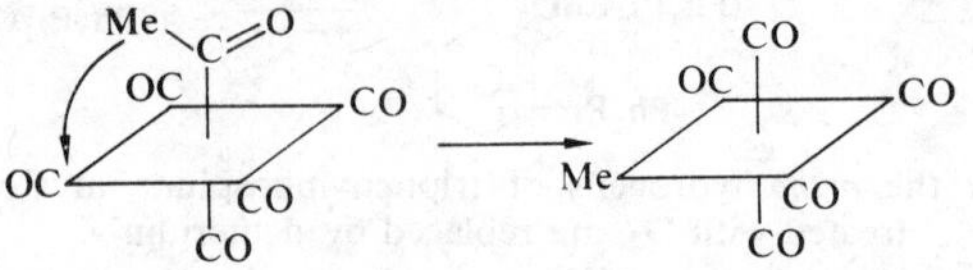

Addition of acetyl chloride to a solution of the cyclo-octeneiridium(I) complex $Ir_2Cl_2(CO)_2(C_8H_{14})_4$ in benzene gives a chlorine bridged methyliridium(III) complex, $Ir_2Cl_4Me_2(CO)_4$, the structure of which has been determined by X-ray diffraction. The reaction probably goes via addition of the acid chloride and spontaneous migration of the methyl group on to the iridium. Other acyl chlorides react similarly.

Elimination of carbon monoxide may be promoted by ultraviolet light, e.g. the benzoyl-iron complex $Fe(COPh)(CO)_2Cp$ gives $FePh(CO)_2Cp$. The acrylyl complex similarly gives a vinyliron complex.

In some cases, by using $RhCl(PPh_3)_3$ as a decarbonylating agent the conversion of acyl to alkyl goes under much milder conditions than with pyrolysis[25].

The *p*-toluensulphonyl chloride adduct of *trans*-$IrCl(CO)(PPh_3)_2$, viz.

$$IrCl_2(SO_2C_6H_4CH_3)(CO)(PPh_3)_2,$$

loses sulphur dioxide on heating to give the *p*-tolyliridium derivative

$$IrCl_2(\textit{p}\text{-tolyl})(CO)(PPh_3)_2{}^{16}.$$

The protonated phenylazoplatinum complex $[PtCl(NNHPh)(PEt_3)_2]BF_4$ prepared from diazonium fluorborate and *trans*-$PtHCl(PEt_3)_2$ loses nitrogen and HBF_4 on being passed through an alumina column giving $PtClPh(PEt_3)_2$ [26].

1.1.1f. *Miscellaneous Methods*

Cyclopropane reacts with H_2PtCl_6 to give a brown solid of composition $PtCl_2C_3H_6$ which reacts with pyridine to give a white complex, $PtCl_2C_3H_6py_2$, probably with the structure [1.1.10]. This reacts with chloroform to give the quaternary pyridinium complex [1.1.11][27].

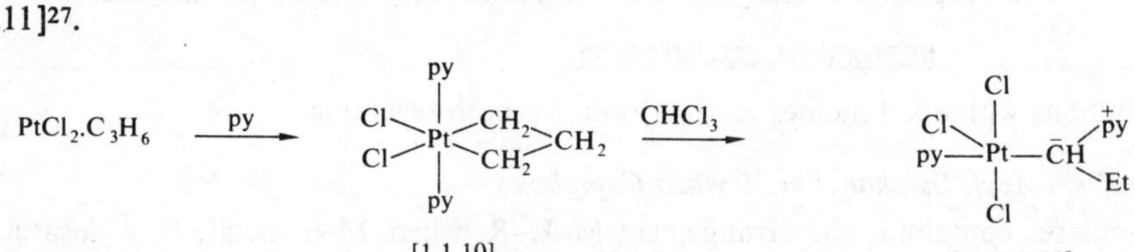

1.1.2. Some Examples of Alkyl– and Aryl–Transition Metal Complexes (Not of Fluorcarbons)

Because the number of types of alkyl and aryl complexes is so large it is difficult to discuss all of them under such headings as methods of synthesis, structure and bonding, etc. Hence for the convenience of the reader we summarize the main types of alkyl- and

25 J. J. Alexander and A. Wojcicki, *J. Organomet. Chem.* **15** (1968) 23.

26 G. W. Parshall, *J. Am. Chem. Soc.* **87** (1965) 2133.

27 N. A. Bailey, R. D. Gillard, M. Keeton, R. Mason and D. R. Russell, *Chem. Commun.* (1966) 396.

aryl-derivatives of the transition metals in this subsection metal by metal. Many of the complexes are discussed in some detail elsewhere in section 1.1.

Titanium

TiMe$_4$	Prepared from TiCl$_4$ and MeLi in Et$_2$O at $-80°$. Decomposes $> -78°$ but forms more stable adducts with diamines. Volatile at $-80°$ *in vacuo*[28].
TiPh$_4$	Decomposes at $-10°$ to (TiPh$_2$)$_n$[29].
TiX$_3$R	X = Cl or Br. R = Me, Et, Ph. Very unstable thermally and towards oxidation. TiCl$_3$Me is a violet solid melting to a yellow liquid[30].
Ti(OPr^i)$_3$Ph	The first to be prepared and one of the more stable of organotitanium complexes[31].
TiR$_2$Cp$_2$, TiClEtCp$_2$ TiMe$_3$Cp [33]	R = Me or Ph. Stable thermally[32].

Zirconium

ZrMe$_4$, Li$_2$[ZrMe$_6$] [34]	
ZrClRCp$_2$	R = Et. Bright yellow. Stable under nitrogen at $0°$ in the dark. R = Me also known[35].
PhCp$_2$ZrOZrCp$_2$Ph [36]	

Vanadium

VRCp$_2$	R = Me or Ph. Black crystals, dark green solutions in dimethyl cellusolve[37].
Li$_4$[VPh$_6$]	Paramagnetic (μ_{eff} = 3.85 BM). Permanganate colour[38].

Niobium

Nb(σ-C$_5$H$_5$)$_2$Cp$_2$	Very air sensitive, paramagnetic[39].
NbCl$_2$Me$_3$	Golden yellow crystals, sublimes readily *in vacuo* at $20°$. Slowly decomposes at $20°$. Highly reactive towards air and water[40].
Li$_2$[NbPh$_7$] [41]	

[28] H. J. Berthold and G. Groh, *Z. anorg. allgem. Chem.* **319** (1963) 230.
[29] V. N. Latjaeva, G. A. Razuvaev, A. V. Malisheva and G. A. Kiljakova, *J. Organomet. Chem.* **2** (1964) 388.
[30] C. Beermann and H. Bestian, *Angew. Chem.* **71** (1959) 618.
[31] D. F. Herman and W. K. Nelson, *J. Am. Chem. Soc.* **75** (1953) 3877, 3882.
[32] T. S. Piper and G. Wilkinson, *J. Inorg. Nucl. Chem.* **3** (1956) 104.
[33] U. Giannini and S. Cesca, *Tetrahedron Letters* **14** (1960) 19.
[34] H. J. Berthold and G. Groh, *Angew. Chem.* **78** (1966) 495.
[35] H. Sinn and G. Oppermann, *Angew. Chem. Int. Edn.* **5** (1966) 962.
[36] E. M. Brainina, G. G. Dvoryantseva and R. Kh. Freidlina, *Dokl. Akad. Nauk SSSR* **156** (1964) 1375.
[37] H. J. de Liefde Meijer, M. J. Janssen and G. J. M. van der Kerk, *Chem. Ind. (London)* (1960) 119.
[38] E. Kurras, *Angew. Chem.* **72** (1960) 635.
[39] E. O. Fischer and A. Treiber, *Chem. Ber.* **94** (1961) 2193.
[40] G. L. Juvinall, *J. Am. Chem. Soc.* **86** (1964) 4202.
[41] B. Sarry and V. Dobrusskin, *Angew. Chem.* **74** (1962) 509.

Tantalum

TaCl$_2$Me$_3$ Similar to the niobium compound but less stable thermally[40].

Chromium

CrMeCp(CO)$_3$, CrRCp(NO)$_2$	R = Me, Et, Ph [32].
Cr^{III}(o-dimethylaminophenyl)$_3$	Red crystals. μ_{eff} = 3.88 BM [42].
Cr^{II}(o-anisyl)$_2$	Yellow. μ_{eff} = 0.54 BM. Other related complexes[42].

Li$_3$[CrR$_6$]
Li$_2$[Cr$_2$Ph$_6$]3Et$_2$O   }   R = Me or Ph. Several derivatives of similar type have
Li$_2$[CrMe$_4$(THF)$_2$]   been made[43].

CrR$_3$L$_3$ R = many different aryl groups, vinyl, methyl or benzyl. L = ether or nitrogen ligand. With R = Ph, L = THF deep red needles, m.p. 85° (decomp.)[44].

CrCl$_{3-n}$R$_n$ (solvent)   R = benzyl, Ph, Me, etc. n = 1 or 2 [44].

[CrR(H$_2$O)$_5$]$^{2+}$ R = benzyl, CHCl$_2$, CHBr$_2$, CHI$_2$, CH$_2$(C$_5$H$_4$N). The benzyl complex cation is yellow to brownish red, and solutions of it in dilute perchloric acid have a half-life of 1$\frac{1}{2}$ days[45].

Molybdenum

MoR(CO)$_3$Cp   R = Me, Et, Pr^i [32].

Tungsten

WR(CO)$_3$Cp   R = Me, Et, *iso*-Pr, allyl, etc. Similar to molybdenum but appear to be slightly less stable[46].

Li$_2$[WPh$_6$]   Red violet in benzene. Pyrophoric[47].

Manganese

MnR(CO)$_5$ R = Me, Et, Pr^n, Ph, etc. Stable to oxidation and thermally stable at 20°. The ethyl and n-propyl derivatives are less stable than the methyl derivative. The complex with R = CH$_2$COOH is a very weak acid (pK_a = 6.1) [48].

MnR(CO)$_4$L   R = Me, Et. L = P(OCH$_2$)$_3$CCH$_3$ [49].

[MnR$_2$]$_x$, LiMnMe$_3$ R = Me or Ph. [MnMe$_2$]$_x$ is an explosive yellow powder[50].

[42] Fr. Hein and D. Tille, *Z. anorg. allgem. Chem.* **329** (1964) 72.
[43] E. Kurras and K. Zimmermann, *J. Organomet. Chem.* **7** (1967) 348.
[44] H. H. Zeiss and R. P. A. Sneeden, *Angew. Chem. Int. Edn.* **6** (1967) 435.
[45] D. Dodd and M. D. Johnson, *J. Chem. Soc.* A (1968) 34.
[46] M. L. H. Green and A. N. Stear, *J. Organomet. Chem.* **1** (1965) 230.
[47] B. Sarry, M. Dettke and H. Grossmann, *Z. anorg. allgem. Chem.* **329** (1964) 218.
[48] W. Hieber and G. Braun, *Z. Naturforsch.* **14b** (1959) 132.
[49] M. Green, R. I. Hancock and D. C. Wood, *J. Chem. Soc.* A (1968) 2718.
[50] C. Beermann and K. Clauss, *Angew. Chem.* **71** (1959) 627.

Rhenium

ReR(CO)$_5$ R = Me, Ph, etc. Similar to the manganese complexes[51].
ReR$_3$(PR$_3'$)$_2$, [ReR$_2$(PR$_3'$)$_2$]$_n$ [52]

Iron

FeR(CO)$_2$Cp   Many compounds of this type known. For R = alkyl the stability falls off in the order R = Me > Et > Pr^n > Pr^i. Compounds with R = Ph, CH$_2$Cl, CH$_2$CHO, CH$_2$CN known. With R = CH$_2$COOH the complex is a very weak acid pK_a = 6.7 [53].

trans-Fe(C$_6$Cl$_5$)$_2$(PEt$_2$Ph)$_2$ Golden yellow. μ_{eff} = 3.6 BM [54].

Ruthenium

RuR(CO)$_2$Cp   Similar to iron complexes[55].

RuXR(P—P)$_2$ P—P = chelating ditertiary phosphine. R = Me, Et or Ph. X = halogen[56].

RuH(aryl)(P—P)$_2$ Aryl = phenyl, 2-naphthyl, anthryl or phenanthryl[22].

Osmium

OsXR(P—P)$_2$ X = H, halogen or alkyl group. R = alkyl or aryl group. Similar to the ruthenium complexes[56].

Cobalt

CoR(CO)$_4$ R = Me or Et. Very unstable and labile[57].

CoCH$_2$Ph(CO)$_3$PPh$_3$ Yellow crystals, m.p. 135° [58].

trans-CoR$_2$(PEt$_2$Ph)$_2$ R = mesityl, C$_6$Cl$_5$, 2-biphenylyl. *Trans*-planar from dipole moment and X-ray studies. μ_{eff} = 2.5 BM. The aryl groups are vertical to the plane of the complex[59].

[CoR(CN)$_5$]$^{3-}$ R = Me, Et, CH$_2$Ph, allyl, vinyl, etc. Usually stable in aqueous solution in the absence of acid or oxygen. With acid decompose to give organic nitriles[24].

CoHEt$_2${(PPh$_2$CH$_2$)$_3$CCH$_2$PPh$_2$}   Formed from Co(acac)$_3$, AlEt$_2$OEt and C{CH$_2$PPh$_2$}$_4$ [60].

CoMe$_2$CpPPh$_3$ [61]

Co{*o*-C$_6$H$_4$CH$_2$NMe$_2$}$_3$ Red crystalline. The *p*-tert-butyl analogue also made[62].

Co$_3$R(CO)$_9$ [Co(CO)$_4$]$^-$ reacts with carbon tetrachloride to give the trinuclear species with R = CCl. Compounds with R = CH$_3$, H also known[63].

[51] W. Beck, W. Hieber and H. Tengler, *Chem. Ber.* **94** (1961) 862.
[52] J. Chatt, J. D. Garforth and G. A. Rowe, *J. Chem. Soc.* A (1966) 1834.
[53] M. L. H. Green, M. Ishaq and R. N. Whiteley, *J. Chem. Soc.* A (1967) 1508.
[54] J. Chatt and B. L. Shaw, *J. Chem. Soc.* (1961) 285.
[55] A. Davison, J. A. McCleverty and G. Wilkinson, *J. Chem. Soc.* (1963) 1133.
[56] J. Chatt and R. G. Hayter, *J. Chem. Soc.* (1963) 6017.
[57] W. Beck and R. E. Nitzschmann, *Chem. Ber.* **97** (1964) 2098.
[58] Z. Nagy-Magos, G. Bor and L. Markó, *J. Organomet. Chem.* **14** (1968) 205.
[59] P. G. Owston and J. M. Rowe, *J. Chem. Soc.* (1963) 3411.
[60] J. Ellermann and W. H. Gruber, *Angew. Chem. Int. Edn.* **7** (1968) 129.
[61] R. B. King, *Inorg. Chem.* **5** (1966) 82.
[62] A. C. Cope and R. N. Gourley, *J. Organomet. Chem.* **8** (1967) 527.
[63] G. Bor, L. Markó and and B. Markó, *Chem. Ber.* **95** (1962) 333.

CoIMeCp(PPh$_3$) [64],
CoMe$_2$Cp(PPh$_3$) [64]

CoRL(DMG)$_2$ $\left\{ \right.$ R = alkyl. L = py or tertiary phosphine. DMG = di-methylglyoximato. These compounds have a marked analogy to the alkyls of vitamin B$_{12}$ derivatives[65].

CoR(BAE)   R = alkyl. BAEH$_2$ is the condensation product of ethylenediamine and acetylacetone[66].

Vitamin B$_{12}$ alkyls[67]

Rhodium

RhX$_2$Me(CO)L$_2$ X = halogen. L = PR$_3$ or AsR$_3$ [68].
RhI$_2$Me(PPh$_3$)$_2$ [69]   see p. 797
RhBr(1-naphthyl)$_2$(PEt$_2$Ph)$_2$ [70]
Rh$_2$I$_2$Me$_4$(SMe$_2$)$_3$ [71]
RhCl$_2$(CHMeCN)py$_3$ Formed by adding acrylonitrile and pyridine to RhCl$_3$ in ethanol[72].

RhCl$_2$R(PPh$_3$)$_2$ R = Et or vinyl[10].
RhR(PPh$_3$)$_3$ R = Me or Ph [73].
K$_2$[Rh(CN)$_4$Me(H$_2$O)] [74]
Rh^{IV}ClPh(tetraphenylporphine)   X-ray structure shows Rh in plane of four nitrogen atoms[75].

Iridium

IrXYR(CO)L$_2$ X, Y halogen. R = Me or Ph. L = tertiary phosphine or tertiary arsine[16].

IrX$_y$Me$_{3-y}$L$_3$ X = halogen. y = 1, 2 or 3. L = PR$_3$, AsR$_3$. Very stable series of compounds[7].

[IrCl$_2$R(CO)$_2$]$_2$, IrCl$_2$R(CO)$_2$py   R = Me, Et, Pr^i or Ph [76].
IrCl$_2$C$_{15}$H$_{13}$O(Me$_2$SO) [77]

Nickel

Ni(CPh$_3$)$_2$ [78]
NiXRL$_2$, NiR$_2$L$_2$ X = halogen, NO$_2$, SCN. R = o-substituted aryl group such as 2-methylphenyl, 1-naphthyl, 9-anthryl, 9-phenanthryl, etc.; also trichlorovinyl. L = tertiary phosphine, phosphite or arsine. These complexes are often very stable[5].

[64] H. Yamazaki and N. Hagihara, *Bull. Chem. Soc. Japan* **38** (1965) 2212.
[65] G. N. Schrauzer and R. J. Windgassen, *J. Am. Chem. Soc.* **89** (1967) 1999.
[66] G. Costa, G. Mestroni, G. Tauzher and L. Stefani, *J. Organomet. Chem.* **6** (1966) 181.
[67] R. Bonnett, *Chem. Revs.* **63** (1963) 573.
[68] A. J. Deeming and B. L. Shaw, *J. Chem. Soc.* A (1969) 597.
[69] D. N. Lawson, J. A. Osborn and G. Wilkinson, *J. Chem. Soc.* A (1966) 1733.
[70] J. Chatt and A. E. Underhill, *J. Chem. Soc.* (1963) 2088.
[71] H. P. Fritz and K. E. Schwarzhans, *J. Organomet. Chem.* **5** (1966) 283.
[72] K. C. Dewhirst, *Inorg. Chem.* **5** (1966) 319.
[73] W. Keim, *J. Organomet. Chem.* **14** (1968) 179.
[74] J. P. Maher, *Chem. Commun.* (1966) 785.
[75] E. B. Fleischer and D. Lavallee, *J. Am. Chem. Soc.* **89** (1967) 7132.
[76] B. L. Shaw and E. Singleton, *J. Chem. Soc.* A (1967) 1683.
[77] J. Trocha-Grimshaw and H. B. Henbest, *Chem. Commun.* (1967) 544.
[78] G. Wilke and H. Schott, *Angew. Chem.* **78** (1966) 592.

NiRLCp   R = alkyl, aryl. L = phosphine, arsine or stibine[79].
NiR$_2$(bipy)   R = Me, Et [80].

Palladium

PdXRL$_2$, PdR$_2$L$_2$,   X = halogen. R = alkyl, aryl. L = tertiary phosphine.
   PdR$_2$ (chelating diene),   Some of these compounds are stable but others give
   PdR$_2$(bipyridyl)   palladium metal easily[81].
Pd$_2$Cl$_2$(C$_6$H$_4$CH$_2$NMe$_2$)$_2$ From Li$_2$PdCl$_4$ and *N,N*-dimethylbenzylamine[82].
Pd$_2$Cl$_2${2-(phenylazo)phenyl}$_2$ From azobenzene and Li$_2$PdCl$_4$ [6].
Pd$_2$Cl$_2$(C$_6$H$_5$C$_5$H$_4$N)$_2$ From 2-phenylpyridine and palladium chloride[83].
Pd$_2$Cl$_2$(CR:CCl·CR'R"NMe$_2$)$_2$ From RC:CCR'R"NMe$_2$ and palladium chloride/lithium
                                       chloride[12].
Pd$_2$Cl$_2$(CH$_2$CHOMeCH$_2$NMe$_2$)   From Li$_2$PdCl$_4$ and *N,N*-dimethylallylamine in metha-
                                       nol[84].

Platinum

PtXRL$_2$ ⎫ X = halogen, SCN, NO$_2$, CN. R = alkyl (including
PtR$_2$L$_2$ ⎬ vinyl) or aryl with many examples. L = tertiary phos-
PtX$_{4-y}$R$_y$L$_2$ (y = 1–4)   ⎭ phine or tertiary arsine. L$_2$ = chelating diphosphine or
                                       diarsine. Many stereoisomers known[85].

PtR$_2$ (chelating diolefin)   Chelating diolefin = COD, cyclo-octatetraene, norborna-
                                       diene, etc. Less stable than compounds containing
                                       tertiary phosphines[86].

[PtXMe$_3$]$_4$, PtMe$_3$C$_5$H$_5$ ⎫ Many other stable complexes containing the trimethyl-
[Pt(acac)Me$_3$]$_2$,   ⎬ platinum grouping are known[87].
   [Pt(acac)Me$_3$bipyridyl]   ⎭

K[Pt(acac)$_3$], K[PtCl(acac)$_2$] ⎫ Have chelated, i.e. *O*-bonded acac groups or the acac
K[PtCl$_2$(acac)]   ⎬ groups can be bonded to the platinum through the
                                    ⎭ central carbon atom, i.e. *C*-bonded (or γ-bonded)[88, 89].

Copper

(CuR)$_x$ The stability falls off Ph > Me > Et. CuMe is an insoluble
                                       bright yellow powder. It reacts with MeLi to give the
                                       red [CuMe$_2$]$^-$ [90].

[79] H. Yamazaki, T. Nishido, Y. Matsumoto, S. Sumida and N. Hagihara, *J. Organomet. Chem.* **6** (1966) 86.
[80] T. Saito, Y. Uchida, A. Misono, A. Yamamoto, K. Morifuji and S. Ikeda, *J. Am. Chem. Soc.* **88** (1966) 5198.
[81] G. Calvin and G. E. Coates, *J. Chem. Soc.* (1960) 2008.
[82] A. C. Cope and E. C. Friedrich, *J. Am. Chem. Soc.* **90** (1968) 909.
[83] A. Kasahara, *Bull. Chem. Soc. Japan* **41** (1968) 1272.
[84] A. C. Cope, J. M. Kliegman and E. C. Friedrich, *J. Am. Chem. Soc.* **89** (1967) 287.
[85] R. J. Cross, *Organomet. Chem. Rev.* **2** (1967) 97.
[86] C. R. Kistner, J. H. Hutchinson, J. R. Doyle and J. C. Storlie, *Inorg. Chem.* **2** (1963) 1255.
[87] A. Robson and M. R. Truter, *J. Chem. Soc.* (1965) 630.
[88] D. Gibson, J. Lewis and C. Oldham, *J. Chem. Soc. A* (1967) 72.
[89] G. T. Behnke and K. Nakamoto, *Inorg. Chem.* **7** (1968) 330.
[90] C. E. H. Bawn and F. J. Whitby, *Discussions Faraday Soc.* **2** (1947) 228.

$Cu(\sigma\text{-}C_5H_5)(PEt_3)$	Stable colourless crystals, m.p. 127–128°. Temperature dependent n.m.r. spectrum[91].

Silver

$(AgR)_2AgNO_3$	Formed from alkyl– or aryl–lead or –tin compounds and an excess of silver nitrate[92].
$(AgR)_x$	Compounds with R = Me, Et, Ph or CH_2Ph. The methyl compound is formed from silver nitrate and an excess of $PbMe_4$. It decomposes, giving ethane and silver[90].

Gold

$AuMe_3$	Very unstable[93].
$R_3P\text{-}AuMe_3$	More stable[93].
$[AuXR_2]_2$, $AuMe_2acac$	R = Me or Et (less stable). X = bridging group = Br, I, SCN, SPh[94].
$[AuMe_2(en)]I$	
$[AuR_2CN]_4$	R = Et or Pr. The X-ray structure of Pr compound shows that the Au–CN–Au units form the sides of a square[95].
$AuR(PR_3')$	R, R′ = Me, Et or Ph[96].

1.1.3. Structures, Stability and Bonding of Alkyl and Aryl Complexes

1.1.3a. *Structures*

There can be little doubt that the great majority of octahedral (d^6) and square planar (usually d^8) complexes have been assigned the correct stereochemistry and generally there has been no need for confirmation by X-ray diffraction. So far little systematic attempt has been made to study factors affecting metal–carbon (alkyl) bond lengths in simple systems. From structural determinations on more complex molecules it has been difficult to conclude anything definite about σ- and π-contributions to alkyl– or aryl–metal bonding.

One would expect that pentafluorophenyl–metal bonds would be shorter than in the corresponding phenyl complexes, because they are so much more stable. However, the two compounds $NiC_6H_5Cp(PPh_3)$ and $NiC_6F_5Cp(PPh_3)$ show no significant difference[97].

$$Ni\text{—}C_6H_5 \quad 1.919 \pm 0.013 \text{ Å}$$
$$Ni\text{—}C_6F_5 \quad 1.914 \pm 0.014 \text{ Å}$$

These distances are less than 0.1 Å longer than the Ni–C bond in nickel tetracarbonyl, suggesting a lot of double-bond character and that the bond lengths are relatively insensitive to changes in bond order.

In the two molybdenum complexes $Mo(C_2H_5)Cp(CO)_3$ and $Mo(C_3F_7)Cp(CO)_3$ the molybdenum–alkyl bond lengths are respectively 2.40 ± 0.03 Å and 2.28 ± 0.02 Å. The difference suggests a correlation of stability with shortness of the bond, but the errors are

91 G. M. Whitesides and J. S. Fleming, *J. Am. Chem. Soc.* **89** (1967) 2855.
92 G. Semerano and L. Riccoboni, *Chem. Ber.* **74** (1941) 1089.
93 G. E. Coates and C. Parkin, *J. Chem. Soc.* (1963) 421.
94 F. H. Brain and C. S. Gibson, *J. Chem. Soc.* (1939) 762.
95 R. F. Phillips and H. M. Powell, *Proc. Roy. Soc.* A, **173** (1939) 147.
96 G. Calvin, G. E. Coates and P. S. Dixon, *Chem. and Ind. (London)* (1959) 1628.
97 M. R. Churchill, T. A. O'Brien, M. D. Rausch and Y. F. Chang, *Chem. Commun.* (1967) 992.

large and the difference may not be significant[98]. The length of the Mo–C bond in the ethyl complex is identical with the sum of an sp^3 carbon radius and the molybdenum radius, deduced from the Mo–Mo bond length in $Mo_2Cp_2(CO)_6$.

The X-ray structures of several other alkyl– or aryl–transition metal complexes have been determined. Some show features of special interest.

In the cobalt complex *trans*-Co(mesityl)$_2$(PEt$_2$Ph)$_2$ the mesityl groups are vertical to the plane of the ligand atoms, and the cobalt–carbon distance of only 1.96 ± 0.01 Å suggests some multiple bond character[59].

The structure of the complex [Pt$_2$Cl$_2$(dicyclopentadieneOMe)$_2$] shows that the OMe group is *exo* and that the Pt–Cl bond *trans* to the σ-bonded carbon is longer than the one *trans* to the olefinic double bond, i.e. 2.51 Å and 2.34 Å respectively[99]. Similarly, in the structure of the complex [Ir$_2$Cl$_4$Me$_2$(CO)$_4$] [1.1.12] the bridging chlorine system is asymmetric, with the Ir–Cl bond length in *trans*-position to methyl longer (2.52 Å) than the one *trans* to carbon monoxide (2.38 Å) [100].

$$
\begin{array}{c}
\text{O} \\
\text{C} \\
\text{Me} \quad | \quad \overset{2.38\text{Å}}{\diagdown}\text{Cl} \quad \text{Cl} \\
\text{Ir} \quad \quad \text{Ir} \quad \text{CO} \\
\text{OC} \quad | \quad \underset{2.52\text{Å}}{\diagup}\text{Cl} \quad \text{Cl} \quad \text{Me} \\
\text{Cl} \quad \text{C} \\
\text{O}
\end{array}
$$

[1.1.12]

The asymmetry in these two structures correlates well with the known large *trans*-bond weakening effect of alkyl groups, as shown by i.r. and kinetic measurements.

The X-ray structure of RhI$_2$CH$_3$(PPh$_3$)$_2$ shows the square pyramidal structure [1.1.13], which is unusual for a complex with the metal in d^6-electron configuration. The Rh–CH$_3$ distance is 2.08 Å [101].

$$
\begin{array}{c}
\text{Me} \\
| \\
\text{Ph}_3\text{P} \diagdown \quad \diagup \text{I} \\
\text{Rh} \\
\text{I} \diagup \quad \diagdown \text{PPh}_3
\end{array}
$$

[1.1.13]

The product formed from azobenzene and sodium chloropalladite is thought to have the structure [1.1.14]. This product reacts with triethylphosphine to give the complex *trans*-PdCl(2-phenylazophenyl)(PEt$_3$)$_2$, which has the expected structure[102]; hence the structure [1.1.14] is virtually certain to be correct.

$$
\begin{array}{c}
\text{Ph} \\
\text{N} \diagdown \text{N} \quad \diagup \text{Cl} \\
\text{Pd} \\
\diagup \bigg]_2
\end{array}
$$

[1.1.14]

[98] M. R. Churchill and R. Mason, *Advances Organomet. Chem.* **5** (1967) 93.
[99] W. A. Whitla, H. M. Powell and L. M. Venanzi, *Chem. Commun.* (1966) 310.
[100] N. A. Bailey, C. J. Jones, B. L. Shaw and E. Singleton, *Chem. Commun.* (1967) 1051.
[101] P. G. H. Troughton and A. C. Skapskii, *Chem. Commun.* (1968) 575.
[102] R. W. Siekman and D. L. Weaver, *Chem. Commun.* (1968) 1021.

There have been several X-ray structural investigations on complexes containing the trimethylplatinum group. Trimethylplatinum chloride has the tetrameric structure [1.1.15]; each platinum is in an essentially octahedral environment and has the expected bond lengths to carbon. The dimeric nonane-4,6-dionato complex Pt_2Me_6(nonane-4,6-dionato)$_2$

[1.1.15]

[1.1.16]

has the unusual structure [1.1.16] (propyl groups omitted for clarity) with a Pt–CH(acac) distance of 2.39 Å. A more recent neutron diffraction study on this complex shows that the central hydrogen is pushed away from the platinum with a Pt–C–H angle of 98° and a C–H distance of 1.10 ± 0.06 Å [103].

The complex Pt(acac)Me$_3$(dipyridyl) has the structure [1.1.17] with a monodentate C-bonded acetylacetonato group {as in K[PtCl(acac)$_2$], etc.}.

[1.1.17]

Other trimethylplatinum complexes whose structures have been determined by X-ray diffraction are [PtMe$_3$(CH$_3$COCHCOOEt)]$_2$, [PtMe$_3$(8-hydroxyquinoline)]$_2$, [PtMe$_3$(salicylaldehydato)]$_2$ and [Pt$_2$Me$_6$(en)].

1.1.3b. Stability and Bonding

Practically nothing is known about the strength of transition metal–alkyl or –aryl bonds, with the exception of trans-PtPh$_2$(PEt$_3$)$_2$, for which the Pt–C bond strength has been estimated at 60 kcal mole^{-1} [104]. It seems likely that they are often relatively weak, e.g. the force constant of the Ti–C bond in TiCl$_3$Me is low (1.85×10^5 dynes cm^{-1}) and simple alkyl–titanium compounds are thermally unstable, e.g. TiMe$_4$ decomposes at $-40°$. Simple alkyl derivatives of other transition metals are usually even more unstable. In general one would expect alkyl– or aryl–transition metal complexes to be thermodynamically unstable since C–C or C–H bonds could be formed in the decomposition and both these bonds are strong. Typical products from the decomposition of transition metal–alkyls (M–R) are alkanes (RH), alkenes (formed by loss of H from R), or coupled products {R·R, e.g. (AgMe)$_x$ decomposes to give ethane}[90].

103 R. N. Hargreaves and M. R. Truter, Chem. Commun. (1968) 473.
104 S. J. Ashcroft and C. T. Mortimer, J. Chem. Soc. A (1967) 930.

In general not much is known about the mechanism of thermal decomposition, but the initial step in the process is likely to be one of the following:

(i) Homolytic fission $M—R \rightarrow M + R\cdot$.

(ii) Heterolytic fission $M—R \rightarrow M^+ + R^-$ or $M^- + R^+$.

(iii) A bond-breaking and bond-forming process occurring within the coordination sphere of the metal such as a reductive elimination, i.e.

$$M—R \rightarrow M + R_2$$
$$\diagdown$$
$$R$$

(iv) A β-hydrogen elimination giving an alkene and an alkane (RH).

The alkyl– or aryl–transition metal complexes differ considerably in their thermal stability and inertness towards chemical attack. Even though a complex is thermodynamically unstable with respect to decomposition products, the process may be slow and the complex may be kept virtually unchanged for years, i.e. it is kinetically stable. Factors which are important in conferring "stability" on an alkyl– or aryl–transition metal complex will now be discussed.

(i) *Dependence on the other ligands.* For the Group VIII metals, rhenium and gold, tertiary phosphines and tertiary arsines are excellent ligands for conferring stability, the complexes frequently being relatively inert towards air, water, ethanol, dilute acids and thermally stable. As a class, compounds containing these ligands are by far the most numerous and the most stable of all alkyl and aryl derivatives of transition metals. For the Group VIA and Group VIIA metals and for iron, ruthenium or osmium, carbonyl, either alone or in association with the π-cyclopentadienyl group, will also stabilize alkyl or aryl derivatives, e.g. in $MnMe(CO)_5$, $FeMe(CO)_2Cp$, etc. π-Cyclopentadienyl stabilizes derivatives of vanadium, titanium or zirconium, e.g. $TiPh_2Cp_2$ and many alkyl or aryl derivatives of cobalt(III) containing the cyanide ion are known. All the above-mentioned ligands are strong field ligands. One can make the generalization that such ligands are best for stabilizing the alkyl or aryl derivatives of the transition metals, especially in the latter half of a transition series. There are several alkyl- or aryl-derivatives not containing such strong field ligands, however, such as $[CrCH_2Ph(H_2O_5)]^{2+}$, and alkyls related to vitamin B_{12}. The stability of these is probably due mainly to kinetic factors, chromium(III) and cobalt(III) complexes often being particularly inert to attack. Alkyl and aryl groups themselves can act as strong field ligands (see section 1.1.4) and this may be an important factor conferring stability in trimethylplatinum(IV) and gold(III) complexes.

(ii) *Dependence on electron configuration.* Some of the most stable alkyl or aryl complexes have metals with d^6- or d^8-"non-bonding" electron configurations. Many obey the inert gas rule, i.e. they have eighteen electrons in the valence shell, in which case all the metal valence orbitals are used in the bonding. Four-coordinate planar complexes usually have sixteen electrons and a vacant p_z-orbital. Other common electron configurations are d^0, d^3 and d^{10}, e.g. in $TiPh_2Cp_2$, $CrPh_3(THF)_3$ and $AuMe(PPh_3)$ respectively. The electron configurations d^0, d^3, d^6 (low spin), d^8 (low spin) and d^{10} have a special stability since they are associated with empty, half-filled, or filled electronic shells or subshells.

(iii) *Dependence on the alkyl group.* Stabilities seem to correlate well with the carbanion stability of the alkyl, aryl or acetylide group, i.e. the smaller the pK_a value of the corresponding alkane or arene (R–H) the greater the stability of the alkyl– or aryl–metal derivative. Thus the pK_a values of methyl–H, ethyl–H, n-propyl–H and isopropyl–H are respectively

39, 40.5, 41 and 41.5 [105], and this is the order of decreasing stability of the alkyl–metal derivatives. Phenyl–H is more acidic (pK_a 37) and phenyl–transition metal complexes are usually more stable than alkyl–transition metal complexes. Phenylacetylides are even more stable and this correlates well with the lower pK_a value (18.5). These stabilities also agree with Jaffé's [106] theoretical conclusions on metal–carbon bond strengths, namely that they should increase with increasing electro-negativity of the alkyl group.

(iv) *Dependence on the transition series.* For analogous alkyl- or aryl-complexes of a triad of metals from the three transition series the order of stability is usually third transition series > second transition series > first transition series. This certainly applies for complexes of the type $MXR(PR_3')_2$ and $MR_2(PR_3')_2$, i.e. the stabilities decrease in the order M = Pt > Pd > Ni (R, R' = alkyl or aryl) and also very probably for complexes of the type MXR(chelating diphosphine)$_2$, M = Os > Ru > Fe. Alkyl– or aryl–iridium(III) complexes are much more stable than analogous rhodium complexes; the corresponding cobalt complexes are not usually known. For complexes of the type $MR(CO)_3Cp$, M = Cr, Mo or W, the chromium complexes are the least stable although the molybdenum complexes seem to be more stable than the tungsten ones.

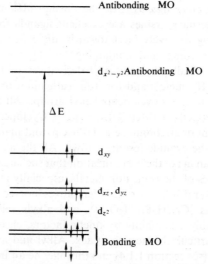

FIG. 4 Part of the energy level diagram for a square planar complex with d^8-electron configuration. ΔE is the energy gap between the lowest antibonding MO and the highest occupied level.

An explanation for the marked instability of some alkyl derivatives of transition metals has been given [1, 5, 107]. It was suggested that factors which weaken the metal–carbon bond decrease the stability and that decomposition (which could occur in a variety of ways) was preceded either (1) by promotion of an electron into an antibonding orbital of the M–C bond or (2) by promotion of an electron from the M–C bonding orbital into an unfilled orbital. Either process would weaken the M–C bond. Because transition metals have penultimate d-levels close to the valence s- and p-levels such electron promotions can often

[105] R. E. Dessy, W. Kitching, T. Psarras, R. Salinger, A. Chen and T. Chivers, *J. Am. Chem. Soc.* **88** (1966) 460.

[106] H. H. Jaffé and G. O. Doak, *J. Chem. Phys.* **21** (1953) 1118.

[107] J. Chatt and B. L. Shaw, *J. Chem. Soc.* (1959) 705.

occur readily and the compound decomposes. It was suggested[107] that for stability an alkyl- or aryl-derivative should have more than a certain minimum energy (ΔE) between the highest occupied level and the lowest energy orbital which is vacant.

The arguments have been discussed in most detail for square planar complexes containing many d-electrons, but could be applied to other systems. Figure 4 shows part of the energy level diagram for a square planar alkyl derivative of nickel(II), palladium(II) or platinum(II), i.e. with eight non-bonding d-electrons.

Decomposition in such complexes would occur on promoting an electron into an antibonding MO; this would be most likely to occur from the d_{xy} into the antibonding $d_{x^2-y^2}$ orbital and the ease with which it could occur would depend on ΔE. There are several factors which could increase ΔE and hence increase stability:

(1) Since ligand field splittings (ΔE) increase in the order Ni < Pd < Pt, one would also expect this to be the order of increasing stability, as is generally observed.

(2) Strongly bonding strong field ligands such as PR_3, AsR_3, and to a lesser degree SR_2, CO, diolefins, etc., would give large ΔE. These ligands could also π-bond to the metal, i.e. they have vacant orbitals which can overlap with the filled d_{xy} orbital and hence lower its energy. This explains why such ligands are so good at stabilizing alkyl and aryl derivatives of this type.

(3) Aryl groups and particularly ethynyl (acetylide) groups would lower the energy of the d_{xy} orbital since they have vacant π^*-orbitals of the correct symmetry. Hence one would expect these to form more stable derivatives than alkyl groups. The observed order of stability is ethynyl > aryl > alkyl. In particular *ortho*-substituted aryl groups form much more stable derivatives with nickel(II) than do aryl groups not substituted in the *ortho* position. In such compounds the *ortho*-substituent holds the plane of the aryl group vertical to the plane of the four ligand atoms, hence π-bonding interaction is restricted to the d_{xy} orbital, giving maximum ΔE and stability. Steric effects are also important in giving kinetic stability to these *o*-substituted aryl complexes.

Although these arguments are qualitative they do explain some of the important factors causing stability or instability in alkyl– or aryl–transition metal complexes.

1.1.4. Physical Properties

1.1.4a. *Nuclear Magnetic Resonance Spectra*

There are a lot of n.m.r. results for alkyl and aryl derivatives of transition metals, but so far there have not been many systematic attempts to correlate the data with electronic structural or other features. However, some ^{1}H and ^{31}P n.m.r. results on platinum complexes correlate well with each other and with other properties. For tertiary phosphine–platinum complexes J(Pt–P) falls by a factor of 0.60–0.68 in going from platinum(II) to platinum(IV) due to a decrease in the s-character of the Pt–P bond in going from dsp^2 to d^2sp^3 hybridization[108]. For *cis*-[PtClMe(PEt$_3$)$_2$], J(Pt–P) is 4.18 KHz for the P *trans* to Cl but only 1.72 KHz for the P opposite the strongly *trans*-bond weakening methyl group. Similarly for complexes of the type *trans*-PtXMe(PEt$_3$)$_2$, ^2J(Pt–CH_3) decreases as the electro-negativity of X decreases. Some ^{1}H n.m.r. results for methylplatinum complexes containing dimethylphenylphosphine as stabilizing ligand are given in Table 1. Many other compounds of similar type have been studied[14] and one can make the following generalizations.

[108] F. H. Allen and A. Pidcock, *J. Chem. Soc.* A (1968) 2700.

TABLE 1. NUCLEAR MAGNETIC RESONANCE DATA FOR METHYLPLATINUM COMPLEXES CONTAINING PMe₂Ph (= L) AS STABILIZING LIGAND (IN CHCl₃ SOLUTION)

$\tau \pm 0.01$. J in Hz ± 0.5 Hz. The data are presented in the order τ-value [J(PH) in parentheses] J(PtH). d = doublet, t = triplet, q = quartet, c = complex resonance

Complex	Configuration	Methyl groups (phosphorus) trans-ligand atom			Methyl groups (platinum)		
		Halogen	P	C	Halogen	P	C
PtClMeL₂	cis	8.44d (10.4) 47		8.35 (9.0) 19		9.22q (4.6) 56 (7.7) 57	
PtClMeL₂	trans A		8.15t (7.0) 31.5		9.82t (7.5) 85		
PtCl₂Me₂L₂	B		7.90t (8.2) 17.6		9.48t (5.6) 68		
PtMe₄L₂	cis		8.51d (9.3) 10.5 8.60d (8.4) 11.5			8.61c (complex) 57 9.61c (−) 57	10.23t (6.4) 44

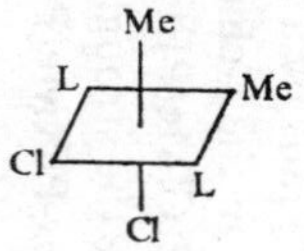

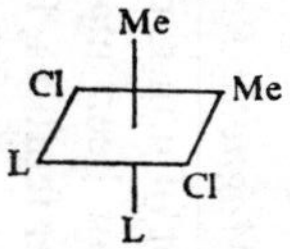

(i) ^2J(Pt–H) falls markedly in going from platinum(II) to platinum(IV) {as found for J(Pt–P)}.

(ii) ^2J(Pt–H) depends mainly on the nature of the *trans*-ligand, e.g. for platinum(IV) complexes the ranges are 67–73 Hz (*trans*-halogen), 56–59 Hz (*trans*-P) and *ca.* 44 Hz (*trans*-C), i.e. the behaviour parallels that of J(Pt–P).

(iii) There are similar trends in ^3J(Pt–H), i.e. for the phosphorus methyls (see Table 4).

In a series of trimethylplatinum(IV) complexes of type *fac*-PtMe$_3$L$_3$, ^2J(Pt–H) depends on the *trans*-ligand L varying from 60.8 Hz (L = ⁻CN) to 79.7 Hz (L = H$_2$O). In water/ pyridine mixtures the various species PtMe$_3$(H$_2$O)$_x$py$_{3-x}$ can be observed by n.m.r.[109]. Many other trimethylplatinum complexes have been studied. In the complex [PtCl(C$_2$H$_5$)$_3$]$_4$, ^2J(Pt–H) (86 Hz) and ^3J(Pt–H) (72 Hz) are opposite in sign.

There are a lot of ^{1}H n.m.r. data on methyl complexes of ruthenium, osmium, rhodium and iridium. Most of the complexes contain tertiary phosphines or tertiary arsines as ligands. Generally for the methyl bonded to the metal ^3J(P–H) is within the range 5–10 Hz, and there is not much difference in the size of J for phosphines *cis* to the methyl and those *trans* {contrast with transition metal complex hydrides where ^2J(P–H) *trans* $\gg$ ^2J(P–H) *cis*}. Couplings of methyl groups to ^{103}Rh are much smaller than to ^{195}Pt, partly because of the smaller nuclear moment.

τ-Values for methyl groups bonded to transition metals vary from 7 to 12.5. Sometimes in ethyl complexes the methyl group is at lower τ than the methylene, e.g. in NiEt$_2$(bipyridyl). The high field shift of coordinated methyls has been correlated with a low electro-negativity for the metal. However, paramagnetic contributions towards chemical shift are probably very important in such complexes, and so one cannot use τ-values to estimate electro-negativities with any confidence. In the methyltitanium complexes TiXMeCp$_2$, τ_{CH_3} increases in the order Cl < Br < I, but for the compounds of type [PtXMe$_3$]$_4$ the reverse order is found.

In some platinum or rhodium acetylacetonate complexes, spin–spin coupling between the metal and the central CH of the acetylacetonate indicates the presence of a metal–carbon bond i.e. M—CH(COCH$_3$)$_2$ [110].

1.1.4b. *Infrared and Raman Data*

Metal–carbon stretching frequencies have been identified for several alkyl-transition metal complexes. The bands are frequently weak, especially for groups other than methyl. As with n.m.r., methylplatinum complexes have been the most widely studied and give the most information[14]. For compounds of the type *trans*-PtXMe(PR$_3$)$_2$ or *cis*-PtMe$_2$(PR$_3$)$_2$, X = anionic ligand, R = alkyl group, v(Pt–C) and δ(Me) tend to increase with increasing electro-negativity of the *trans*-ligand as do values of v(Pt–H) in complexes of type *trans*-PtHX(PR$_3$)$_2$. There is a mutual electronic interaction between two *trans*-groups, e.g. for *trans*-PtClMe(PEt$_3$)$_2$, v(Pt–C) at 551 cm^{-1} is high but v(Pt–Cl) at 274 cm^{-1} is low. Similar mutual interactions are found in methyliridium(III) complexes[7].

In trimethylplatinum(IV) complexes, e.g. [PtClMe$_3$]$_4$, bands in the region of 570 cm^{-1} have been assigned to v(Pt–C), and in the ranges 1230–1250 cm^{-1} to δ_s(Me) and 1260–1280 cm^{-1} to δ_{as}(Me). For [PtClEt$_3$]$_4$ no absorption due to v(Pt–C) could be observed. The Raman spectrum of *fac*-PtMe$_3$(NH$_3$)$_3$ shows a band at 584 cm^{-1} {v(Pt–C)} and one at 390

109 D. E. Clegg and J. R. Hall, *Australian J. Chem.* **20** (1967) 2025.
110 G. W. Parshall, *J. Am. Chem. Soc.* **88** (1966) 704.

cm^{-1} $\{\nu(\text{Pt-N})\}$; in the corresponding deuterammonia complex PtMe$_3$(ND$_3$)$_3$, the frequencies are at 585 cm^{-1} and 364 cm^{-1} [111].

Infrared has been used to identify O-bonded (chelated) and C-bonded or γ-acetylacetonatoplatinum complexes[89].

1.1.4c. Dipole Moments

These are useful for determining stereochemistry and for studying polarizations across metal complexes[107]. Thus the moment of trans-PtClMe(PEt$_3$)$_2$ is 4.3 D whilst cis-PtClMe(PEt$_3$)$_2$ has a moment of 8.4 D. The value of 3.4 D indicates a strong electron drift $\overrightarrow{\text{Me-Pt-Cl}}$; this drift is associated with a weak Pt-Cl bond {low $\nu(\text{Pt-Cl})$} and a strong Pt-Me bond {high $\nu(\text{Pt-C})$}. Similar effects are found in palladium(II), ruthenium(II), osmium(II) and iridium(III) complexes.

1.1.4d. Magnetic Moments

Several paramagnetic alkyl or aryl derivatives are known, but usually only their room temperature magnetic moments have been measured. For aryl derivatives of chromium(III), with three unpaired spins, the moments are 3.6–3.9 BM, close to the spin-only value of 3.82 BM. In contrast, aryl derivatives of chromium(II), e.g. CrAr$_2$ or LiCrAr$_3$, have very low moments, 0.35–0.54 BM, which could arise from Cr–Cr-interaction. Aryl–cobalt complexes of type trans-Co(aryl)$_2$(PEt$_2$Ph)$_2$ have moments of ca. 2.5 BM, much higher than the spin-only value of 1.73 BM, indicating significant orbital contribution.

1.1.4e. Electronic Spectra—The Position of Alkyl and Aryl Groups in the Spectrochemical Series

Studies on the electronic spectra (d–d transitions) of various alkyl– or aryl–metal complexes put the alkyl or aryl groups high in the spectrochemical series for ruthenium(II) and cobalt(III) but lower for chromium(III) or titanium(III) [4]. For RuXCl(Me$_2$PCH$_2$CH$_2$PMe$_2$)$_2$ the order is X = CN > Ph > Me > H $\gg$ Cl. For [CoX(CN)$_5$]$^{3-}$ X = Me approaches that of CN. For [CrX$_6$]$^{3-}$ X = CN $\gg$ NH$_3$ > Me > H$_2$O > F. For TiCl$_2$Xpy$_3$ X = Cl > Ph > Me.

It appears therefore that the ligand field effects of alkyl or aryl groups are large only when the number of non-bonding d-electrons is large.

1.1.5. Reactions[111a]

1.1.5a. Cleavage of the Metal–Carbon Bond

This can be effected by reagents with active hydrogens, halogens, alkyl halides, hydrogen, LiAlH$_4$, metal halides and by homolytic cleavage.

Alkyl or aryl derivatives of transition metals usually have a low anionic reactivity; this is because of the very low electro-positive character of the metal atom. Thus many of them

[111] D. E. Clegg and J. R. Hall, Spectrochim. Acta 23A (1967) 263.

[111a] J. P. Candlin, K. A. Taylor and D. T. Thompson, Reactions of Transition–Metal Complexes, Elsevier, Amsterdam (1968).

can be recrystallized from alcohols and are unaffected by dilute acids. Anhydrous hydrogen halides usually bring about rapid cleavage, however, e.g.

$$\text{cis-PtPh}_2(\text{PEt}_3)_2 \xrightarrow[-C_6H_6]{\text{HCl (1 mol)}} \text{cis-PtClPh(PEt}_3)_2 \xrightarrow{\text{HCl}} \text{cis-PtCl}_2(\text{PEt}_3)_2$$

This is the most convenient route to cis-PtClPh(PEt$_3$)$_2$. The rate of fission depends on the trans-ligand, e.g. cis-PtClPh(PEt$_3$)$_2$ reacts a lot faster with HCl than does trans-PtClPh(PEt$_3$)$_2$. Similar cleavage reactions of methylplatinum complexes with methanolic HCl obey the rate expression

$$\text{Rate} = k_1[\text{H}^+] + k_2[\text{H}^+][\text{Cl}^-][\text{Pt complex}]$$

and a two-step mechanism involving a labile platinum(IV) hydride intermediate is proposed[112]. Other similar cleavage reactions include:

$$\text{PdMe}_2(\text{PEt}_3)_2 + \text{PhSH} \rightarrow \text{Pd(SPh)}_2(\text{PEt}_3)_2$$
$$\text{NiMe(Cp)PPh}_3 + \text{HBr} \rightarrow \text{NiBr(Cp)PPh}_3$$

Halogens also cleave the metal–carbon bond, e.g.

$$[\text{AuBrEt}_2]_2 \xrightarrow{\text{Br}_2} [\text{AuBr}_2\text{Et}]_2$$
$$\text{TiPh}_2\text{Cp}_2 \xrightarrow{\text{Cl}_2} \text{TiCl}_2\text{Cp}_2$$

Fission may go via an oxidative addition[113], e.g.

$$\text{PtPh}_2(\text{PEt}_3)_2 \xrightarrow[\text{fast}]{\text{I}_2} \text{PtI}_2\text{Ph}_2(\text{PEt}_3)_2 \xrightarrow[\text{EtOH}]{\text{slow in}} \text{PtIPh(PEt}_3)_2 + \text{PhI}$$

but the reaction may be complex, e.g.

$$\text{NiCH}_3\text{Cp(PPh}_3)_2 + \text{I}_2 \rightarrow \text{NiICp(PPh}_3)_2 + \text{CH}_3\text{I} + \text{some ethane}$$

Organic halides can effect fission, e.g. Pt(o-tolyl)$_2$py$_2$ reacts with a variety of organic iodides. The reactions involve an intermediate platinum(IV) adduct:

$$\text{Pt(o-tolyl)}_2\text{py}_2 + \text{PhI} \rightarrow \text{PtI(o-tolyl)py}_2 + p\text{-methylbiphenyl}$$

Mercuric chloride can cause fission, e.g.

$$\text{TiMeCl}_3 + \text{HgCl}_2 \rightarrow \text{TiCl}_4 + \text{MeHgCl}$$

as can anhydrous magnesium iodide.

Hydrogen will cleave metal–carbon bonds, e.g. Li$_3$[CrPh$_6$]xEt$_2$O readily reacts to give the complexes Li$_3$[CrHPh$_5$]3Et$_2$O or [Li$_2$Cr$_2$H$_3$Ph$_6$]EtO and benzene[114]. There are several other examples. LiAlH$_4$ or LiAlD$_4$ have also been used to cleave metal–carbon bonds.

The products of homolytic cleavage may be complex. TiPh(OR)$_3$ at 80° gives Ti(OR)$_3$, benzene and biphenyl. TiPh(OBu)$_3$ polymerizes styrene at a rate comparable with free radical initiation.

Olefins may be eliminated, e.g. [Co(CH$_2$CHMe$_2$)(CN)$_5$]$^{3-}$ slowly loses isobutene at 20° and trans-PtClEt(PEt$_3$)$_2$ on heating gives PtHCl(PEt$_3$)$_2$ and ethylene.

112 U. Belluco, M. Guistiniani and M. Graziani, J. Am. Chem. Soc. 89 (1967) 6494.
113 R. Ettorre, Inorg. Nucl. Chem. Letts. 5 (1969) 45.
114 F. R. Hein and R. Weiss, Naturwissenschaften 46 (1959) 321.

1.1.5b. *Oxidative Addition Reactions*

Alkyl ligands have low electro-negativity and enhance the reactivity of the central metal atom towards oxidative addition reactions, e.g. *cis*-PtMe$_2$(PMe$_2$Ph)$_2$ readily adds on methyl halides, acid chlorides, etc., to give platinum(IV) adducts, but *cis*-PtCl$_2$(PMe$_2$Ph)$_2$ is unreactive (see section 1.1.1d).

1.1.5c. *Insertion Reactions*

These are very important and occur widely in catalysis. In the reaction an unsaturated molecule (U) becomes inserted into a metal–alkyl or –aryl bond (M–R) to give the "insertion" product M–U–R. Examples of unsaturated molecules which will react in this way include CO, SO$_2$, nitriles, isonitriles, olefins, diolefins, acetylenes, ketones and possibly N$_2$.

The reaction of MnMe(CO)$_5$ with CO to give Mn(COMe)(CO)$_5$ has already been mentioned. This and related reactions have been studied extensively. In the reaction the methyl group moves from the metal to one of the carbonyl groups which is already coordinated. For the conversion of MnR(CO)$_5$ to Mn(COR)(CO)$_5$ the rates decrease in the order of R = *n*-Pr > Et > Ph > Me > PhCH$_2$ or CF$_3$. If R is an optically active radical it retains its optical activity during the conversion. In the conversion of MnMe(CO)$_5$ by PPh$_3$ to Mn(COMe)(CO)$_4$PPh$_3$ the *cis*-isomer is formed first but may isomerize to the *trans*-isomer by dissociation of PPh$_3$ and recombination in a first-order process[115].

Alkyl derivatives of many other metals will insert CO to give acyls, e.g. of molybdenum, iron, cobalt, nickel, palladium and platinum. In the conversion of MoR(CO)$_3$Cp by PR$_3'$ to Mo(COR)(CO)$_2$PR$_3'$Cp the rates decrease in the order of

$$R = Et > Me > PhCH_2 > CH_2{:}CHCH_2$$

and the rates are very solvent dependent, e.g. the reaction is 10^3 × faster in MeCN than in THF.

A very important example of CO insertion into a metal–alkyl bond occurs in the industrial Oxo-process (hydroformylation of olefins) (see section 2).

Sulphur dioxide insertions can occur very rapidly[116], e.g. FeMe(CO)$_2$Cp reacts with SO$_2$ at −40° to give Fe(SO$_2$Me)(CO)$_2$Cp. Sulphur dioxide insertion occurs with other compounds of this type, i.e. FeR(CO)$_2$Cp, and the rates decrease in the order of R = Me ~ Et > Ph ~ CH$_2$Ph ≫ C$_6$F$_5$. Sulphur dioxide inserts very rapidly into σ-allylic complexes of manganese or iron. A cyclic mechanism is probably involved since the crotyl-manganese complex [1.1.18] gives the α-methylallyl sulphone derivative [1.1.19].

The propargyl complex Mn(CH$_2$C$\,$:$\,$CH)(CO)$_5$ and SO$_2$ react at −75° to give a quantitative yield of the allenyl sulphinate complex

$$Mn(\overset{O}{\underset{}{S}}\!\!-OCH{:}C{:}CH_2)(CO)_5.$$

[115] K. Noack, M. Ruch and F. Calderazzo, *Inorg. Chem.* **7** (1968) 345.
[116] J. P. Bibler and A. Wojcicki, *J. Am. Chem. Soc.* **88** (1966) 4862.
[117] F. A. Hartman, P. J. Pollick, R. L. Downs and A. Wojcicki, *J. Am. Chem. Soc.* **89** (1967) 2493.

The cyclic attack is thought to occur on the oxygen because the linearity of the $CH_2C\!:\!CH$ system prevents attack on the sulphur of the SO_2.

The decomposition of alkylpentacyanocobaltate ions to alkyl cyanides by acid is an example of a migration of an alkyl group from a metal to a protonated cyanide ligand[118]:

$$[CoR(CN)_5]^{3-} \xrightarrow{H^+} [CoR(CN)_4CNH]^{2-} \xrightarrow{\text{"insertion"}} [CoCR(CN)_4]^{2-} \xrightarrow{CN^-} [Co(CN)_6]^{3-} + RCN + H^+.$$
$$\underset{NH}{\overset{\|}{}}$$

The methyl complex $NiMeCpPPh_3$ reacts with cyclohexyl isonitrile to give an isonitrile-inserted product

$$\underset{Ni(C\!:\!NC_6H_{11})(NCC_6H_{11})Cp}{\overset{Me}{\underset{|}{}}}\ ^{119}$$

Olefin insertion into a metal–alkyl bond is very important in catalysis, but the products usually react further and have not been isolated. One example, however, is the insertion of isobutene into the titanium–methyl bond of $TiCl_3Me$ to give the neopentyl complex $TiCl_3CH_2C(CH_3)_3$ [120].

A convenient synthesis of $Pd_2Cl_2(\pi$-1-benzylallyl$)_2$ consists of treating the chloropalla-date(II) ion with phenylmercuric chloride and butadiene. Insertion of butadiene into a transient phenylpalladium complex is clearly a key step in this synthesis. Butadiene insertion into a σ-allyl–metal bond probably occurs in many of the butadiene oligomeriza-tion reactions catalysed by transition metal complexes (nickel, cobalt, palladium, etc.) (see section 3).

When a mixture of $TiCl_2Cp_2$ and PhLi (fivefold excess) in ether is treated with nitrogen, ammonia and aniline are obtained on hydrolysis. With 100 atm nitrogen the yields of aniline and ammonia are increased to 0.15 and 0.65 mole respectively. With p-tolyl-lithium, p-toluidine is formed. It is suggested that insertion of N_2 into a titanium–aryl bond is involved[121], i.e.

$$\text{Ti—Ph} \rightarrow \text{Ti—N}\!:\!\text{N} \rightarrow \underset{|}{\text{Ti—N}\!:\!\text{NPh}} \xrightarrow[\text{(2) hydrolysis}]{\text{(1) reduction by LiPh}} PhNH_2 + NH_3$$
$$\underset{Ph}{}$$

1.1.5d. *Conversion to Coordinated π-Complexes*

Some simple alkyl–metal complexes can be converted into olefin–metal complexes either by hydride abstraction (with Ph_3C^+) or by protonation, e.g. for iron(II) complexes[122].

$$Cp(CO)_2Fe—\underset{\diagdown CH_3}{\overset{\diagup CH_3}{CH}}$$

$$Cp(CO)_2Fe—CH_2CH\!=\!CH_2 \xrightarrow{H^+} Cp(CO)_2\overset{+}{Fe}—\underset{CH_2}{\overset{CH_3}{\underset{\|}{CH}}}$$

$$Cp(CO)_2Fe—CH_2CH_2CH_3$$

118 M. D. Johnson, M. L. Tobe and L. Y. Wong, *Chem. Commun.* (1967) 298.
119 Y. Yamamoto, H. Yamazaki and N. Hagihara, *Bull. Chem. Soc. Japan* **41** (1968) 532.
120 H. de Vries, *Rec. trav. Chim.* **80** (1961) 866.
121 M. E. Volpin, V. B. Shur, R. V. Kudryavtsev and L. A. Prodayko, *Chem. Commun.* (1968) 1038.
122 M. L. H. Green and P. L. I. Nagy, *J. Chem. Soc.* (1963) 189.

Treatment of the corresponding crotyl complex with D^+ shows that attack occurs at carbon atom 3:

$$Cp(CO)_2FeCH_2CH{=}CHCH_3 \xrightarrow{D^+} Cp(CO)_2Fe{-}{-}\begin{array}{l} CH_2 \\ \| \\ CH \\ CHDCH_3 \end{array}$$

Other ligands can be protonated in this way, e.g. $-CH_2COCH_3$, $-CH_2CN$ and $-CH_2C\vdots CH$, and similar reactions have also been observed with molybdenum and tungsten complexes.

The conversion of aryls to π-arenes is an exceedingly complex process and has been studied in detail for chromium[44, 123]. Thus $CrPh_3(THF)_3$ in benzene for 2–3 days gives an insoluble black solid and some soluble products. The soluble products on hydrolysis give some diphenyl and $Cr(PhPh)_2$. The black solid on hydrolysis gives $Cr(arene)_2$, Cr_{aq}^{3+} and hydrogen; the arene can be biphenyl or benzene.

1.1.5e. *Miscellaneous Reactions*

The COD complex $PtCl_2COD$ reacts with isopropylmagnesium bromide at $-40°$ to give $Pt(iso{-}C_3H_7)_2(COD)$; this with more COD in ultraviolet light gives the platinum(0) complex $Pt(COD)_2$.

$RhMe(PPh_3)_3$ on heating evolves methane and gives $Rh\{o{-}C_6H_4PPh_2\}(PPh_3)_2$. With $P(C_6D_5)_3$ as the ligand, CH_3D is evolved.

trans-$NiCl_2(PEt_3)_2$ with 2-allylphenylmagnesium chloride gives *trans*-$NiCl(2$-allyl-phenyl$)(PEt_3)_2$. This when heated in tetrachloroethylene gives some indene (up to 58% yield), other organic products and the very stable *trans*-$NiCl(CCl\vdots CCl_2)(PEt_3)_2$ [21].

1.2. VITAMIN B₁₂ AND ITS ANALOGUES

1.2.1. Vitamin B_{12} Chemistry[67, 124–126]

Vitamin B_{12} or cyanocobalamin is a six-coordinate cobalt(III) species with a porphyrin-type ring system and coordinated 5,6-dimethylbenzimidazole (L) and cyanide (R) groups [1.2.1]. A closely related coenzyme, normally known as DBC, 5,6-dimethylbenzimidazolyl-cobamide, has a C5′-deoxyadenosyl group bonded to the cobalt, and is thus the first naturally occurring organometallic compound.

[1.2.1]

Aquocobalamin [1.2.1], $R = H_2O$, when reduced with mild reducing agents, gives a cobalt(II) species, vitamin B_{12r}, and stronger reducing agents such as KBH_4 give the grey–green vitamin B_{12s}, an air-sensitive cobalt(I) species. When vitamin B_{12s} is treated with

123 J. Hähle and G. Stolze, *J. Organomet. Chem.* **8** (1967) 311.
124 G. N. Schrauzer, *Accounts Chem. Res.* **1** (1968) 97.
125 E. L. Smith, *Vitamin B₁₂*, 3rd edn., Methuen, London and New York (1965).
126 D. Crowfoot-Hodgkin, *Proc. Roy. Soc. (London)* A, **288** (1965) 294.

alkylating agents, acetylenes or some olefins, much more stable cobalt(III) compounds containing Co–C (alkyl) bonds are produced. Some of these reactions are summarized in Fig. 5. In two of these transformations a proton from the solvent is involved. Conversion of vitamin B_{12s} to DBC is by treatment with 2',3'-isopropylidine-5'-tosyladenosine followed by removal of the isopropylidine group. These alkyl derivatives of cobalamin are stable to cleavage by acids but are light-sensitive, giving vitamin B_{12r} and various organic products by fission of the cobalt–alkyl bond.

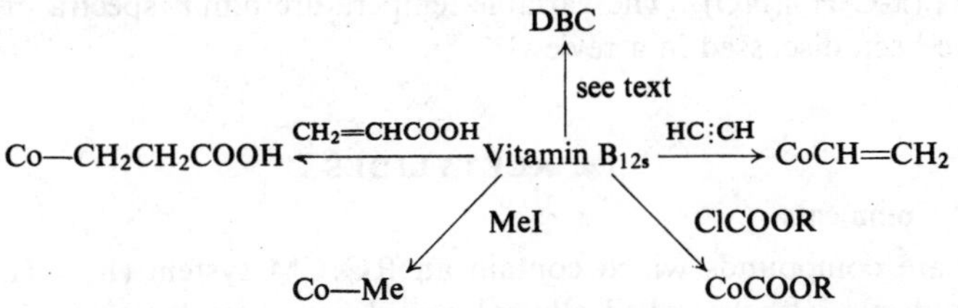

$$\text{DBC}$$
$$\uparrow \text{ see text}$$

$$\text{Co–CH}_2\text{CH}_2\text{COOH} \xleftarrow{\text{CH}_2=\text{CHCOOH}} \text{Vitamin } B_{12s} \xrightarrow{\text{HC}\vdots\text{CH}} \text{CoCH}=\text{CH}_2$$

$$\swarrow \text{MeI} \qquad \text{ClCOOR} \searrow$$

$$\text{Co–Me} \qquad\qquad \text{CoCOOR}$$

FIG. 5. Some reactions of vitamin B_{12s} in which Co–C bonds are formed:[67, 124, 125]. The other ligands are omitted.

1.2.2. Cobaloximes[124] and Bis(acetylacetone)ethylenediamine Complexes[66]

By reacting cobalt(II) species with dimethylglyoxime in the presence of a base L, e.g. pyridine or PPh_3, complexes of the type [1.2.2] are formed. X = anionic ligand, e.g. Cl, L = py or PPh_3, etc. Such compounds are known as cobaloximes and show a similar

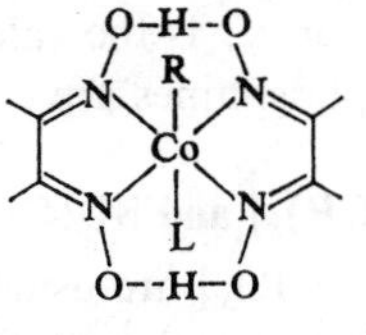

[1.2.2]

chemistry to the vitamin B_{12} derivatives. Cobaloxime compounds can be reduced by strong reducing agents to a cobalt(I) species which will undergo all the reactions outlined in Fig. 5. The alkylcobaloximes are stable to acids and bases but, as with the vitamin B_{12} alkyls, light causes fission of the Co–C bond, e.g. methylcobaloxime in benzene gives toluene.

The tetradentate ligand bis(acetylacetone)ethylenediamine ($BAEH_2$) also stabilizes cobalt–alkyl complexes including some pentacoordinate derivatives, e.g.CoR(BAE) [66]. These are synthesized in a similar way to the alkyl derivatives of vitamin B_{12} or cobaloxime.

1.3. σ-CYCLOPENTADIENYL COMPLEXES

The cyclopentadienyl anion can act as a six-electron donor, as in π-cyclopentadienyl–metal complexes, but there are some compounds in which it acts as an alkyl (monodentate) group. These compounds are usually prepared by treating a metal halide with cyclopentadienylsodium. Two isomers,

$$\text{Fe}(\sigma\text{-C}_5\text{H}_4\text{CH}_3)(\pi\text{-C}_5\text{H}_5)(\text{CO})_2 \quad \text{and} \quad \text{Fe}(\sigma\text{-C}_5\text{H}_5)(\pi\text{-C}_5\text{H}_4\text{CH}_3)(\text{CO})_2$$

are known. The most interesting features of σ-cyclopentadienyls are their variable temperature n.m.r. spectra.

Fe(σ-C$_5$H$_5$)(π-C$_5$H$_5$)(CO)$_2$ shows a variable temperature ^{1}H n.m.r. spectrum with two sharp singlets at 30° indicating a rapidly rotating σ- and a π-cyclopentadienyl group. At −85°, however, the rotation of the σ-C$_5$H$_5$ group has stopped and a complex AA′BB′X pattern is obtained together with a sharp singlet for the π-C$_5$H$_5$ group. Other σ-cyclopentadienyls show similar effects, e.g. Cu(σ-C$_5$H$_5$)PEt$_3$ shows one C$_5$H$_5$ resonance at 0°, but at −70° three resonances corresponding to a "freezing out" of the rotation are obtained[91]. Other complexes showing rapid rotation of a σ-cyclopentadienyl group are Au(σ-C$_5$H$_5$)PPh$_3$ and Cr(σ-C$_5$H$_5$)(π-C$_5$H$_5$)(NO)$_2$. The variable-temperature n.m.r. spectra of such fluxional molecules have been discussed in a review[127].

1.4. ACETYLIDES

1.4.1. General Comments

Acetylides are compounds which contain an RC⦂CM system (R = H, alkyl or aryl; M = metal) and may also be called alkynyl complexes. Acetylenes are more acidic than alkanes or arenes and the metal–acetylide bond seems to be less reactive (and probably stronger) than metal–alkyl or metal–aryl bonds. There is probably considerable back donation from filled d-orbitals on the metal into the antibonding π^*-orbitals on the acetylide, giving a metal–carbon bond which is stronger than metal–alkyl bonds. Many acetylides can be coordinated to one metal atom, e.g. K$_3$[Mn(C⦂CH)$_6$]. The small volume taken up by the acetylide grouping in the region of the valence orbitals of the metal makes this possible.

It is convenient to classify acetylides into (1) those of types A$_w$[M(C⦂CR)$_x$] and M$_y$(C⦂CR)$_z$ (A = alkali metal, $x-w$ or z/y is the valence of the metal M), and (2) those in which other ligands such as tertiary phosphines are coordinated to the metal.

1.4.2. Acetylides of Types A$_w$[M(C⦂CR)$_x$] and M$_y$(C⦂CR)$_z$ [128]

The compounds of type A$_w$[M(C⦂CR)$_x$] are usually made by treating a thiocyanate or nitrate of the metal M with the alkali metal acetylide AC⦂CR in liquid ammonia. The formulae and colours of compounds of this type are shown in Table 2. Many such acetylides

TABLE 2. COMPLEX ACETYLIDES OF TYPE A$_w$[M(C⦂CR)$_x$] [128]
(A = Na or K)

K$_3$[Cr(C⦂CH)$_6$]	orange	K$_4$[Ni(C⦂CH)$_4$]	yellow
A$_2$(Mn(C⦂CR)$_4$]	pink	K$_2$[Pd(C⦂CR)$_2$]	yellow
A$_2$[Mn(C⦂CH)$_6$]	dark brown	K$_2$[Pt(C⦂CR)$_4$]	white
A$_4$[Fe(C⦂CR)$_6$]	yellow	K$_2$[Pt(C⦂CR)$_2$]	yellow
A$_3$[Fe(C⦂CH)$_6$]	brown–violet	A$_2$[Cu(C⦂CR)$_2$]	white
A$_4$[Co(C⦂CR)$_6$]	green	K$_2$[Cu(C⦂CR)$_3$]	white
A$_3$[Co(C⦂CR)$_6$]	yellow	K[Ag(C⦂CR)$_2$]	white
A$_2$[Ni(C⦂CR)$_4$]	yellow	K[Au(C⦂CR)$_2$]	white

are similar in formula, colour and magnetic properties to complex cyanides (CN and C⦂CH are isoelectronic). They are usually much more readily hydrolysed than complex cyanides and are frequently explosive (this may be a consequence of the energy-rich C⦂C bond, not the M–C bond). Their stability falls off in the order of R = Ph > H > alkyl.

[127] F. A. Cotton, *Accounts Chem. Res.* **1** (1968) 257.
[128] R. Nast, *Abstracts, Fifth International Conference, Coordination Chemistry, London,* Chemical Society (London) Special Publication No. 13 (1959), p. 103.

There are several uncharged acetylides known of type $M_y(C\vdots CR)_z$, where M = Mn, Ni, Cu, Ag or Au. These are made by direct action of the acetylene with the metal salt. They are polymeric with the acetylide group bonded to more than one atom, e.g. the X-ray structure of copper phenylacetylide shows the acetylide to be unequally bonded to three copper atoms[1]. The tert-butylethynyl complex of copper(I) is octameric, i.e. $[CuC\vdots CBu^t]_8$, whilst the gold(I) complex is tetrameric: each acetylide group is probably σ-bonded to one gold atom and π-bonded (through its triple bond) to another.

Silver or copper acetylides, M_2C_2, are made from acetylene and a silver or copper salt. They are very insoluble and explosive when dry. They probably have each acetylide group associated with several metal atoms.

1.4.3. Acetylide Complexes with Tertiary Phosphines or Other Ligands

The nickel triad forms many complexes of types $trans$-$M(C\vdots CR')_2(PR_3)_2$ or $MX(C\vdots CR')(PR_3)_2$, where M = Ni, Pd or Pt, X = halide, etc., R = alkyl or aryl, R' = alkyl, aryl or H. The diacetylides can be made by metathesis from the corresponding metal halides[5], e.g.

$$NiCl_2(PEt_3)_2 + 2NaC\vdots CPh \xrightarrow[\text{NH}_3]{\text{in liquid}} Ni(C\vdots CPh)_2(PEt_3)_2$$

$PtBr(C\vdots CPh)(PPh_3)_2$ is formed from $Pt(PPh_3)_3$ and $BrC\vdots CPh$ [19]. An unsual platinum(IV) hydride acetylide is formed by treating $Pt(PPh_3)_4$ with ethynylcyclohexanol; it has been assigned the structure $PtH_2(C\vdots C\cdot C_6H_{10}OH)_2(PPh_3)_2$ [129]. $trans$-$Ni(C\vdots CPh)_2(PEt_2Ph)_2$ readily takes up another mole of PEt_2Ph to give the five-coordinate species $Ni(C\vdots CPh)_2(PEt_2Ph)_3$ [5]. Two groups have determined the X-ray structure of $trans$-$Ni(C\vdots CPh)_2(PEt_3)_2$ and find an Ni–C bond length of 1.87 Å.

Tertiary phosphine derivatives of copper, silver or gold acetylides are also made by metathesis from the corresponding halides. The complexes are usually polymeric. The X-ray structure of $[Cu(C\vdots CPh)PMe_3]_4$ [130] shows two different types of copper atoms in the molecule and that the phenylethnynyl groups are associated with two or three copper atoms. $[AgC\vdots CPh(PMe_3)]_x$ is dimeric in benzene but with a complex polymeric structure in the crystal[131]. Gold(I) forms a series of complexes of the type $Au(C\vdots CR)L$ (L = tertiary phosphine, tertiary arsine or ammonia).

Iridium forms an acetylide hydride $IrHCl(C\vdots CPh)CO(PPh_3)_2$. There are some gold(I) acetylide–amine complexes known. Chromium and molybdenum form carbonyl acetylides, e.g. $K_3[Mo(C\vdots CH)_3(CO)_3]$, and iron forms the complex $Fe(C\vdots CPh)(CO)_2Cp$.

1.5. ACYL–METAL COMPLEXES

These have been mentioned already (in section 1.1). They may be formed (1) by the insertion of carbon monoxide into an alkyl– or aryl–metal bond, e.g. in manganese, iron, palladium, platinum and rhodium complexes[132]; (2) by reacting an acid halide with a carbonylate ion[7a, 7b]; (3) by the oxidative addition of an acid halide to a metal complex in a low valence state, e.g. $trans$-$IrCl(CO)L_2 + RCOCl$ gives $IrCl_2(COR)COL_2$, where L = PR'_3

129 D. M. Roundhill and H. B. Jonassen, *Chem. Commun.* (1968) 1233.
130 P. W. R. Corfield and H. M. M. Shearer, *Acta Cryst.* **21** (1966) 957.
131 P. W. R. Corfield and H. M. M. Shearer, *Acta Cryst.* **20** (1966) 502.
132 K. Noack, U. Schaerer and F. Calderazzo, *J. Organomet. Chem.* **8** (1967) 517.

or AsR_3' [16]; (4) carbalkoxy complexes can be formed by attack of alkoxide on a carbonyl group, e.g.

$$[Ir(CO)_3(PPh_3)_2]^+ \underset{H^+}{\overset{OMe^-}{\rightleftharpoons}} [Ir(COOMe)(CO)_2(PPh_3)_2]$$

Methyl chloroformate adds to trans-$IrCl(CO)(PMe_2Ph)_2$ as in (3) above, and the product —when treated with hydrogen chloride followed by water—gives a carboxylic acid complex $IrCl_2(COOH)(CO)(PMe_2Ph)_2$. This readily loses carbon dioxide on heating[133].

Acyl–metal complexes show an intense infrared absorption band due to $\nu(C=O)$. The frequency is usually 1600–1700 cm^{-1}. The acetyl group shows a large trans-bond weakening effect in platinum metal complexes similar to that of the methyl group, i.e. in $IrClBr(COCH_3)CO(PMe_2Ph)_2$ $\nu(Ir–Cl)$ for chlorine trans to acetyl is very low (244 cm^{-1}).

The acetyl group in $Fe(COCH_3)(CO)_2Cp$ is readily protonated, the protonated species could be regarded as a carbene complex[134], i.e.

$$Cp(CO)_2Fe\underset{\underset{O}{\|}}{CMe} \overset{H^+}{\longrightarrow} \left[Cp(CO)_2Fe\overset{Me}{\underset{OH}{\cdots C}}\right]^+$$

1.6. FLUOROALKYL, FLUOROARYL, FLUOROACYL AND RELATED COMPLEXES

This large and rapidly expanding field has been reviewed[135, 136]. Fluoroalkyl– and fluoroaryl–transition metal complexes are usually much more stable than their hydrocarbon analogues, and many hundreds have been prepared.

1.6.1. Preparations

1.6.1a. From Perfluoro-lithium or -Grignard Reagents

This method is particularly useful for trifluorovinyl- and pentafluorophenyl-derivatives of titanium, cobalt, nickel, palladium and platinum.

$$NiCl_2(PEt_3)_2 + CF_2:CFMgCl \rightarrow NiBr(CF:CF_2)(PEt_3)_2$$
$$TiCl_2Cp_2 + C_6F_5Li \rightarrow Ti(C_6F_5)_2Cp_2$$

Complexes of $C_6F_5C:C-$ and $CF_3C:C-$ are prepared similarly.

1.6.1b. From Carbonylate Anions and Fluoroacyl Halides or Fluoro-olefins

Per- or poly-fluoroacyl halides (or acid anhydrides) react to give acyl–metal complexes readily, e.g.

$$R_fCOCl + [Mn(CO)_5]^- \rightarrow R_fCOMn(CO)_5$$

where $R_f = CF_3$, C_2F_5, C_3F_7, C_6F_5, etc.; and fluoroacyl complexes decarbonylate either spontaneously or on heating:

$$C_3F_7CORe(CO)_5 \overset{heat}{\longrightarrow} C_3F_7Re(CO)_5 + CO$$
$$CF_3COCo(CO)_4 \overset{-3°}{\longrightarrow} CF_3Co(CO)_4 + CO$$

133 A. J. Deeming and B. L. Shaw, J. Chem. Soc. A (1969) 443.

134 M. L. H. Green and C. R. Hurley, J. Organomet. Chem. 10 (1967) 188.

135 P. M. Treichel and F. G. A. Stone, Advances Organomet Chem. 1 (1964) 143.

136 M. I. Bruce and F. G. A. Stone, "Fluorocarbon complexes of transition metals" in Preparative Inorganic Reactions (ed. W. L. Jolly), Interscience (1968).

Replacement of one or more of the carbonyl groups on the metal by tertiary phosphines makes decarbonylation difficult. Ultraviolet light promotes the decarbonylation, however.

Fluoro-olefins and fluoroaromatics are highly susceptible to nucleophilic attack by some carbonylate anions, e.g. $[Mn(CO)_5]^-$, $[Re(CO)_5]^-$, $[Mo(CO)_3Cp]^-$, $[Fe(CO)_2Cp]^-$ and $[Ni(CO)Cp]^-$.

The reaction can take several courses:

$$CF_3CF:CF_2 + [Mn(CO)_5]^- \longrightarrow (CO)_5MnCF:CFCF_3$$

$$CF_2:CFCl + [Mn(CO)_5]^- \xrightarrow{H^+} (CO)_5MnCF_2CHFCl \qquad \{H^+ \text{ comes from the solvent}\}$$

$$CF_2:CF\cdot CF_2Cl + [Mn(CO)_5]^- \longrightarrow (CO)_5MnCF_2CF:CF_2$$

Many polyfluoro-vinylic, -aromatic and -heterocyclic compounds have been made by similar reactions.

1.6.1c. *Insertion Reactions*

Fluoro-olefins or fluoroacetylenes will insert into metal–hydride, metal–alkyl or metal–metal bonds:

$$CF_2:CF_2 + FeH_2(CO)_4 \longrightarrow (HCF_2CF_2)_2Fe(CO)_4$$

$$CF_2:CFCl + MnCH_3(CO)_5 \xrightarrow{u.v.} CH_3CF_2CFClMn(CO)_5$$

$$CF_2:CF_2 + Co_2(CO)_8 \longrightarrow (CO)_4CoCF_2CF_2Co(CO)_4$$

$$CF_3C:CCF_3 + Me_3SnFe(CO)_2Cp \longrightarrow Me_3SnC(CF_3):C(CF_3)Fe(CO)_2Cp$$

1.6.1d. *Oxidative Additions*

Perfluoralkyl iodides will add to metals in low valence states:

$$Pt(PPh_3)_4 + C_3F_7I \rightarrow cis\text{-}PtI(C_3F_7)(PPh_3)_2$$

$$Fe(CO)_5 + CF_3I \rightarrow FeI(CF_3)(CO)_4$$

1.6.1e. *Additions of Fluoro-olefins*

For example

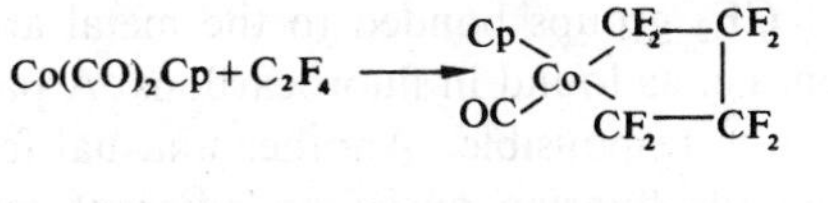

Perfluorobutadiene similarly adds to pentacarbonyliron

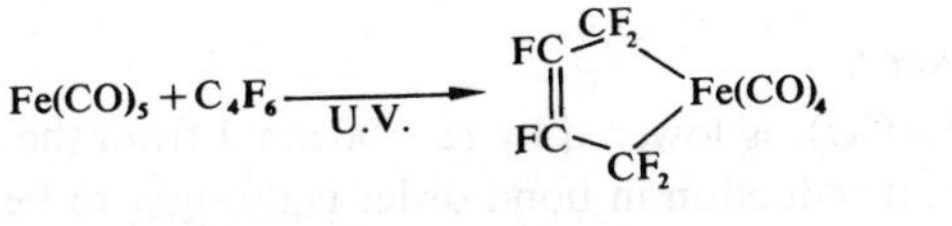

1.6.1f. *Cleavage of Alkyl–Metal Bonds by Perfluoroalkyl Iodides*

For example

$$MoMe(CO)_2Cp + C_3F_7I \rightarrow Mo(C_3F_7)(CO)_2Cp$$

[137] M. I. Bruce and F. G. A. Stone, *Angew. Chem. Int. Edn.* **7** (1968) 747.
[138] R. L. Hunt, D. M. Roundhill and G. Wilkinson, *J. Chem. Soc.* A (1967) 982.

1.6.2. Structure, Stability and Bonding

The metal–carbon bond lengths in $NiC_6F_5Cp(PPh_3)$ and $Mo(C_3F_7)(CO)_2Cp$ have already been discussed (section 1.1.3). The structure of the difluorovinyl complex cis-CFH:CFMn(CO)$_5$ has some interesting features[139]. Although the Mn–CF bond length of 1.95 Å is close to the sum of the single bond radii, the C–F bonds are long $(1.48 \pm 0.02$ Å) compared with the parent fluorocarbon (1.33 Å). Possibly the canonical forms [1.6.1] and [1.6.2] are contributing to the structure[139].

[1.6.1]   [1.6.2]

The large C–C–Mn angle of 130° lends support to this hypothesis[2].

The structures of $Fe(CF_2CF_2H)_2(CO)_4$, $Co(C_2F_4H)(CO)_3PPh_3$, $RhI(C_2F_5)(CO)Cp$ and $[(CN)_5CoCF_2CF_2H]^{3-}$ have also been determined by X-ray diffraction.

Fluoroalkyl– and fluoroaryl–transition metal complexes are much more stable than their hydrocarbon analogues (if they exist). This high thermal stability and inertness possibly come from a metal–carbon bond strength which is higher than in the hydrocarbon analogue. The highly electro-negative perfluoro group would be expected to induce a more positive charge on the metal and hence contract the metal orbitals. These could then overlap better with the orbitals on the carbon. There is also the possibility of some double bonding between the metal and the carbon ligand atom due to overlap of the filled d-orbitals on the metal with the C–F antibonding orbitals or, in valence bond terms, contributions by structures such as [1.6.1]. However, these arguments are speculative and very little is definitely known about factors controlling metal–carbon bond strengths.

1.6.3. Physical Properties

1.6.3a. *Nuclear Magnetic Resonance Spectroscopy*

^{19}F n.m.r. spectroscopy is extremely useful in this field because the large chemical shifts often allow first-order interpretation. A feature for perfluoroalkyl–transition metal complexes is that CF_2 or CF_3 groups bonded to the metal absorb 30–70 ppm downfield from the "normal" region, i.e. as found in fluorocarbons. A paramagnetic contribution to the shielding is thought to be responsible. Another unusual feature is the small spin–spin coupling frequently found for fluorine nuclei on adjacent carbon atoms. In perfluoro-phenyl complexes the *ortho*-fluorines are also strongly deshielded.

1.6.3b. *Infrared Spectra*

ν(C–F) in $CF_3Mn(CO)_5$ is lowered by *ca.* 100 cm^{-1} from the values for CF_3X (X = Cl, Br or I). This apparent reduction in bond order is thought to be due to a drift of electrons from the filled d-orbitals on the metal into the C–F antibonding orbitals[140] (see section 1.6.2). Perfluoroalkyl–metal carbonyls have higher values of ν(CO) than the corresponding hydrocarbon alkyl–metal carbonyls because the electro-negative perfluoroalkyl group will reduce back-donation from the metal to the π^*-orbitals of the carbon monoxide. The frequencies are comparable with those found for the carbonyl halides.

[139] F. W. B. Einstein, H. Luth and J. Trotter, *J. Chem. Soc.* A (1967) 89.
[140] F. A. Cotton and R. M. Wing, *J. Organomet. Chem.* 9 (1967) 511.

1.6.4. Reaction

The perfluoroalkyl- or perfluoroaryl-group have a very low order of reactivity, and in the great majority of the reactions of their complexes they remain bonded to the metal. Thus in reactions where a halide ligand is displaced (e.g. by pyridine, PPb_3, ClO_4, OEt, SCN, Me, etc.) the perfluoroalkyl (aryl) group is unaffected, e.g.

$$TiClC_6F_5Cp_2 \xrightarrow{\text{NaOEt}} Ti(OEt)C_6F_5Cp_2$$

Similarly, carbon monoxide can be displaced from perfluoroalkyl–metal carbonyls without affecting the perfluoroalkyl group:

$$(CF_3)_2Fe(CO)_4 + diphos \rightarrow (CF_3)_2Fe(CO)_2diphos$$

1.7. CARBENE COMPLEXES

1.7.1. Preparation

Treatment of the Group VI metal carbonyls with methyl- or phenyl-lithium gives anions $[(CO)_5MCOR]^-$, and these react with diazomethane to give "carbene" complexes.

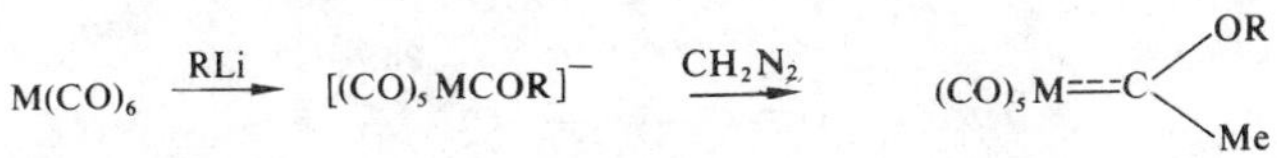

$$M(CO)_6 \xrightarrow{\text{RLi}} [(CO)_5MCOR]^- \xrightarrow{\text{CH}_2\text{N}_2} (CO)_5M{=}\!\!=\!\!C\!\!\begin{smallmatrix} \diagup OR \\ \diagdown Me \end{smallmatrix}$$

where M = Cr, Mo or W.

An improved method for the chromium complex is to treat the intermediate anion with trimethyloxonium fluoroborate in water, when the methoxy-methylcarbene complex is obtained in 85% yield[141]. Similar methods are used to make carbene complexes of manganese and rhenium, e.g.

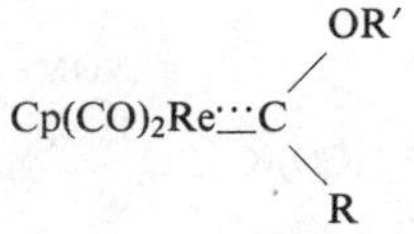

$$Cp(CO)_2Re{:}{:}C\!\!\begin{smallmatrix} \diagup OR' \\ \diagdown R \end{smallmatrix}$$

from $CpRe(CO)_3$ (R' = Me, Et or H).

A chromium carbene complex [1.7.2] is formed (80% yield) on heating 1,3-dimethylimidazolinium hydrogen pentacarbonylchromate [1.7.1]:

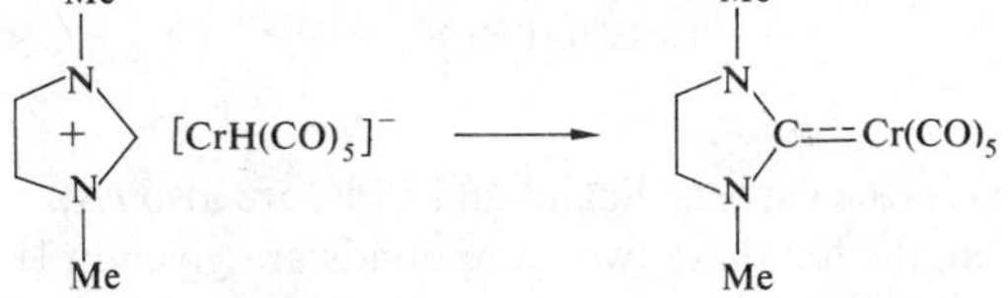

[1.7.1]   [1.7.2]

Treatment of disodium pentacarbonylchromate with 1,2-diphenyl-3,3-dichlorocyclopropene gives the unusual carbene complex [1.7.3].

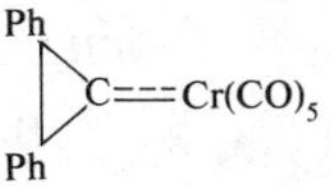

[1.7.3] 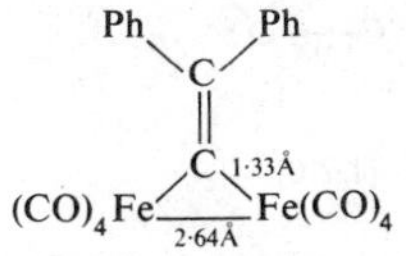   [1.7.4]

(The first example of a carbene ligand not containing a hetero-atom.)

141 R. Aumann and E. O. Fischer, *Chem. Ber.* **101** (1968) 954.

A carbene-bridged complex [1.7.4] is formed from iron pentacarbonyl and diphenyl-ketene in ultraviolet light; its structure has been determined by X-ray diffraction[142].

The preparation of other types of carbene complexes is described under "Reactions" (section 1.7.3). The possible formation of an iron–carbene complex on protonating $Fe(COCH_3)(CO)_2Cp$ has already been mentioned (section 1.5).

1.7.2. Structures and Bonding

The tungsten complex

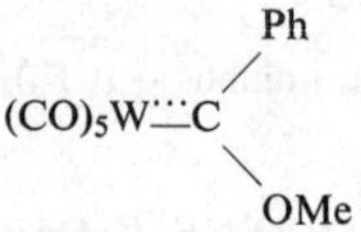

$$(CO)_5W\text{═}C\begin{smallmatrix}\nearrow Ph \\ \\ \searrow OMe\end{smallmatrix}$$

has octahedral coordination; the W–C bond to the carbene is longer (2.05 Å) than that to the carbonyls (1.89 Å). Other distances (Å) are shown in [1.7.5]. The Me is *trans* to the Ph.

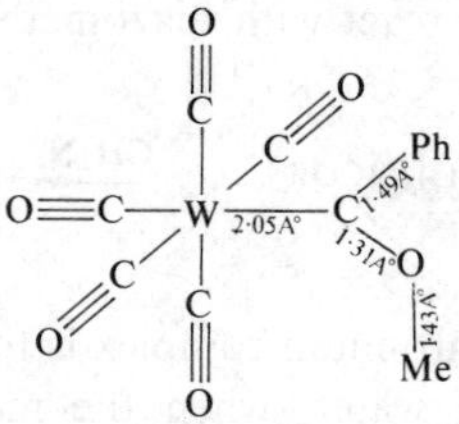

[1.7.5]

Similarly, in

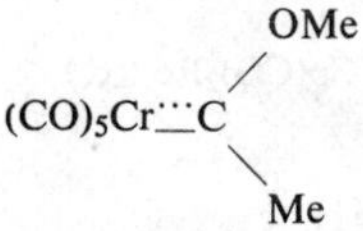

$$(CO)_5Cr\text{═}C\begin{smallmatrix}\nearrow OMe \\ \\ \searrow Me\end{smallmatrix}$$

the Me (of the OMe) is *trans* to the other Me, but in

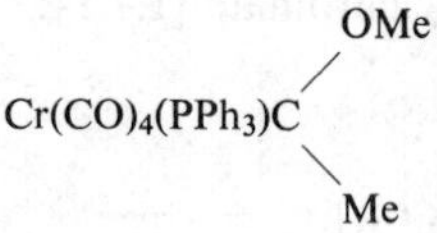

$$Cr(CO)_4(PPh_3)C\begin{smallmatrix}\nearrow OMe \\ \\ \searrow Me\end{smallmatrix}$$

the OMe and Me are *cis* (the carbene ligand and PPh₃ are also *cis*).

Important bond lengths for these two compounds are given in [1.7.6] and [1.7.7].

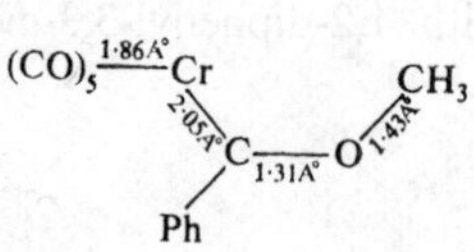

[1.7.6]

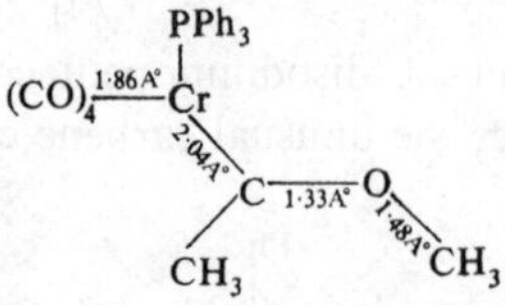

[1.7.7]

There is n.m.r. evidence for restricted rotation about the C–O (carbene) bond in these complexes[143]. It is therefore suggested that there is some double-bond character in this

142 O. S. Mills and A. D. Redhouse, *J. Chem. Soc.* A (1968) 1282.
143 E. Moser and E. O. Fischer, *J. Organomet. Chem.* 13 (1968) 209.

bond due to overlap of a lone pair on the oxygen with the vacant p_z-orbital of the carbene carbon, i.e. in VB terms there is some contribution from the structure [1.7.8]:

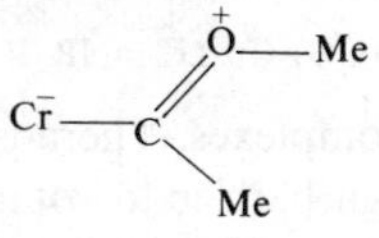

$$\overset{+}{\underset{Me}{Cr\!-\!\overset{\displaystyle \overset{O-Me}{\|}}{C}}}$$

[1.7.8]

As can be seen from the three structures [1.7.5], [1.7.6] and [1.7.7] the metal–carbene distances (2.04–2.05 Å) are much longer than the metal–CO distances (1.86 Å). The radii of sp^2- and sp-hybridized carbon differ by 0.07 Å, which accounts for part of this difference, but the results suggest that the metal–carbene bonds do not have a lot of double-bond character. The C–O distances of 1.31, 1.31 and 1.33 Å are much shorter than single-bond distances.

1.7.3. Reactions

Carbene ligands show a high degree of reactivity. Some of the reactions of $(CO)_5CrC(OMe)Me$ are outlined in Fig. 6.

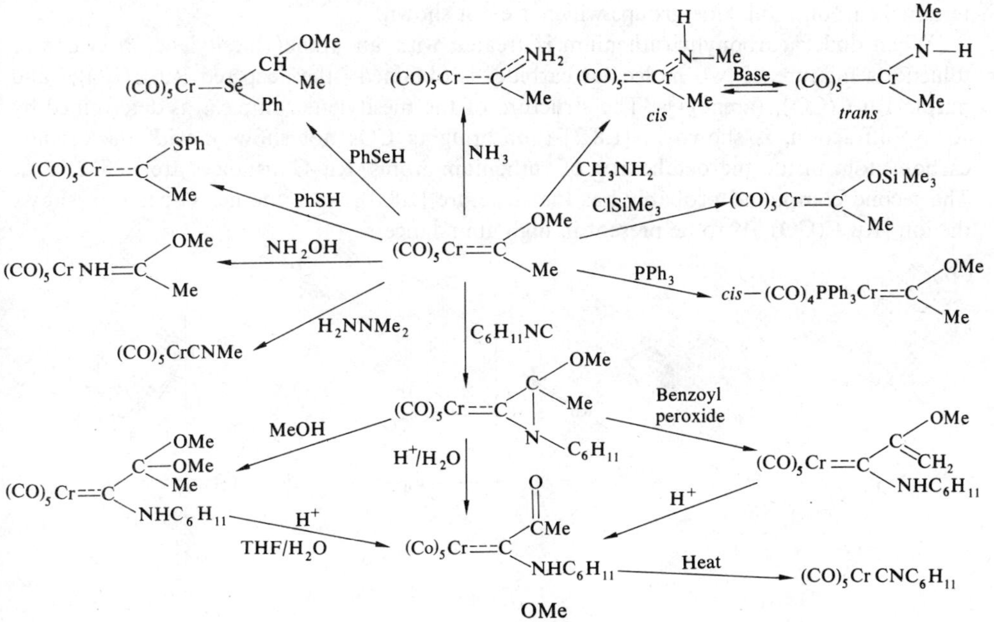

FIG. 6. Some reactions of $(CO)_5Cr\!=\!C\overset{\displaystyle OMe}{\underset{\displaystyle Me}{}}$ and its derivatives[141, 144, 145].

The kinetics of reactions with primary amines, i.e. to give $(CO)_5CrC(Me)\!=\!NHR$, show a first-order dependence in complex but a dependence on $[NH_2R]^3$, in n-decane as solvent.

144 E. Moser and E. O. Fischer, *J. Organomet. Chem.* **15** (1968) 147.
145 E. Moser and E. O. Fischer, *J. Organomet. Chem.* **12** (1968) 1.

There have been extensive ^1H n.m.r. studies on compounds of the type

$$\underset{\underset{(CO)_5Cr \doteq C=NR_1R_2,}{|}}{\overset{Me}{}}$$

including studies with ^{15}N labelled complexes. There is much double bond character in the $C=NR_1R_2$ bond and therefore a high barrier to rotation[144, 146].

The ethoxymethylcarbene complex $(CO)_5CrC(OEt)Me$ reacts with pyridine to give ethyl vinyl ether and $Cr(CO)_5py$.

1.8. MOLECULAR CARBIDES

The first example of these remarkable compounds was $Fe_5C(CO)_{15}$, obtained in small yield by heating $Fe_3(CO)_{12}$ with methylphenylacetylene or pent-1-yne in light petroleum. The structure, determined by X-ray diffraction[147], shows an approximately equilateral square pyramid of iron atoms with an Fe–Fe distance of 2.64 Å. The carbide atom, which is located slightly below the centre of the basal plane of iron atoms, is at an average distance of 1.89 Å from the four basal iron atoms and 1.96 Å from the apical iron atom. It probably originates from reduction of a carbon monoxide. Each iron atom in [1.8.1] has three terminal carbon monoxide groups which are not shown.

When dodecacarbonyltriruthenium is treated with an arene (mesitylene, *m*-xylene or toluene) a mixture of two molecular carbides is obtained—the deep red $Ru_6C(CO)_{17}$ and purple $Ru_6C(CO)_{14}(arene)$ [148]. The structure of the mesitylene complex, as determined by X-ray diffraction, is shown in [1.8.2] (non-bridging COs not shown); with the carbide carbon atom inside the octahedron of ruthenium atoms, Ru–C distances are 1.88–2.12 Å. The second compound probably has the structure [1.8.3], since the mass spectrum shows the ion $[Ru_6C(CO)_{17}]^+$ to be present in high abundance.

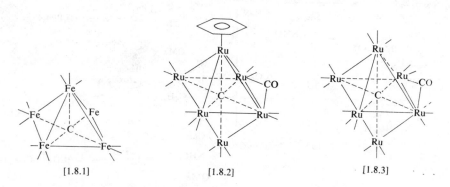

[1.8.1]   [1.8.2]   [1.8.3]

[146] E. Moser and E. O. Fischer, *J. Organomet. Chem.* **13** (1968) 387.
[147] E. H. Braye, L. F. Dahl, W. Hübel and D. L. Wampler, *J. Am. Chem. Soc.* **84** (1962) 4633.
[148] B. F. G. Johnson, R. D. Johnston and J. Lewis, *J. Chem. Soc.* A (1968) 2865.

2. OLEFIN, ALLENE AND ACETYLENE COMPLEXES

2.1. COMPLEXES WITH NON-CONJUGATED OLEFINS[149-151]

Olefin complexes are usually prepared by treating a metal complex with an olefin, although there are a few other methods of formation. It is convenient to discuss methods of synthesis in terms of the ligands being displaced.

2.1.1a. Formation by Displacement of Halide Ligands, Solvent or Weakly Coordinating Ligands such as Perchlorate or Nitrate

The famous Zeise's salt, $K[PtCl_3(C_2H_4)],H_2O$, first prepared in 1827, is made by shaking an aqueous solution of potassium chloroplatinate(II), containing hydrochloric acid, in an atmosphere of ethylene for several days. The yellow salt $K[PtCl_3(C_2H_4)],H_2O$ crystallizes out. Propylene reacts similarly, but the reaction takes even longer. The reaction is catalysed by stannous chloride, which cuts down the reaction time from days to hours[152]. Butadiene gives $[PtCl_3C_4H_6PtCl_3]^{2-}$, in which each platinum atom is thought to be bonded to one of the double bonds of butadiene[153].

Evaporation of an acid solution of Zeise's salt gives ethyleneplatinous chloride, $Pt_2Cl_4(C_2H_4)_2$, which has the structure [2.1.1]:

[2.1.1]

A good method of preparing olefin–palladium complexes is to treat a solution of palladium chloride in 50% acetic acid at 20° with the olefin. This gives complexes of type $Pd_2Cl_4(olefin)_2$, analogous to [2.1.1], with a variety of olefins such as isobutene, cyclohexene, α-methylstyrene, etc.[154]. The same reaction mixtures at 90° give π-allylicpalladium complexes (see section 3.1). An alternative route to olefin–palladium complexes is to treat palladium chloride with the olefin in benzene, often under pressure. Other solvents seem to give better results than benzene, e.g. palladium chloride suspended in either ethyl chloride or t-butyl chloride is converted completely to $Pd_2Cl_4(C_2H_4)_2$ by 10 atm of ethylene in less than a minute[155].

The (ethylene)rhodium(I) complex $Rh_2Cl_2(C_2H_4)_4$ [2.1.2] is made from rhodium trichloride and ethylene, some of the ethylene being converted into acetaldehyde. Cyclo-octene

$$2RhCl_3 + 6C_2H_4 + 2H_2O \rightarrow Rh_2Cl_2(C_2H_4)_4 + 2CH_3CHO + 4HCl$$

[2.1.2]

[149] M. A. Bennett, *Chem. Rev.* **62** (1962) 611.
[150] E. O. Fischer and H. Werner, *Angew. Chem. Int. Edn.* **2** (1963) 80.
[151] R. G. Guy and B. L. Shaw, *Advances Inorg. Chem. Radiochem.* **4** (1962) 77.
[152] R. Cramer, *Inorg. Chem.* **4** (1965) 445.
[153] M. J. Grogan and K. Nakamoto, *Inorganica chimica acta* **1** (1967) 228.
[154] R. Hüttel, J. Kratzer and M. Bechter, *Chem. Ber.* **94** (1961) 766.
[155] A. D. Ketley, L. P. Fisher, A. J. Berlin, C. R. Morgan, E. H. Gorman and T. R. Steadman, *Inorg. Chem.* **6** (1967) 657.

reacts with rhodium trichloride in cold ethanol to give $Rh_2Cl_2(C_8H_{14})_4$ but in hot ethanol $Rh_2Cl_2(CO)_2(C_8H_{14})_2$ is formed together with methane (from the breakdown of the ethanol).

Chelating diolefins react with halides to give complexes. Typical chelating diolefins include *cis,cis*-cyclo-octa-1,5-diene (COD), bicyclo[2,2,1]heptadiene (norbornadiene, NBD), dipentene, dicyclopentadiene, hexa-1,5-diene, diallyl ether, and Dewar hexamethylbenzene (DHMB). Sodium chloroplatinate in *n*-propanol reacts with COD to give $PtCl_2(COD)$ [2.1.3]. Other chelating diolefins react similarly. A convenient route to the palladium

[2.1.3] [2.1.4]

complexes of type $PdCl_2(diolefin)$ is from $PdCl_2(PhCN)_2$ and the diolefin. Although rhodium trichloride reacts with COD in ethanol to give the rhodium(I) complex $Rh_2Cl_2(COD)_2$, [2.1.4], iridium(III) or iridium(IV) chlorides give the hydro(COD)iridium(III) complex $Ir_2H_2Cl_4(COD)_2$ [2.1.5]. However, this loses hydrogen chloride reversibly in water to give the iridium(I) species $Ir_2Cl_2(COD)_2$ [156] [2.1.6], analogous to [2.1.4]. Hydrated iridium trichloride in ethanol reacts with duroquinone to give a complex $[IrHCl_2(duroquinone)]_2$, which is analogous in many ways to the complex [2.1.5].

[2.1.5] [2.1.6]

Chelating diolefins such as COD or NBD react with ruthenium trihalides in ethanol to give very insoluble halide bridged complexes of composition $[RuX_2(diolefin)]_n$. Osmium forms a similar complex with COD.

There is an extensive series of complexes of chelating mono-olefins, in which a double bond and a tertiary phosphorus or tertiary arsenic atom form a chelating system. Typical chelating mono-olefins include pent-4-enyl(dimethyl)arsine, pent-4-enyl(diphenyl)arsine, *o*-allylphenyl(diphenyl)phosphine, *o*-styryl(dimethyl)arsine. Complexes of palladium, platinum, copper, silver and gold have been prepared[157] (also of chromium, molybdenum and tungsten carbonyls; see later). The shift in frequency of the band due to $C=C$ stretch from 1640 to *ca.* 1500 cm^{-1} indicates that the C–C double bond is coordinated.

Olefins may also split halogen bridges; e.g. complexes of the type $Pt_2X_4(PR_3)_2$ react with olefins to give the complexes *cis*-$PtX_2(olefin)PR_3$ (X = Cl or Br; olefin = ethylene, propylene or but-1-ene; R = Et, Pr or Bu).

Replacement of a chloride ligand by an olefin can be promoted by aluminium chloride, which takes up the chloride as $AlCl_4^-$. Examples are:

$$MClCp(CO)_3 + C_2H_4 + AlCl_3 \xrightarrow[\text{pressure}]{\text{under}} [MCp(CO)_3(C_2H_4)]^+ \quad (M = Cr, Mo \text{ or } W)$$

$$ReCl(CO)_5 + C_2H_4 + AlCl_3 \longrightarrow [Re(CO)_5C_2H_4]^+$$

$$FeClCp(CO)_2 + C_2H_4 + AlCl_3 \longrightarrow [FeCp(CO)_2C_2H_4]^+$$

[156] G. Winkhaus and H. Singer, *Chem. Ber.* **99** (1966) 3610.
[157] M. A. Bennett, W. R. Kneen and R. S. Nyholm, *Inorg. Chem.* **7** (1968) 556.
[158] J. Chatt, N. P. Johnson and B. L. Shaw, *J. Chem. Soc.* (1964) 1662.

Examples of olefin–metal complex formation with perchlorate, nitrate, fluoroborate or solvent displacement occur mainly with silver. Silver nitrate, perchlorate or fluoroborate solutions absorb olefins to give labile complexes. Strong silver fluoroborate solutions in water will absorb large volumes of ethylene. Solid complexes with diolefins can often be isolated, but usually they easily dissociate, are low melting and are variable in composition (see section 2.1.3). The following types have been made: $Ag(mono\text{-}olefin)_2X$, $Ag(diolefin)X$, and $Ag_2(diolefin)X_2$, $X = NO_3$, ClO_4 [159]. Natural rubber seems to form complexes with silver salts.

2.1.1b. *From Metal Carbonyls*

The bonding of an olefin to a metal is similar to that of carbon monoxide, and olefins will frequently displace carbon monoxide to give an olefin complex, e.g.

$$Rh_2Cl_2(CO)_4 + COD \rightarrow Rh_2Cl_2(COD)_2$$

The displacement of carbon monoxide may be promoted by ultraviolet light:

$$MnCp(CO)_3 + C_2H_4 \xrightarrow{u.v.} MnCp(CO)_2(C_2H_4)$$

$$Cr(CO)_3(mesitylene) + C_2H_4 \xrightarrow{u.v.} Cr(CO)_2(C_2H_4)(mesitylene)$$

Similarly, iron pentacarbonyl reacts with acrylonitrile in sunlight to give

$$Fe(CO)_4(CH_2\!:\!CHCN)$$

This product can also be produced thermally from the more reactive $Fe_2(CO)_9$. The X-ray structure of $Fe(CO)_4(CH_2\!:\!CHCN)$ shows that the C:C double bond is coordinated to the iron. $Fe_2(CO)_9$ will react with many olefins to give products of the type $Fe(CO)_4(olefin)$. With butadiene the product $Fe(CO)_4(C_4H_6)$ has only one of the double bonds coordinated. Olefin acids such as acrylic acid, fumaric acid, maleic acid, cinnamic acid, etc., give products of this type, i.e. $Fe(CO)_4(olefin\ acid)$. The fumaric acid and acrylic acid complexes have been resolved into D- and L-forms, using the brucine salts[160]; the two mirror images from fumaric acid are shown in [2.1.7] and [2.1.8].

[2.1.7]   [2.1.8]

Maleic acid gives only a *meso*-form.

Chelating diolefins such as COD or NBD react with Group VI metal carbonyls to give compounds of the type $M(CO)_4(diolefin)$. Substitution is slow but may be promoted by ultraviolet light. The light induces loss of carbon monoxide and the diolefin can then attack the coordinatively unsaturated product. An interesting displacement of carbon monoxide

159 E. O. Fischer and K. Fichtel, *Chem. Ber.* **94** (1961) 1200.
160 G. Paiaro and R. Palumbo, *Gazz. chim. ital.* **97** (1967) 265.

occurs when the allyl ether derivative [2.1.9] is irradiated to give the cyclized olefin complex [2.1.10].

[2.1.9]   [2.1.10]

Group VI metal carbonyls, $M(CO)_6$, react with the chelating mono-olefin o-styryl-(diphenyl)phosphine (SP) to give complexes $M(CO)_4SP$. Alternatively, the acetonitrile $M(CO)_3(CH_3CN)_3$ or norbornadiene complexes $M(CO)_4(NBD)$ may be treated with SP to give the same product[161]. With o-allylphenyl(diphenyl)phosphine the products are of the rearranged o-propenylphenyl(diphenyl)phosphine (PP).

Similar complexes have been made from manganese- or rhenium-carbonyl halides[162].

2.1.1c. *Displacement of Other Ligands Including Other Olefins*

The displacement of ethylene from Zeise's salt or from ethylene-platinous chloride is an important method of making the corresponding complexes of higher olefins. The higher olefin will have less affinity than ethylene for platinum, but the ethylene can be displaced from the system and complete replacement brought about[163]. The displacement of ethylene from $Rh_2Cl_2(C_2H_4)_4$ by olefins can be used for making other olefin–rhodium complexes.

The displacement of coordinated nitriles is also a good method for making olefin complexes of palladium, chromium, molybdenum or tungsten. Thus the benzonitrile ligands of $PdCl_2(PhCN)_2$ are readily displaced by olefins, especially by chelating diolefins, to give olefin–palladium complexes, e.g.

$$PdCl_2(PhCN)_2 + NBD \rightarrow PdCl_2(NBD)$$

The acetonitrile derivatives of molybdenum or tungsten carbonyls, $M(CO)_3(MeCN)_3$, are easily prepared, and when reacted with chelating diolefins give complexes of the type $M(CO)_4(diolefin)$ [164].

Tertiary phosphines may be displaced by olefins to give olefin complexes. Thus platinum(0)– or palladium(0)–triphenylphosphine complexes of the type $M(PPh_3)_x$ ($x = 2$–4) react with olefins containing electro-negative substituents, to give olefin complexes of the type $M(olefin)(PPh_3)_2$ [165, 166].

$$Pd(PPh_3)_4 + olefin \rightarrow Pd(olefin)(PPh_3)_2 \quad (olefin = diethyl\ fumarate,\ tetracyanoethylene,\ etc.)$$

161 M. A. Bennett, R. S. Nyholm and J. D. Saxby, *J. Organomet. Chem.* **10** (1967) 301.
162 L. V. Interrante and G. V. Nelson, *Inorg. Chem.* **7** (1968) 2059.
163 R. A. Alexander, N. C. Baenziger, C. Carpenter and J. R. Doyle, *J. Am. Chem. Soc.* **82** (1960) 535.
164 R. B. King and A. Fronzaglia, *Inorg. Chem.* **5** (1966) 1837.
165 S. Cenini, R. Ugo, F. Bonati and G. laMonica, *Inorg. Nucl. Chem. Letters* **3** (1967) 191.
166 P. Fitton and J. E. McKeon, *Chem. Commun.* (1968) 4.

π-Allylic ligands can also be displaced by olefins, e.g.

$$Ni(C_3H_5)_2 + COD \rightarrow Ni(COD)_2 + hexa\text{-}1,5\text{-}diene$$
$$Rh_2Cl_2(C_3H_5)_4 + COD \rightarrow Rh_2Cl_2(COD)_2 + hexa\text{-}1,5\text{-}diene$$

2.1.1d. *Miscellaneous Methods of Forming Olefin Complexes, Including Reductive Olefination*

Reduction of a metal complex in the presence of an olefin can be a convenient route to an olefin complex of a metal in a low valency state. The reaction is analogous to reductive carbonylation, which is the most important method of making metal carbonyls. So far the method has been applied to the synthesis of only a few olefin complexes. Examples are given below.

Complexes of the type $Pt(olefin)(PPh_3)_2$ mentioned in section 2.1.1c can be made by reducing *cis*-$PtCl_2(PPh_3)_2$ with hydrazine in the presence of the olefin. Di(cyclo-octa-1,5-diene)nickel(0) is formed from nickel acetylacetonate and cyclo-octa-1,5-diene using an alkyl derivative of aluminium as reducing agent. The complex $Fe(COD)(cyclo\text{-}octa\text{-}1,3,5\text{-}triene)$ is prepared by the reduction of an iron halide with a Grignard reagent in the presence of the two olefins. Finally, treatment of $PtCl_2(COD)$ with isopropylMgBr gives $Pt(C_3H_7)_2(COD)$, and this with more COD in ultraviolet light gives $Pt(COD)_2$ [167].

2.1.2. Some Examples of Transition Metal–Olefin Complexes

A list of the most important types of metal–olefin complexes is now given, metal by metal (as in section 1.1.2 for alkyls and aryls). The majority of these compounds are discussed further in section 2.1.

Chromium

$Cr(CO)_5(tetracyanoethylene)$ [168]

$Cr(CO)_4(chelating\ diolefin)$ Diolefin = COD [169], NBD [170] or DHMB [171].

$Cr(CO)_4(chelating\ mono\text{-}olefin)$ Chelating mono-olefin = *o*-styryl(diphenyl)phosphine or *o*-propenylphenyl(diphenyl)phosphine [161].

Molybdenum

$Mo(CO)_4(chelating\ diolefin)$ Diolefin = NBD [170], COD [169], bicyclo[2,2,2]octa-2,5-diene, DHMB [171].

$Mo(CO)_4(chelating\ mono\text{-}olefin)$ *o*-Styryl(diphenyl)phosphine [161], *o*-propenylphenyl(diphenyl)phosphine.

$Mo(CO)_2(acrylonitrile)_2$ [172]

$Mo(MeCOCH:CH_2)_3$ From $Mo(CO)_3(MeCN)_3$ and methyl vinyl ketone [173].

$[MoX(benzoquinone)_3][Et_4N]$ and   X = halogen.
    $Mo(benzoquinone)_3$ [174]

[167] J. Müller and P. Goser, *Angew. Chem.* **79** (1967) 380.
[168] M. Herberhold, *Angew. Chem. Int. Edn.* **7** (1968) 305.
[169] E. O. Fischer and W. Fröhlich, *Chem. Ber.* **92** (1959) 2995.
[170] M. A. Bennett, L. Pratt and G. Wilkinson, *J. Chem. Soc.* (1961) 2037.
[171] E. O. Fischer, W. Berngruber and C. G. Kreiter, *Chem. Ber.* **101** (1968) 824.
[172] A. G. Massey and L. E. Orgel, *Chem. and Ind. (London)* (1961) 436.
[173] R. B. King, *J. Organomet. Chem.* **8** (1967) 139.
[174] F. Calderazzo and R. Henzi, *J. Organomet. Chem.* **10** (1967) 483.

Tungsten

W(CO)$_4$(chelating diolefin)   Diolefin = COD [169], NBD [170], DHMB [171], dimethyl(divinyl)silane[175].

W(CO)$_4$(chelating mono-olefin)   *o*-Styryl(diphenyl)phosphine[161] or *o*-propenylphenyl(diphenyl)phosphine.

[WX(benzoquinone)$_3$][Et$_4$N]   X = halogen[174].

Manganese

Mn(CO)$_2$(olefin)Cp   Olefin = C$_2$H$_4$, propylene, pent-1-ene, cycloheptene, acrylonitrile, etc. The kinetics of the displacement of the olefins by ligands such as Ph$_3$P, Ph$_2$S, *n*-Bu$_3$P, etc., have been studied[176].

{MnCp(CO)$_2$}$_2$(butadiene)   The butadiene forms a bridge[177].
[Mn(CO)$_5$(C$_2$H$_4$)]$^+$ [178]

Rhenium

ReCp(CO)$_2$(olefin)   Olefin = cyclopentene or cyclopentadiene[179].
[Re(CO)$_4$(C$_2$H$_4$)$_2$]$^+$ [180]

Iron

Fe(CO)$_4$(olefin)   Olefin = maleic acid, fumaric acid, acrylic acid[181], acrylonitrile[182], vinyl chloride, ethyl vinyl ether, styrene[183], β-chlorovinyl ketones[184].

Fe(CO)$_3$(acrylonitrile)$_2$ [185]
Fe(CO)$_3$(duroquinone)   From Fe(CO)$_5$ + duroquinone[186].
Fe(diphos)$_2$C$_2$H$_4$ From Fe(acac)$_3$, AlEt$_2$OEt, diphos and ethylene[187].

Ruthenium

Ru(CO)$_3$(chelating diolefin)   COD or various substituted COD [188].
[RuX$_2$(chelating diolefin)]$_n$ X = Cl, Br or I; chelating diolefin = COD or NBD [189].

[175] T. A. Manuel and F. G. A. Stone, *Chem. and Ind. (London)* (1960) 231.
[176] R. J. Angelici and W. Loewen, *Inorg. Chem.* 6 (1967) 682.
[177] M. Ziegler, *Z. anorg. allgem Chem.* 355 (1967) 12.
[178] M. L. H. Green and P. L. I. Nagy, *J. Organomet. Chem.* 1 (1963) 58.
[179] M. L. H. Green and G. Wilkinson, *J. Chem. Soc.* (1958) 4314.
[180] E. O. Fischer and K. Öfele, *Angew. Chem.* 74 (1962) 76.
[181] E. Weiss, K. Stark, J. E. Lancaster and H. D. Murdoch, *Helv. chim. Acta* 46 (1963) 288.
[182] S. F. A. Kettle and L. E. Orgel, *Chem. and Ind. (London)* (1960) 49.
[183] E. Koerner von Gustorf, M. C. Henry and D. J. McAdoo, *Justus Liebigs Ann. Chem.* 707 (1967) 190.
[184] A. N. Nesmeyanov, K. Ahmed, L. V. Rybin, M I. Rybinskaya and Y. A. Ustynyuk, *J. Organomet. Chem.* 10 (1967) 121.
[185] G. N. Schrauzer, *Chem. Ber.* 94 (1961) 642.
[186] G. N. Schrauzer and H. Thyret, *Angew. Chem.* 75 (1963) 641.
[187] G. Hata, H. Kondo and A. Miyake, *J. Am. Chem. Soc.* 90 (1968) 2278.
[188] R. J. H. Cowles, B. F. G. Johnson, P. L. Josty and J. Lewis, *Chem. Commun.* (1969) 392.
[189] E. W. Abel, M. A. Bennett and G. Wilkinson, *J. Chem. Soc.* (1959) 3178.

Ru(chelating diolefin)-
(cycloheptatriene)[190]

$RuCl_2(COD)(PEt_2Ph)_2$	From $[Ru_2Cl_3(PEt_2Ph)_6]Cl$ and COD [191].
$[RuCl_2(CO)(COD)]_2$	From $RuCl_3 + COD + CO$ in EtOH [192].
$RuHCl(NBD)(PPh_3)_2$	From $RuHCl(PPh_3)_3$ and NBD [193].

Osmium

$[OsCl_2(COD)]_n$ [194]	
$OsCl_2(COD)(PEt_2Ph)_2$	Similar to Ru complex[191].

Cobalt

CoCp(chelating diolefin)	Diolefin = COD or various substituted CODs [195], COT [196] or duroquinone[186].

Rhodium

$Rh_2Cl_2(mono\text{-}olefin)_4$	C_2H_4, propylene, etc.[197], cyclo-octene, methyl maleate, acrylonitrile[198], etc.
$Rh(acac)(mono\text{-}olefin)_2$	Mono-olefin = C_2H_4, propylene or vinyl chloride[199].
$RhCp(C_2H_4)_2$ [200]	
$RhCl(mono\text{-}olefin)(MPh_3)_2$	Olefin = C_2H_4, propylene, C_2F_4; M = P or As [201].
$Rh_2Cl_2(chelating\ diolefin)_2$ $RhCl(chelating\ diolefin)L$	Chelating diolefin = COD, dicyclopentadiene, COT [202], NBD, hexa-1,5-diene, 2,3-dimethyl-hexa-1,5-diene[200], *cis, trans*-cyclo-deca-1,6-diene[203]. The chlorine bridge may be split by amines, phosphines, arsines (L).
$RhCl(CO)_2(cyclic\ olefin)$	Olefin = cycloheptene, norbornene[204] or substituted fulvenes[205].
$Rh(SnCl_3)(NBD)_2$ [206]	
$Rh_2Cl_2(C_7H_2Cl_6)_2$	$C_7H_2Cl_6$ = 1,2,3,4,7,7-hexachlorobicyclo-[2,2,1]hepta-2,5-diene[207].

[190] B. D. Beregin and G. V. Sennikova, *Proc. Acad. Sci. USSR* **59** (1964) 1127.
[191] J. Chatt and R. G. Hayter, *J. Chem. Soc.* (1961) 896.
[192] S. D. Robinson and G. Wilkinson, *J. Chem. Soc.* A (1966) 300.
[193] P. S. Hallman, B. R. McGarvey and G. Wilkinson, *J. Chem. Soc.* A (1968) 3143.
[194] G. Winkhaus, H. Singer and M. Kricke, *Z. Naturforsch.* **21b** (1966) 1109.
[195] J. Lewis and A. W. Parkins, *J. Chem. Soc.* A (1967) 1150.
[196] A. Nakamura and N. Hagihara, *Bull. Chem. Soc. Japan* **32** (1959) 880.
[197] R. Cramer, *Inorg. Chem.* **1** (1962) 722.
[198] L. Porri and A. Lionetti, *J. Organomet. Chem.* **6** (1966) 422.
[199] R. Cramer, *J. Am. Chem. Soc.* **89** (1967) 4621.
[200] R. Cramer, *J. Am. Chem. Soc.* **86** (1964) 217.
[201] J. T. Mague and G. Wilkinson, *J. Chem. Soc.* A (1966) 1736.
[202] J. Chatt and L. M. Venanzi, *J. Chem. Soc.* (1957) 4735.
[203] J. C. Trebellas, J. R. Olechowski, H. B. Jonassen and D. W. Moore, *J. Organomet. Chem.* **9** (1967) 153.
[204] G. Winkhaus and H. Singer, *Chem. Ber.* **99** (1966) 3602.
[205] J. Altman and G. Wilkinson, *J Chem. Soc.* (1964) 5654.
[206] J. F. Young, R. D. Gillard and G. Wilkinson, *J. Chem. Soc.* (1964) 5176.
[207] G. Winkhaus, M. Kricke and H. Singer, *Z. Naturforsch.* **22b** (1967) 893.

Iridium

Ir_2Cl_2(chelating diolefin)$_2$ COD [156], 1,2,3,4,7,7-hexabicyclo[2,2,]-hepta-2,5-diene[207].

$IrCl(COD)PPh_3$ [208]
$Ir(acac)COD$ [209], $IrCp(COD)$ [209]
$Ir_2H_2X_4(COD)_2$ X = Cl, Br or I [209].
$Ir(SnCl_3)$(chelating diolefin)$_2$ COD or NBD [206].
$IrCl(CO)(olefin)_3$
$IrCl_2(CO)_2(olefin)_4$ }   Cycloheptene, cyclo-octene[76, 156].

$IrX(PPh_3)_2(CO)(olefin)$ X = Cl, Br, or I. Olefin = C_2H_4. Stability of Cl > I [210]. Olefin = $C_2(CN)_4$ [211], C_2F_4 [212].

$IrMe(PPh_3)_2COD$ From $Ir_2Cl_2(COD)_2 + PPh_3 + MeMgI$ [213].
$Ir(SnCl_3)(MPh_3)_2COD$ M = P or As [206].

Nickel

$Ni(COD)_2$ [214]
Ni(cyclododeca-1,5,9-triene)[214]
$Ni_x(olefin)_{2x}$ The olefins must have electro-negative substituents, e.g. acrylonitrile, fumaronitrile, acrolein, duroquinone[215].

Ni(duroquinone)(chelating diolefin)   COD, COT, dicyclopentadiene[216].
$Ni(acrylonitrile)_2PPh_3$,
   $Ni(acrylonitrile)_2(PPh_3)_2$ [216]
$Ni^IX(COD)$ X = Br or I, from $Ni_2X_2(C_3H_5)_2 + COD$ [217].
$Ni(olefin)(PR_3)_2$ Olefin = ethylene, styrene, methylstyrene, stilbene; R = Ph or Et [214].

Palladium

$Pd(olefin)(PPh_3)_2$ Olefin = maleic anhydride, diethyl fumarate, $C_2(CN)_4$ [166].

$Pd_2Cl_4(olefin)_2$ Many different olefins used (see text)[154].
$PdCl_2(C_2H_4)_2$? [155]
PdX_2(chelating diolefin)   X = halogen; chelating diolefin = COD, NBD, COT, dicyclopentadiene[218], DHMB [219].

[208] H. C. Volger, K. Vrieze and A. P. Praat, *J. Organomet. Chem.* **14** (1968) 429.
[209] S. D. Robinson and B. L. Shaw, *J. Chem. Soc.* (1965) 4997.
[210] L. Vaska and R. E. Rhodes, *J. Am. Chem. Soc.* **87** (1965) 4970
[211] J. A. McGinnety and J. A. Ibers, *Chem. Commun.* (1968) 235.
[212] G. W. Parshall and F. N. Jones, *J. Am. Chem. Soc.* **87** (1965) 5356.
[213] H. Yamazaki, M. Takesada and N. Hagihara, *Bull. Chem. Soc. Japan* **42** (1969).275.
[214] G. Wilke, *Angew. Chem. Int. Edn.* **2** (1963) 105.
[215] G. Schrauzer, *Advances Organomet. Chem.* **2** (1964) 2.
[216] G. N. Schrauzer and H. Thyret, *J. Am. Chem. Soc.* **82** (1960) 6420.
[217] L. Porri, G. Vitulli and M. C. Gallazzi, *Angew. Chem. Int. Edn.* **6** (1967) 452.
[218] J. Chatt, L. M. Vallarino and L. M. Venanzi, *J. Chem. Soc.* (1957) 2496.
[219] H. Dietl and P. M. Maitlis, *Chem. Commun.* (1967) 759.

Pd_2X_2(diene-Y), PdX(diene-Y)am	X = halogen or OAc; diene-Y = substituted diene such as COD or NBD, where Y = OMe, OAc; or conjugate base of acetylacetone, diethyl malonate or ethylacetoacetate; am = amine[13, 220].
PdX_2(chelating mono-olefin)	X = Cl or Br; chelating mono-olefin = o-allylphenyl(dimethyl)arsine or o-allyl-phenyl(diphenyl)phosphine[157].

Platinum

Pt(olefin)(PPh$_3$)$_2$	Olefin = stilbene, 4,4'-dinitrostilbene[221], acrylonitrile, maleic anhydride, methyl vinyl ketone, etc.[165].
Pt(COD)$_2$ [167]	
[PtX$_3$(olefin)]$^-$	X = halogen, olefin = ethylene, propylene, etc., allyl alcohol, various substituted styrenes[222, 223].
PrClBr(C$_2$H$_4$)NH$_3$	Using the strong *trans*-effect of ethylene, etc., all three possible isomers have been prepared[224].
PtCl(C$_2$H$_4$)py(NH$_3$)NO$_3$ [225]	
trans-PtCl$_2$(C$_2$H$_4$)$_2$	From Pt$_2$Cl$_4$(C$_2$H$_4$)$_2$ and C$_2$H$_4$ in acetone at $-70°$. Decomposes at $-10°$ [226].
cis-PtCl$_2$(C$_2$H$_4$)$_2$ [1]	
trans-PtCl$_2$(mono-olefin)L	Ethylene, propylene, substituted styrenes, etc. L = amine, pyridine N-oxides[222, 227].
cis-PtCl$_2$(mono-olefin)L	Ethylene, propylene, but-1-ene, L = tertiary phosphine[158].
PtX$_2$(chelating diolefin)	Chelating diolefin = COD, NBD, dipentene, dicyclopentadiene, DHMB [228], hexa-1,5-diene, COT. X = Cl, Br or I [13].
Pt$_2$X$_2$(diene-Y), PtX(diene-Y)am	A very extensive series, similar to Pd-compounds, but usually more stable[13].
PtX$_2$(chelating mono-olefin)	X = Cl, Br or I. Chelating mono-olefin = o-allylphenyl(diphenyl)phosphine, o-allyl-phenyldimethylarsine, pent-4-enyl(dimethyl)-arsine or pent-4-enyl(diphenyl)arsine, o-styryl(dimethyl)arsine[157, 229].

[220] C. B. Anderson and B. J. Burreson, *J. Organomet. Chem.* 7 (1967) 181.
[221] J. Chatt, B. L. Shaw and A. A. Williams, *J. Chem. Soc.* (1962) 3269.
[222] R. N. Keller, *Chem. Rev.* 28 (1941) 229.
[223] J. R. Joy and M. Orchin, *J. Am. Chem. Soc.* 81 (1959) 305.
[224] A. D. Gelman, *Compt. Rend. Acad. Sci. URSS* 38 (1944) 310; *Chem. Abs.* 38 (1944) 526.
[225] A. D. Gelman and E. A. Meilakh, *Compt. Rend. Acad. Sci. URSS* 51 (1946) 207; *Chem. Abs.* 40 (1946) 7042.
[226] J. Chatt and R. G. Wilkins, *J. Chem. Soc.* (1952) 2622.
[227] L. Garcia, S. I. Shupack and M. Orchin, *Inorg. Chem.* 1 (1962) 893.
[228] R. Mason, G. B. Robertson, P. O. Whimp, B. L. Shaw and G. Shaw, *Chem. Commun.* (1968) 868.
[229] M. A. Bennett, H. W. Kouwenhoven, J. Lewis and R. S. Nyholm, *J. Chem. Soc.* 4570 (1964).

$Pt_2H_2(SnCl_3)_2(PPh_3)_4$(diolefin)

Diolefin = COD or NBD. There are other related compounds[230].

Copper

[CuX(olefin)]$_n$

X = Cl or Br, olefin = ethylene, propylene, but-2-ene, acrolein or NBD and COT acting as monodentate ligands. Readily dissociate[231, 232].

Cu_2X_2(diolefin)

Diolefin = COD and other cyclopolyolefins, more stable than with mono-olefins[233].

Cu_2X_2(chelating mono-olefin)

Chelating mono-olefin = *o*-allylphenyl-(dimethyl)arsine, *o*-allylphenyl(diphenyl)-phosphine[234].

Silver

[Ag(olefin)$_2$]X

X = NO_3, ClO_4 or BF_4; olefin = α- or β-pinene, cyclohexene, etc. COT [151].

[Ag(diolefin)]X

Diolefin = dicyclopentadiene, COD, NBD, COT [151, 235, 236].

[Ag$_2$(diolefin)]X$_2$
[Ag$_2$(NBD)$_3$](BF$_4$)$_2$ [236]
[Ag(chelating mono-olefin)]NO_3

Cyclo-octa-1,4-diene, NBD [151].

o-Allylphenyl(diphenyl)phosphine [234]

Gold

AuCl(olefin)

Cyclopentene, cyclohexene, *cis*-cyclo-octene, long-chain terminal olefins[237, 238].

Au_2Cl_2(diolefin)

Hexa-1,4-diene, hexa-1,5-diene, COD, NBD and others[238].

Au_2Cl_4NBD, Au_2Cl_4(NBD)$_2$,
Au_2Cl_4(NBD)$_3$ [239]

2.1.3. Structures, Bonding and Stability

2.1.3a. *Structures*

It was postulated by Chatt and Duncanson from infrared and other studies that the ethylene of Zeise's salt is vertical to the plane of the complex. This was confirmed by two X-ray structural determinations on Zeise's salt and one on ethylene-platinous chloride. These determinations showed the overall geometry but did not determine the C–C or Pt–C distances accurately. The structure of *trans*-$PtCl_2(C_2H_4)NHMe_2$ also shows the ethylene to be vertical to the plane of the other ligand atoms[240].

230 H. A. Tayim and J. C. Bailar, *J. Am. Chem. Soc.* **89** (1967) 4330.
231 H. L. Haight, J. R. Doyle, N. C. Baenziger and G. F. Richards, *Inorg. Chem.* **2** (1963) 1301.
232 S. Kawaguchi and T. Ogura, *Inorg. Chem.* **5** (1966) 844.
233 J. H. van den Hende and W. C. Baird, *J. Am. Chem. Soc.* **85** (1963) 1009.
234 M. A. Bennett, W. R. Kneen and R. S. Nyholm, *Inorg. Chem.* **7** (1968) 552.
235 J. G. Traynham and J. R. Olechowski, *J. Am. Chem. Soc.* **81** (1959) 571.
236 H. W. Quinn, *Can. J. Chem.* **46** (1968) 117.
237 A. J. Chalk, *J. Am. Chem. Soc.* **86** (1964) 4733.
238 R. Hüttel and H. Reinheimer, *Chem. Ber.* **99** (1966) 2778.
239 R. Hüttel, H. Reinheimer and K. Nowak, *Tetrahedron Letters* (1967) 1019.
240 P. R. H. Alderman, P. G. Owston and J. M. Rowe, *Acta Cryst.* **13** (1960) 149.

TABLE 3. SOME OLEFIN COMPLEXES WHOSE STRUCTURES HAVE BEEN DETERMINED BY X-RAY DIFFRACTION
(C–C and M–C distances are given where known)

	C–C (Å)	M–C (Å)	Reference
$Cr(CO)_4HMDB$	1.36	2.33	G. Huttner and O. S. Mills, *Chem. Commun.* (1968) 344
$\{MnCp(CO)_2\}_2C_4H_6$	1.43		M. Ziegler, *Z. anorg. allgem. Chem.* **355** (1967) 12
$Mo(CO)_4\{\text{prop-1-enylphenyl(diphenyl)phosphine}\}$	1.40	2.45, 2.52	H. Luth, M. R. Truter and A. Robson, *J. Chem. Soc. A* (1969) 28
$Fe(\text{acrylonitrile})(CO)_4$			A. R. Luxmoore and M. R. Truter, *Acta Cryst.* **15** (1962) 1117
$Fe(\text{fumaric acid})(CO)_4$ [a]	1.42	2.04	P. Corradini, C. Pedone and A. Sirigu, *Chem. Commun.* (1968) 275
$Rh_2Cl_2(COD)_2$			J. A. Ibers and R. G. Snyder, *Acta Cryst.* **15** (1962) 923
$IrBr(CO)(PPh_3)_2\{C_2(CN)_4\}$	1.507	2.15	J. A. McGinnety and J. A. Ibers, *Chem. Commun.* (1968) 235
$Ir(SnCl_3)(COD)_2$	1.45, 1.32	2.23, 2.28, 2.16, 2.18	P. Porta, H. M. Powell, R. J. Mawby and L. M. Venanzi, *J. Chem. Soc. A* (1967) 455
$Ni(C_2H_4)(PPh_3)_2$ [b] ?	1.46	2.00, 2.02	C. D. Cook, C. H. Koo, S. C. Nyburg and M. T. Shiomi, *Chem. Commun.* (1967) 426
$Ni(C_2H_4)(PPh_3)_2$ [b] ?	1.41		W. Dreissig and H. Dietrich, *Acta Cryst.* **24B** (1968) 1324
$PdCl_2(NBD)$	1.36		N. C. Baenziger, G. F. Richards and J. R. Doyle, *Acta Cryst.* **18** (1965) 924
$Pt\{C_2(CN)_4\}(PPh_3)_2$	1.52	2.10, 2.11	C. Panattoni, G. Bombieri, U. Belluco and W. H. Baddley, *J. Am. Chem. Soc.* **90** (1968) 798
$trans\text{-}PtCl_2(C_2H_4)NHMe_2$	1.47	2.21	P. R. H. Alderman, P. G. Owston and J. M. Rowe, *Acta Cryst.* **13** (1960) 149
$Pt_2(acac)_2(allyl)_2$	1.37, 1.41	2.11, 2.12, 2.15, 2.14	W. S. McDonald, B. E. Mann, G. Raper, B. L. Shaw and G. Shaw, *Chem. Commun.* (1969) 1254
$Pt_2Cl_2(C_{12}H_{17})_2$	1.38, 1.48	2.34, 2.36, 2.16, 2.16	R. Mason, G. B. Robertson, P. O. Whimp, B. L. Shaw and G. Shaw, *Chem. Commun.* (1968) 868
$[CuCl(NBD)]_4$	1.345	2.05, 2.11	N. C. Baenziger, H. L. Haight and J. R. Doyle, *Inorg. Chem.* **3** (1964) 1535
$Cu_2Cl_2(COD)_2$	1.40, 1.41		J. H. van den Hende and W. C. Baird, *J. Am. Chem. Soc.* **85** (1963) 1009
$[CuCl(COT)]_x$	1.39	2.07, 2.095	N. C. Baenziger, G. F. Richards and J. R. Doyle, *Inorg. Chem.* **3** (1964) 1529
$[Ag_2NBD](NO_3)_2$	1.39	2.41, 2.31	N. C. Baenziger, H. L. Haight, R. Alexander and J. R. Doyle, *Inorg. Chem.* **5** (1966) 1399

[a] The absolute configuration (R,R) of the optically active complex was determined.
[b] One of these two may be the O_2 adduct and not the ethylene complex, see *Ann. Rep. Chem. Soc., Part A* (1968) 38.

The structures of several olefin–metal complexes have been determined by X-ray diffraction. Table 3 gives a list of them, together with the C–C (olefin) distances and M–C distances, where reported. The errors in C–C distances may be large, especially for the heavier metals.

As can be seen from Table 3, when the metal is in a low valency state, the C–C distances tend to be significantly longer than in the free olefin (in which it is less than 1.4 Å). This is especially so with the two complexes derived from tetracyanoethylene (TCNE). Both these complexes show interesting structural features which will now be discussed.

The structure of $Pt\{C_2(CN)_4\}(PPh_3)_2$ is shown in [2.1.11].

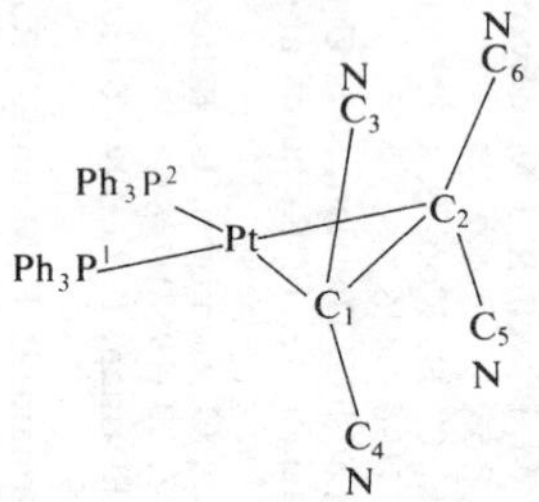

[2.1.11]

The C_1–C_2 distance of 1.52 Å is 0.18 Å longer than in TCNE itself (1.339 Å). Another feature is that C_1–C_2 is tilted 10° with respect to the plane P_1–Pt–P_2. Other similar molecules showing tilts are $Pt(CS_2)(PPh_3)_2$ (6°), $Pt(PhC\mathbin{:}CPh)(PPh_3)_2$ (14°) and, possibly, $Ni(C_2H_4)(PPh_3)_2$ (12°). The angle C_1–Pt–C_2 is only 42°. One can regard the TCNE as functioning as a strong π-acid with very little σ-donor character, i.e. it accepts a lot of electron density from the platinum into its π* orbitals. This lengthens C_1–C_2 to a single bond distance and also has the effect of decreasing the C–CN distances from 1.449 Å (free ligand) to 1.40 Å, in agreement with MO calculations. The alternative VB description of a three-membered ring σ-framework with bent bonds is considered to be a completely equivalent description.

The structure of the iridium complex $IrBr(CO)(TCNE)(PPh_3)_2$ is shown [2.1.12].

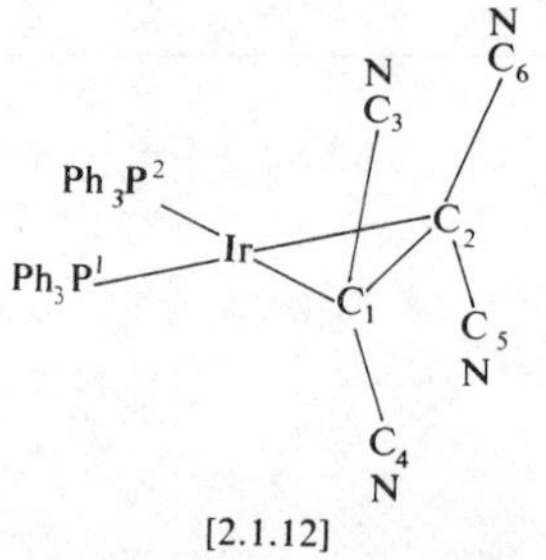

[2.1.12]

Again, C_1–C_2 (1.507 Å) is much longer than in TCNE itself, the C–CN distance (1.430 Å) is shorter and the CN groups are bent back away from the iridium. The dihedral angle between the planes C_3–C_1–C_4 and C_6–C_2–C_5 is 110°. The bonding has been discussed in terms of a three-centre MO scheme. This scheme accounts for many of the features shown by this and similar complexes.

For the complex $Ni(C_2H_4)(PPh_3)_2$ the two structural determinations (see Table 3) give significantly different results, and it has been suggested that one of the determinations was done on the oxygen complex, $Ni(O_2)(PPh_3)_2$, in error.

The structures of some other olefin–metal complexes will now be discussed.

The structure of the Dewar hexamethylbenzene complex $Cr(CO)_4(DHMB)$ is shown in [2.1.13], and confirms that the DHMB framework is retained in the complex

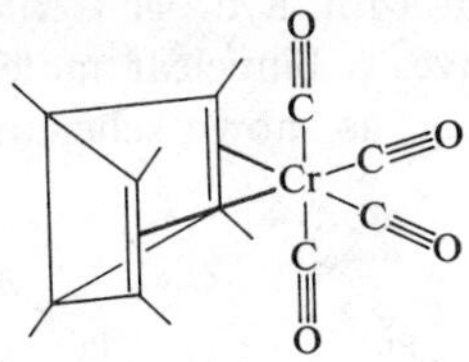

[2.1.13]

The isomerization of the double bond of 2-allylphenyl(diphenyl)phosphine on complexation with molybdenum(0) has been confirmed in the structure of the complex *cis*-(*o*-prop-l-enylphenyl)(diphenyl)phosphinemolybdenum tetracarbonyl [2.1.14].

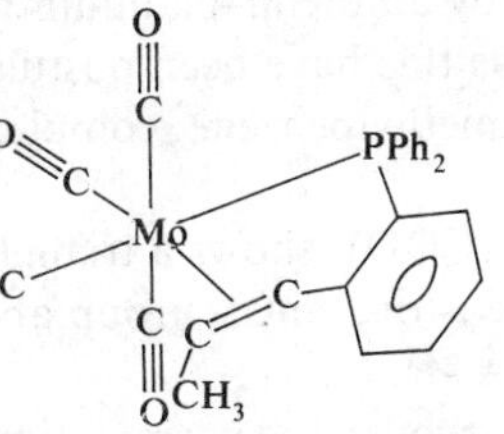

[2.1.14]

The C=C bond lies slightly out of the plane of the Mo the P and the CO's *trans* to P and C=C.

The butadiene complex of manganese, $[MnCp(CO)_2]_2C_4H_6$, has the geometry shown in [2.1.15]. The butadiene forms a bridge between the two manganese atoms.

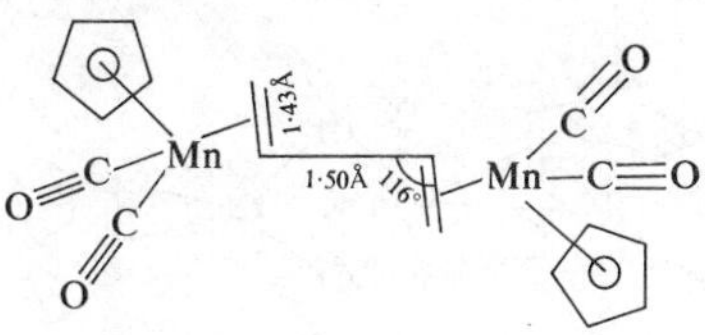

[2.1.15]

Duroquinone react with nickel tetracarbonyl to give di(duroquinone)nickel. This complex reacts with COD to give Ni(duroquinone)COD, which has the structure [2.1.16][241].

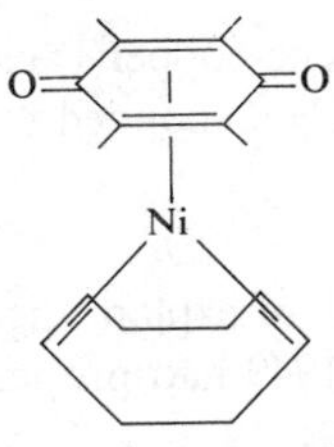

[2.1.16]

241 M. D. Glick and L. F. Dahl, *J. Organomet. Chem.* **3** (1965) 200.

The C=C distance of the duroquinone in the complex, at 1.40 ± 0.01 Å, is considerably longer than in the free ligand (1.322 Å). The nickel atom is 2.11 Å from the centre of the duroquinone C=C bonds and only 2.00 Å from the COD C=C bonds.

Treatment of bis(allyl)platinum with hydrogen chloride gives $[PtCl(C_3H_5)]_x$ which with thallium(I) acetylacetonate gives a binuclear molecule $Pt_2(acac)_2(C_3H_5)_2$. The allyl groups in this molecule are bridging, as shown schematically in [2.1.17]. One (terminal)

[2.1.17]

carbon atom of an allyl group is σ-bonded to one of the platinum atoms and the remaining two carbon atoms are bonded by an olefin–platinum bond to the other platinum atom[242]. Bridging allyl complexes such as this have been postulated as intermediates in reactions.

The dehydro Dewar hexamethylbenzene complex $Pt_2Cl_2(C_{12}H_{17})_2$ shows a similar bridging allylic system.

The structure of $Pt_3(SnCl_3)_2(COD)_3$ shows a triangle of platinum atoms with each platinum chelated by a COD. There is one $SnCl_3$ group above and one below the triangle. The average Pt–Pt distance is 2.58 Å [243].

There are several X-ray structural determinations on silver–olefin complexes, e.g. on $NBD \cdot 2AgNO_3$, a cyclo-octatetraene dimer complex $C_{16}H_{16}AgNO_3$, cyclononatriene$\cdot AgNO_3$, (bullvalene)$_3AgBF_4$, (bullvalene)$AgBF_4$ [244], humulene$\cdot 2AgNO_3$ [245] and others. The structures are often complicated, sometimes with several different silver–(C=C) distances in the same complex.

In the complex $Cu_2Cl_2(COD)_2$ the copper atoms are in a tetrahedral environment, as shown in [2.1.18].

[2.1.18]

In $[CuCl(NBD)]_4$ four coppers and four chlorines form an eight-membered tub-shaped ring in the tetrameric unit and only one double bond of a norbornadiene is coordinated to each copper.

In the structure of cyclo-octatetraene-copper(I) chloride there are continuous chains of copper and chlorine atoms with one C=C bonded to each copper atom.

2.1.3b. The Metal–Olefin Bond

The theories proposed before 1951 to explain the nature of the transition metal–olefin bond were not satisfactory. Chatt in 1949 had proposed that filled d-orbitals on the metal

242 W. S. McDonald, B. E. Mann, G. Raper, B. L. Shaw and G. Shaw, *Chem. Commun.* (1969) 1254.
243 L. J. Guggenberger, *Chem. Commun.* (1968) 512.
244 J. S. McKechnie and I. C. Paul, *J. Chem. Soc.* B (1968) 1445.
245 A. T. McPhail and G. A. Sim, *J. Chem. Soc.* B (1966) 112.

were essential for coordination of the olefin. Dewar then proposed the now famous MO description of the bonding in silver–olefin complexes. This was modified by Chatt and Duncanson[246] to explain the nature of the bonding in platinum–olefin complexes. They also explained why the olefin was vertical to the plane of the complex in terms of a mixing in of the vacant $6p_z$-orbital into the $5d_{xz}$, thereby giving better overlap with the π^*-orbitals of the olefin. The essential feature of the Dewar–Chatt–Duncanson theory of metal–olefin bonding is that in addition to donation of π-electrons from the olefin to a vacant orbital on the metal there is a back donation from the metal into the π^*-orbitals of the olefin. The Dewar bonding description for silver–olefin complexes is summarized in [2.1.19] in which a

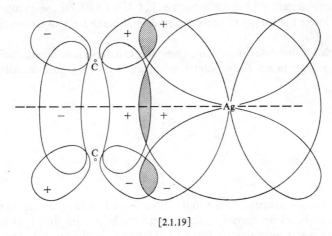

[2.1.19]

σ-type donor bond is formed by overlap of the π-$2p$ MO of the olefin and a vacant $5s$-orbital on the silver and a π-type acceptor bond is formed by overlap of the vacant π^*-$2p$ MO of the olefin and a filled $4d$-orbital of the silver. The Chatt–Duncanson picture of the bonding in platinum–olefin complexes is shown schematically in [2.1.20] in which the σ-type donor bond is formed by overlap of π-$2p$ with a vacant $5d6s6p^2$ hybrid on the platinum and the π-type acceptor bond is formed by overlap of the π^*-$2p$ MO of the olefin with a filled $5d6p$ hybrid of the platinum.

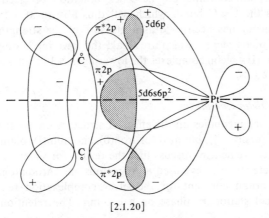

[2.1.20]

246 J. Chatt and L. A. Duncanson, *J. Chem. Soc.* (1953) 2939.

The bond between olefins and metals has also been called a μ-bond[247]. Since there is both donation and acceptance of electrons by the olefin, the polarity of the metal–olefin bond is probably low; an estimate of the polarity of the platinum(II)–olefin bond puts it at *ca.* 2 D. It is thought that donation of electrons from the π-orbitals of the olefin to the metal will enhance back donation from the metal to the π^*-orbitals on the olefin and that there is a synergic effect.

With alkyl substituted olefins one would expect the π-electrons to be more readily donated to the metal (since alkyl substituents lower the ionization potential of the olefin) but such olefins would be less able to accept electrons back into their π^*-orbitals. On the other hand, electro-negative substituents such as F, CN, COOH, etc., will reduce the σ-donor power but will lower the energy of the π^*-orbitals and make back donation from the metal easier.

If back donation becomes dominant, then one can represent metal–olefin bonding in VB terms as two single bonds (bent bonds), as shown in [2.1.21], i.e. a metallacyclopropane structure.

[2.1.21]

The bonding is thought to approximate to this situation when the reacting metal is in a low valency state and the olefin carries electro-negative substituents. In an olefin complex of type [2.1.21] the metal increases its formal oxidation number by $+2$. Thus in the tetra-cyanoethylene complex $Pt\{C_2(CN)_4\}(PPh_3)_2$ the C–C (olefin) distance is that of a single bond, the CN groups are bent away from the platinum and the two phosphorus atoms, the platinum and the two platinum-bonded carbons are nearly coplanar, as in platinum(II) complexes but not as in four-coordinated platinum(0) complexes. The slight deviation ($10°$) from co-planarity might be due to a little participation of the metal-orbital in the bonding.

The Dewar–Chatt–Duncanson description of the metal–olefin bond accounts for the lowering of $\nu(C{=}C)$ on coordination, i.e. back donation of electrons into the olefin π^*-orbitals weakens the C=C bond, which will thus absorb at a lower frequency. The change in this frequency has been used as a measure of the strength of the metal–olefin bond. In silver–olefin complexes, which are weak, the lowering is only 50–60 cm^{-1}, but in the stronger platinum(II)–olefin complexes the lowering is > 100 cm^{-1} and sometimes much greater (see section 2.1.4).

2.1.3c. *Stability of Olefin Complexes*

There have been many studies on the effect of structure on the stability of olefin complexes, particularly for silver(I), but also for iron(0), rhodium(I), platinum(II), platinum(0) and copper(I). It is convenient to discuss first the results for silver(I).

The most complete studies have used an ingenious gas chromatographic method. This measures the equilibrium constant for the silver complexation reaction, using a silver nitrate/ethylene glycol stationary phase on a column. The retention times give a rapid

[247] P. L. Pauson, *Proc. Chem. Soc.* (1960) 297.

method of measuring the equilibrium constant K of the following reaction, occurring within the liquid phase,

$$AgNO_{3_{glycol}} + olefin \rightleftharpoons (olefin)AgNO_{3_{glycol}}.$$

A large number (> 100) of aliphatic, acyclic and alicyclic olefins have been studied in this way. Table 4 gives K-values for a selection of olefins and also K_L-values, where $K_L =$ the partition coefficient for the olefin on a pure ethylene glycol column.

TABLE 4. VALUES OF K (l/mole) FOR THE REACTION
$AgNO_3 + olefin \rightleftharpoons olefin \cdot AgNO_3$,
AS STUDIED BY A g.l.c. METHOD AT 40°
$K_L =$ Partition coefficient for olefin on pure ethylene glycol column [248]

	K	K_L		K	K_L
Ethylene	22.3	0.1	Cyclopentene	7.3	5.8
Propylene	9.1	0.4	Cyclohexene	3.6	14.7
But-l-ene	7.7	0.9	Cycloheptene	12.8	27.0
cis-But-2-ene	5.4	1.1	cis-Cyclo-octene	14.4	56.1
trans-But-2-ene	1.4	1.0	trans-Cyclo-octene	1000	56.1
cis-Hex-3-ene	3.9	4.1	Penta-1, 4-diene	10.2	2.9
trans-Hex-3-ene	1.0	4.1	Hexa-1,5-diene	28.8	5.1
2,3-Dimethylbut-2-ene	0.1	4.5	Hepta-1,6-diene	14.7	9.9

Some of the conclusions from this work are as follows:

(i) Increasing alkyl substitution about the double bond lowers the affinity for all classes of olefins.

(ii) A branched substituent lowers K more than a straight-chain substituent. Thus for olefins of the types $RCH:CH_2$ and $RCH:CHR$, K decreased in the order of

$$R = Me > Et > i\text{-}Pr > t\text{-}Bu,$$

which is the order of increasing bulkiness (Me $\rightarrow$ t-Bu) and increasing electron density on the double bond. Increasing the electron density would strengthen the σ-bond to the metal but weaken the π-acceptor bond. Thus if the π-acceptor bonding is dominant it would explain the observed order. However, one can also explain the observed order in terms of increasing steric hindrance and this is the explanation that has been favoured[248].

(iii) cis-Acyclic olefins form more stable complexes than trans-acyclic olefins. This is attributed to a reduction in steric strain on argentation, i.e. two cis groups become less close. The order of stability is the same as the order of heats of hydrogenation, i.e. cis $>$ trans, for which a similar explanation has been given. trans-Cyclo-octene forms a much more stable complex than cis-cyclo-octene and also has a very much greater heat of hydrogenation (9.3 kcal mole⁻¹ greater). It is easy to separate a mixture of trans- and cis-cyclo-octene using the silver complex.

(iv) K is much higher for a 1,5-diene than for a 1,4-, 1,6-, 1,7- or 1,8-diene. This could be due to the chelate effect, but often potentially chelating diolefins do not give complexes in which the diolefin is chelating (see section 2.1.3a on structures).

[248] M. A. Muhs and F. T. Weiss, J. Am. Chem. Soc. 84 (1962) 4697.

(v) For cyclic cis-olefins the order of affinities is $C_8 > C_7 > C_5 > C_6$.

There have also been several studies on the affinities of olefins for silver(I) using a distribution method. The equilibrium constant for the following reaction

$$(\text{olefin})_{\text{CCl}_4} + \text{AgNO}_{3_{\text{H}_2\text{O}}} \rightleftharpoons (\text{olefin} \cdot \text{AgNO}_3)_{\text{H}_2\text{O}}$$

was measured, either using spectrophotometry or titrimetry. It was found that increasing substitution lowers the K value, e.g. for hex-1-ene K is 860 but for isobutene it is only 61.7. Hexa-1,5-diene gives a very high K value (1850) but conjugated dienes such as 2,3-dimethyl-butadiene ($K = 22.5$) form much less stable complexes. The formation constant for the cis-pent-2-ene complex is approximately twice that for the trans-pent-2-ene complex[249].

The order of affinities of the cyclic olefins for silver(I) in these aqueous/CCl$_4$ systems is $C_5 > C_7 > C_6 > C_8$, i.e. a different order from that found using the g.l.c. method described above. Clearly the solvating ligands round the silver (water or ethylene glycol) affect the relative order of affinities of the olefins for the silver[235].

Cyclopropene, a highly strained olefin, has a very large argentation constant, ca. 10^7, and is quantitatively taken up by aqueous silver nitrate solution.

An interesting extension of the g.l.c. method for studying the affinities of olefins for silver has been a comparison of the relative affinities of deuteriated and non-deuteriated olefins. Remarkably, deuteriated olefins show slightly greater affinity due to a more negative ΔH value on complexation. Ratios of K_D/K_H as high as 1.157 are found (for trans-CD$_3$CD:CDCD$_3$ and CH$_3$CH:CHCH$_3$) [250]. The nature of these secondary isotope effects has not been explained satisfactorily.

Measurements on the stabilities of copper(I)–olefin complexes also show that increasing alkyl substitution on the olefin lowers the stability. With copper(I), conjugated dienes such as butadiene and isoprene form more stable complexes than mono-olefins. Industrial processes to separate dienes from mono-olefins and paraffins are based on this difference in stability[251].

There have also been some valuable and extensive measurements on the affinities of olefins for rhodium(I). Rh(acac)(C$_2$H$_4$)$_2$ is a thermodynamically stable but kinetically labile complex, and when treated with another olefin, some ethylene is displaced. In a closed vessel, equilibration is rapid and the amount of ethylene displaced can be measured by infrared spectroscopy. The affinities of ethylene and propylene have also been compared by an n.m.r. method, with good agreement between the two methods.

The equilibrium constants for the reaction

$$\text{Rh(acac)(C}_2\text{H}_4)_2 + \text{olefin} \rightleftharpoons \text{Rh(acac)(C}_2\text{H}_4)(\text{olefin}) + \text{C}_2\text{H}_4$$

are given in Table 5.

The following conclusions can be drawn.

(i) Increasing alkyl substitution about the double bond lowers the affinity of the olefin, i.e. ethylene > propylene; but-1-ene > but-2-enes. The effects of alkyl substitution are greater than for silver (about ten times as great in K ratios). It was concluded that π-bonding is more important than σ-bonding in rhodium–olefin complexes[199].

(ii) The complex with cis-but-2-ene is more stable than with trans-but-2-ene.

[249] S. Winstein and H. J. Lucas, J. Am. Chem. Soc. 60 (1938) 836.
[250] R. J. Cvetanović, F. J. Duncan, W. E. Falconer and R. S. Irwin, J. Am. Chem. Soc. 87 (1965) 1827.
[251] E. R. Atkinson, D. Rubinstein and E. R. Winiarczyk, Ind. Eng. Chem. 50 (1958) 1553.

TABLE 5. SOME EQUILIBRIUM CONSTANTS FOR THE REACTION
Rh(acac)(ethylene)$_2$ + olefin $\rightleftharpoons$ Rh(acac)(ethylene)(olefin) + ethylene
Errors $\pm 10\%$ [199]

Olefin	K_{eq} (25°)	Olefin	K_{eq} (25°)
Propylene	0.078	CH$_2$:CHCl	0.170
But-l-ene	0.092	CH$_2$:CHF	0.32
cis-But-2-ene	0.0041	CHF:CF$_2$	1.24
trans-But-2-ene	0.002	CF$_2$:CF$_2$	59
Isobutene	0.00035	cis-CHCl:CHCl	0.070

(iii) With the exception of vinyl fluoride, fluoro-olefins form more stable complexes than olefins, but chloro-olefin complexes have very low stability, e.g. CH$_2$:CCl$_2$ or CCl$_2$:CCl$_2$ show no reaction with Rh(acac)(C$_2$H$_4$)$_2$.

Measurements at 0° and at 25° enabled some estimate of enthalpy and entropy terms. The errors were large, but it would appear that the displacement of ethylene by propylene is slightly endothermic and that all ethylene displacements are accompanied by a decrease in entropy.

Ethylene will displace the coordinated ethylene from Rh(acac)(ethylene)$_2$ extremely rapidly (the half-life of coordinated ethylene can be $< 10^{-4}$ sec), but tetrafluoroethylene, in spite of having such a great affinity for rhodium(I), displaces ethylene relatively slowly. The rate of displacement is less than $10^{-6} \times$ that of the ethylene exchange. It is thought therefore that σ-donation into the vacant p-orbital of Rh(acac)(ethylene)$_2$ is an important factor determining the displacement rate.

The affinities of simple olefins for platinum(II) have been compared by measuring the equilibrium constant for the following displacement spectrophotometrically[252]:

$$[(PhCH:CH_2)PtCl_3]^- + olefin \rightleftharpoons [(olefin)PtCl_3]^- + PhCH:CH_2$$

Typical equilibrium constants for this displacement are cis-4-methylpent-2-ene (0.95), trans-4-methylpent-2-ene (0.34) and cis-pent-2-ene (0.14), i.e. as before the complex from the cis-olefin is more stable than that from the trans-olefin. Surprisingly, the complex from cis-4-methylpent-2-ene is more stable than that from cis-pent-2-ene.

There have also been extensive studies on platinum–olefin complex formation, with olefins of the type RRC:CR(CH$_2$)$_n$ER$_3^+$ (E = N, P or As; R = H or alkyl; n = 1 or 2)[253]. Some results are shown in Table 6.

As can be seen, substitution of an olefinic H by CH$_3$ decreases K. Substitution of H by CH$_3$ in the α-position causes a smaller decrease in K. Disubstitution causes further decrease in K.

Since the ionization potentials of olefins decrease as alkyl substitution increases, one would expect the σ-donor power to increase, i.e. ΔH to become more negative on substitution. In fact ΔH becomes less negative. It was therefore concluded that the π-acceptor bonding is more important than the σ-donor bonding and that steric hindrance can weaken the bonding. The stability constants for allyl alcohol and its methyl derivatives as ligands

[252] J. R. Joy and M. Orchin, J. Am. Chem. Soc. **81** (1959) 310.
[253] R. G. Denning, F. R. Hartley and L. M. Venanzi, J. Chem. Soc. A (1967) 328.

TABLE 6. SOME EQUILIBRIUM CONSTANT K; ΔH AND ΔS VALUES FOR THE DISPLACEMENT
$[PtCl_4]^{2-} + olefin^+ = (olefin)PtCl_3 + Cl^-$

"Olefins" are of the type $RRC:CR \cdot CHR\overset{+}{N}H_3$ [253]

Olefin	K (60°)	ΔH (kcal mole^{-1})	ΔS (eV)
$CH_2:CH \cdot CH_2\overset{+}{N}H_3$	1022 (59°)	-7.1 ± 0.2	-7.6 ± 0.7
$CH_2:CH \cdot CHMe\overset{+}{N}H_3$	804	-6.7 ± 0.3	-6.9 ± 1.0
$MeCH:CH \cdot CH_2\overset{+}{N}H_3$	209	-5.1 ± 0.25	-4.6 ± 0.7
$CH_2:CMeCH_2\overset{+}{N}H_3$	3.2		
$Me_2C:CH \cdot CH_2\overset{+}{N}H_3$	2.6		

were also measured. Again, substitution of hydrogen by methyl lowers the stability constant.

The affinities of a large number of substituted styrenes for platinum(II) have been compared by measuring the equilibrium constants for the reaction

$$dodecene + PtCl_2(XC_6H_4CH:CH_2)L \rightleftharpoons PtCl_2(dodecene)L + XC_6H_4CH:CH_2$$

where $XC_6H_4CH:CH_2$ is a 3- or 4-substituted styrene, $X = H$, CH_3O, CH_3, Cl or NO_2, and L is a 4-substituted pyridine N-oxide[254]. The results are difficult to summarize concisely. 4-Nitrostyrene forms more stable complexes than 4-methoxystyrene, but the ratio of their equilibrium constants depends on the pyridine N-oxide, e.g. with 4-carbomethoxy-pyridine N-oxide the ratio is 1.7, but it rises to 150 with pyridine N-oxide itself and to 512 for 4-methylpyridine N-oxide. In fact, substituents on the pyridine N-oxide have a greater effect on the stability constant of the platinum–olefin complex than substituents on the styrene. Attempts to correlate the stability with Hammett parameters were not successful. Clearly the factors which determine the degree of formation of a platinum(II)–olefin are not understood.

With olefin complexes of platinum(0) of the type $Pt(olefin)(Ph_3P)_2$ preparative work shows that electron withdrawing substituents on the olefin increase the stability of the complex, e.g. *trans*-4,4'-dinitrostilbene forms a more stable complex than *trans*-stilbene itself[221]. Similarly, with iron(0)–olefin complexes of the type $Fe(CO)_4(olefin)$ a high thermal stability correlates with a high ionization potential for the olefin, i.e. poor donor but good acceptor properties increase the stability of the complex. Thus the acrylonitrile complex is one of the most stable, the ethylene complex the least stable and the complexes of styrene or vinyl chloride are of intermediate stability[183].

The affinities of some chelating diolefins for rhodium(I) have been compared by measuring the equilibrium constant of the reaction

$$RhCl(diolefin) + diolefin^* \rightleftharpoons RhCl(diolefin^*) + diolefin$$

(the rhodium complexes are actually chloro-bridged dimers). The diolefins studied were NBD (A), bicyclo[2,2,2]octa-2,5-diene (B), 2,3,5,6-tetramethyl-7,8-bis(trifluoromethyl)-

[254] S. I. Shupack and M. Orchin, *J. Am. Chem. Soc.* **86** (1964) 586.

bicyclo[2,2,2]octa-2,5,7-triene (C), hexamethylbicyclo[2,2,0]hexa-2,5-diene (D), COD (E), COT (F), 2,3-dicarbomethoxybicyclo[2,2,1]heptadiene (G) and others[255].

The order of affinities of these diolefins for rhodium(I) is A > B > C ⩾ E > G > D > F. Some of the ratios of the equilibrium constants are A/B = 3, B/C = 2, E/G = 80, G/D = 100 and D/F = 60.

It was also found that norbornadiene (NBD) exchanges extremely rapidly with Rh_2Cl_2(NBD) even at $-40°$, probably via a pentacoordinate species, RhCl(NBD)$_2$. Bicyclo[2,2,2]octa-2,5-diene similarly exchanges rapidly on the n.m.r. time scale, but the other chelating diolefins C → G did not, even at 100°.

2.1.4. Infrared and Nuclear Magnetic Resonance Spectroscopy Including Applications to Olefin Rotation and Exchange

2.1.4a. *Infrared Spectra*

In the infrared absorption spectrum of ethylene-platinous chloride there is a band at 1506 cm^{-1} which Chatt and Duncanson[249] assigned to ν(C=C) lowered by 117 cm^{-1} from the value in free ethylene (in the Raman spectrum). However, others assigned this band to a CH$_2$-deformation mode and interpreted the bonding in terms of a platina-cyclopropane structure[256]. Propylene has an infrared absorption band at 1652 cm^{-1} assigned to ν(C=C) and in the complex K[PtCl$_3$(C$_3$H$_6$)]H$_2$O a band at 1504 cm^{-1} has also been assigned to ν(C=C) lowered by 148 cm^{-1} by coordination. For perdeuteriopropylene, ν(C=C) is at 1588 cm^{-1} and in K[PtCl$_3$(C$_3$D$_6$)]H$_2$O the corresponding band is at 1416 cm^{-1}, i.e. lowered by 172 cm^{-1}. Since complex formation causes similar drops in frequency it was considered that the assignment of the bands at 1504 cm^{-1} and 1416 cm^{-1} as C=C stretches and not CH$_2$- or CD$_2$-deformations, to be essentially correct[257]. However, a more recent analysis[258] of the infrared spectrum of Zeise's salt suggests that the weak band at 1526 cm^{-1} is due to a C=C stretching mode coupling strongly with a CH$_2$-scissoring mode. In the tetradeuterioethylene complex, K[PtCl$_3$(C$_2$D$_4$)], the corresponding band is at 1428 cm^{-1}. In the ethylene complex the CH$_2$-scissoring vibration occurs at 1067 cm^{-1} and this shifts to 978 cm^{-1} in the tetra-deuteriated analogue. A band at 407 cm^{-1} was assigned to platinum–ethylene stretch, and this shifts on deuteriation to 387 cm^{-1}.

Many platinum(II) complexes with the higher olefins show bands at *ca.* 1500 cm^{-1}, which have been assigned to ν(C=C). These bands are still present in platinum complexes of non-terminal olefins, i.e. when there is no CH$_2$-scissoring mode to couple with. Some examples are given in Table 7, and give strong support to the original assignment of Chatt and Duncanson.

Dewar hexamethylbenzene has no hydrogens attached to the double-bonded carbons and has therefore little possibility of coupling of its C=C stretch with hydrogenic vibrations. In its complexes, however, a medium intensity infrared absorption band, within the range 1520–1550 cm^{-1}, is observed {for Cr(0), Mo(0), W(0), Rh(I), Pd(II) and Pt(II)}. It must be due to ν(C=C), lowered in frequency from the value in the free ligand (1681 cm^{-1}).

A successful attempt has been made to correlate changes in ν(C=C) for coordinated ethylene with changes in the electron releasing power of the platinum(II) to which it is

[255] H. C. Volger, M. M. P. Gaasbeck, H. Hogeveen and K. Vrieze, *Inorganica chimica acta* **3** (1969) 145.
[256] A. A. Babushkin, L. A. Gribov and A. D. Gelman, *Russian J. Inorg. Chem.* **4** (1959) 695.
[257] D. M. Adams and J. Chatt, *J. Chem. Soc.* (1962) 2821.
[258] M. J. Grogan and K. Nakamoto, *J. Am. Chem. Soc.* **88** (1966) 5454.

TABLE 7. VALUES OF ν(C=C) IN SOME PLATINUM(II)–OLEFIN COMPLEXES

	ν(C=C) cm^{-1}	Reference
Pt$_2$Cl$_4$(cis-but-2-ene)$_2$	1505m	H. B. Jonassen and W. B. Kirsch, J. Am. Chem. Soc. **79** (1957) 1279
Pt$_2$Cl$_4$(cis-4-methylpent-2-ene)$_2$	1490m ⎫	J. R. Joy and M. Orchin,
Pt$_2$Cl$_4$(trans-4-methylpent-2-ene)$_2$	1513s ⎭	J. Am. Chem. Soc. **81** (1959) 310
K[PtCl$_3$(cis-but-2-ene)]	1505	D. B. Powell and N. Sheppard, J. Chem. Soc. (1960) 2519

coordinated[259]. Thus in a series of substituted pyridine N-oxide (Q) derivatives of type trans-PtCl$_2$(C$_2$H$_4$)Q, ν(C=C) decreases with increasing electron releasing power of the 4-substituted pyridine N-oxide. As can be seen from Table 8, there is excellent correlation between ν(C=C) and the pK_a of the substituted pyridine N-oxide (which is also linearly related to the Hammet σ-value of the 4-substituent). There is, in fact, a linear relationship between pK_a and ν(C=C).

TABLE 8. VALUES OF ν(C=C) (COORDINATED ETHYLENE) FOR A SERIES OF COMPLEXES OF TYPE trans-PtCl$_2$(C$_2$H$_4$)(4-SUBSTI-TUTED PYRIDINE N-OXIDE) AND pK_a VALUES FOR THE CORRES-PONDING 4-SUBSTITUTED PYRIDINE N-OXIDES [259]

4-substituent	pK_a	ν(C=C) cm^{-1}
OMe	2.05	1490
CH$_3$	1.29	1500
H	0.79	1510
Cl	0.36	1515
COOMe	−0.41	1528
NO$_2$	−1.7	1545

Thus the stronger the electron donor power of the 4-substituent the larger the drop in ν(C=C) from the value of free ethylene. It was also considered that the greater the electron-releasing power of the 4-substituent the stronger the platinum–ethylene bonding because of increased back-donation to the π^*-orbitals of the ethylene, i.e. the π-acceptor bonding is more important than the σ-donor bonding.

In the series of complexes formed by potentially chelating mono-olefins such as o-allylphenyl(diphenyl)phosphine, o-styryl(diphenyl)phosphine, etc., with platinum(II), palladium(II), silver(I), copper(I), chromium(0), molybdenum(0) and tungsten(0) the double bond was shown to be coordinated (or otherwise) by a lowering (or otherwise) of ν(C=C) [157]. For platinum(II) a drop of ca. 140 cm^{-1} was found.

2.1.4b. Nuclear Magnetic Resonance Spectra

^{1}H n.m.r. spectroscopy is used a great deal in structural studies on metal–olefin complexes and many recent papers give n.m.r. data. Nuclear magnetic resonance has also been

[259] S. I. Shupack and M. Orchin, J. Am. Chem. Soc. **85** (1963) 902.

used to study olefin rotation and fast olefin exchange in metal complexes. This is discussed later.

Sometimes it is found that the τ-values for olefinic hydrogens are similar in the complexed and uncomplexed olefin, e.g. with Zeise's salt. However, the resonances of olefinic hydrogens are more commonly shifted upfield (sometimes considerably so) on complex formation. There is not the space to discuss all the results here. However, a full analysis of the n.m.r. data for the two rhodium complexes $RhCp(C_2H_4)_2$ and $RhCp(C_2H_4)SO_2$ has been given, and by way of example these results are given in Table 9. The τ-values for the ethylenic hydrogens in the complexes are much higher than in free ethylene. The values of $J_{1,2}$ and $J_{1,3}$ decrease on coordination.

2.1.4c. Olefin Rotation in Metal Complexes

This was first observed with some ethylene–rhodium(I) complexes using ^1H n.m.r. spectroscopy[200]. The systems have been studied more precisely since then[260].

At room temperature $RhCp(C_2H_4)_2$ shows a resonance due to Cp at 4.5τ and two broad peaks at 8.73 and 6.88τ. These broad peaks arise from pairs of non-equivalent ethylene protons undergoing rotation at an intermediate rate on the n.m.r. time scale. The rotation of the ethylenes is around the axes shown in [2.1.22].

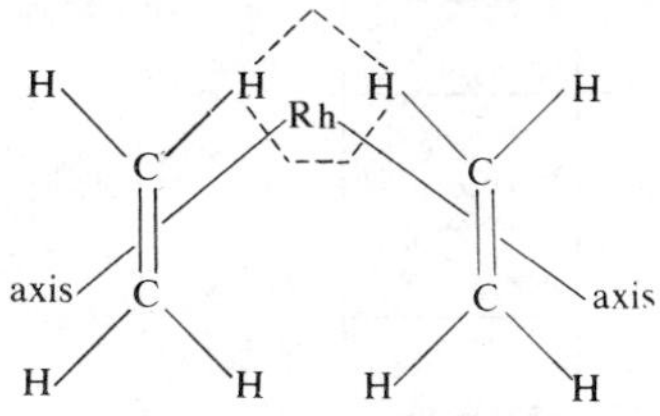

[2.1.22]

At $-25°$ the rotation is slow and a lot of fine structure is observed (see Table 9). As the solution warms up, the ethylenes rotate faster around the axes shown in [2.1.22]. From changes in the n.m.r. spectrum the activation energy for the rotation has been calculated to be 15.0 ± 0.2 kcal mole^{-1}. Similar studies on the ethylene rotations in $RhCp(C_2H_4)SO_2$ and $RhCp(C_2H_4)(C_2F_4)$ give activation energies of 12.2 ± 0.8 and 13.6 ± 0.6 kcal mole^{-1} [260] respectively.

Variable temperature n.m.r. studies on complexes of the type PtCl(acac)olefin also show that the olefin is rotating[261]. The bonding of the olefin in such complexes has been discussed in terms of (1) the olefin lying with its olefinic carbon atoms in the plane defined by the metal–ligand square plane as in [2.1.23], or (2) with these carbon atoms lying above and below the plane as in [2.1.24].

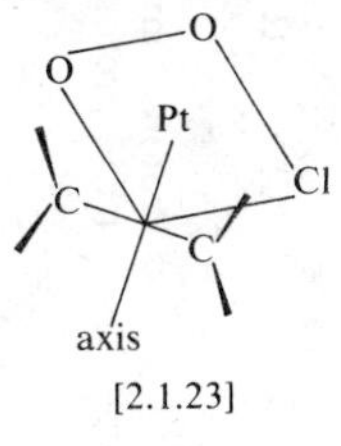

[2.1.23]

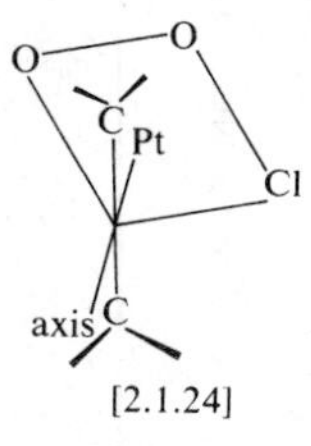

[2.1.24]

[260] R. Cramer, J. B. Kline and J. D. Roberts, J. Am. Chem. Soc. 91 (1969) 2519.
[261] C. E. Holloway, G. Hulley, B. F. G. Johnson and J. Lewis, J. Chem. Soc. A (1969) 53.

TABLE 9. NUCLEAR MAGNETIC RESONANCE PARAMETERS FOR THE COORDINATED ETHYLENE[a] IN RhCp(C₂H₄)₂ AND RhCp(C₂H₄)(SO₂)
In CH₂Cl₂ solution at −25°. J-values in Hz [260]

	$\tau_{1,2}$[b]	$\tau_{3,4}$	$J_{1,2}$	$J_{3,4}$	$J_{1,3}$	$J_{1,4}$	J_{Rh-1}	J_{Rh-3}
RhCp(C₂H₄)₂	8.79	6.92	8.8	8.8	12.2	−0.06	1.8	2.5
RhCp(C₂H₄)(SO₂)	7.32	6.34	8.8	8.8	14.4	−0.07	1.8	2.4
C₂H₄			11.7		19.7	2.4		

[a]

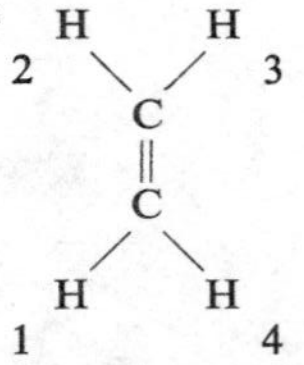

$$\begin{array}{c} {}^{2}\text{H} \quad\quad \text{H}\,{}^{3} \\ \diagdown \quad\quad \diagup \\ \text{C}=\text{C} \\ \diagup \quad\quad \diagdown \\ {}^{1}\text{H} \quad\quad \text{H}\,{}^{4} \end{array}$$

[b] Protons 1 and 2 are adjacent to the other coordinated ethylene or to the sulphur dioxide. The assignment of the resonance at higher τ to protons 1 and 2 is based on the τ-values found for the olefinic protons in [Rh₂Cl₂(hexa-1,5-diene)₂] [200].

In orientation [2.1.23] the olefinic carbon atoms are non-equivalent and therefore the protons bonded directly or indirectly to them would be non-equivalent. In the perpendicular orientation [2.1.24] the protons on different sides of the double bond would be non-equivalent. The complexes studied showed variable temperature n.m.r. spectra due to a rotation about the axis shown. At a sufficiently low temperature the rotation can be slowed down or stopped. When this occurs with the propylene or cis-but-2-ene complex two isomers are present in solution. One would expect two isomers for the propylene complex if either [2.1.23] or [2.1.24] were the most stable orientation, but for the cis-but-2-ene complex only configuration [2.1.24] could give two isomers (with the two methyl groups on the same side as the chlorine or on the opposite side). Thus clearly the perpendicular orientation [2.1.24] is preferred in solution. Table 10 gives the n.m.r. data, including the temperatures at which olefin rotation causes coalescence of the resonances.

TABLE 10. NUCLEAR MAGNETIC RESONANCE DATA (IN $CDCl_3$) FOR COMPLEXES OF TYPE PtCl(acac)(olefin) AT A TEMPERATURE LOW ENOUGH TO STOP OLEFIN ROTATION
Chemical shifts in τ. w = weak

Olefin		Olefin resonances		Coalescence temperature (°K)	Isomer ratio
		$\tau_{=CH}$	τ_{CH_3}		
Ethylene		5.48 }[a] 5.53		245 ± 3 (at 100 MHz)	
Propylene	gem	−w 4.55	8.14w 8.35	260 ± 3 (at 60 MHz)	3:1
	cis	5.44 5.57w			
	trans	5.53w 5.76			
cis-But-2-ene		−w 4.68	8.27 8.49	253 ± 5 (at 60 MHz)	6:1
trans-But-2-ene		5.11 4.86	8.23 8.42	309 (at 60 MHz)	
Tetramethylethylene		—	8.15 8.44	219 (at 100 MHz)	

[a] Most intense lines of an AA′ BB′ multiplet.

The free energies of activation for the olefin rotation are in the order

trans-but-2-ene > propylene ~ cis-but-2-ene ~ ethylene > tetramethylethylene

This order has been explained in terms of changes in steric interactions during rotation.

2.1.4d. Olefin Exchange and Other Exchange Processes

The coordinated ethylene of $K[PtCl_3(C_2H_4)]$ in methanol containing a little hydrogen chloride absorbs at 5.17τ and free ethylene absorbs at 4.63τ. In a solution containing both only one ethylene resonance is observed and its position varies with the ratio of free/coordinated ethylene. Even at −75° the resonance is broadened but not split. These results show that the exchange between free and coordinated ethylene in Zeise's salt is very rapid[152].

There is a similar very rapid exchange between the coordinated ethylenes of $Rh(acac)(C_2H_4)_2$ and free ethylene. It is estimated that in a typical experiment at $25°$ the exchange rate is $> 10^4$ sec^{-1}. In contrast $RhCp(C_2H_4)_2$ exchanges its ethylenes extremely slowly. Thus even after heating $RhCp(C_2H_4)_2$ at $100°$ for 5 hr with C_2D_4 there is negligible exchange. Clearly the presence of a vacant p_z orbital as in $[PtCl_3(C_2H_4)]^-$ and $Rh(acac)(C_2H_4)_2$ is essential for rapid ethylene exchange[200]. Rapid exchange between free and coordinated ethylene occurs in solutions formed from $IrCl(CO)(cyclo\text{-}octene)_2$ and ethylene[76].

There have also been a series of n.m.r. kinetic investigations on systems containing complexes of the type MCl(chelating diolefin)L, where M = Rh [262] or Ir [208], chelating diolefin = COD or NBD and L = PR_3, AsR_3 or SbR_3.

In the COD complexes of rhodium of type RhCl(COD)L [2.1.25]

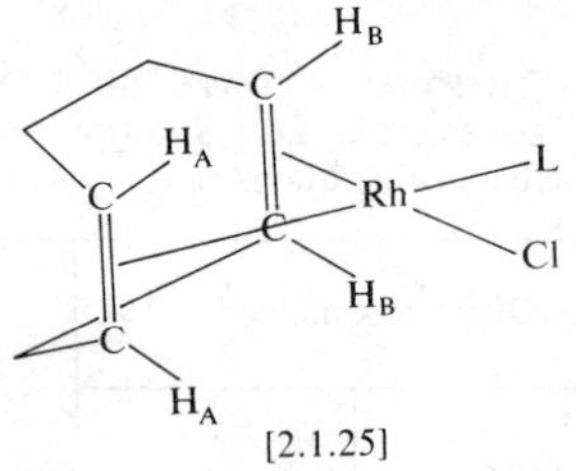

[2.1.25]

(L = PPh_3 or $AsPh_3$), the resonances due to H_B are at higher τ-values than for H_A, e.g. for $RhCl(COD)PPh_3$ in $CDCl_3$ the resonances are at 6.90 and 4.48τ respectively. As solutions of complexes of the type RhCl(COD)L are heated, the signals due to H_A and H_B broaden and coalesce. In the presence of some of the ligand L, equivalence of the protons of types H_A and H_B is caused by the following exchange process [2.1.26] $\rightleftarrows$ [2.1.27]:

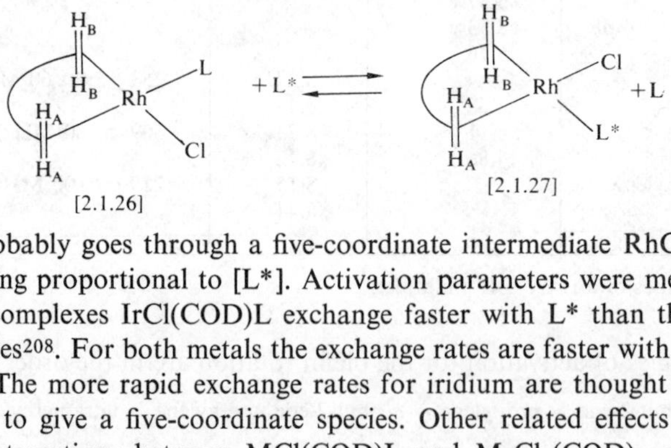

[2.1.26]   [2.1.27]

The exchange probably goes through a five-coordinate intermediate RhCl(COD)LL*, the exchange rate being proportional to [L*]. Activation parameters were measured.

The iridium complexes IrCl(COD)L exchange faster with L* than the corresponding rhodium complexes[208]. For both metals the exchange rates are faster with L = $AsPh_3$ than with L = PPh_3. The more rapid exchange rates for iridium are thought to be due to the greater tendency to give a five-coordinate species. Other related effects were studied by n.m.r. such as interactions between MCl(COD)L and $M_2Cl_2(COD)_2$, and also between MCl(COD)L and Cl$^-$.

2.1.5. Reactions of Olefin Complexes[111a]

Some reactions of olefin complexes have already been mentioned, e.g. olefin exchange, bridge splitting reactions, etc. In this section we shall be concerned almost entirely with reactions of the coordinated olefin.

262 K. Vrieze, H. C. Volger and A. P. Praat, *J. Organomet. Chem.* **14** (1968) 185.

2.1.5a. *Nucleophilic Attack and Related Processes*

Olefins coordinated to palladium(II) or platinum(II) are very susceptible to nucleophilic attack. The first example was the reaction of dicyclopentadiene with K_2PtCl_4 in methanol. This gives the complex $Pt_2Cl_2(C_{10}H_{12}OMe)_2$, formed by attack of OMe^- on one of the double bonds of the dicyclopentadiene and loss of chloride ion. The nature of this compound was not established until 1956. It was subsequently shown that complexes of palladium or platinum halides with chelating dienes [2.1.28] such as COD, NBD or dipentene react with methanol in the presence of a base, e.g. Na_2CO_3, to give chlorine-bridged complexes of the type [2.1.29], M = Pd or Pt,

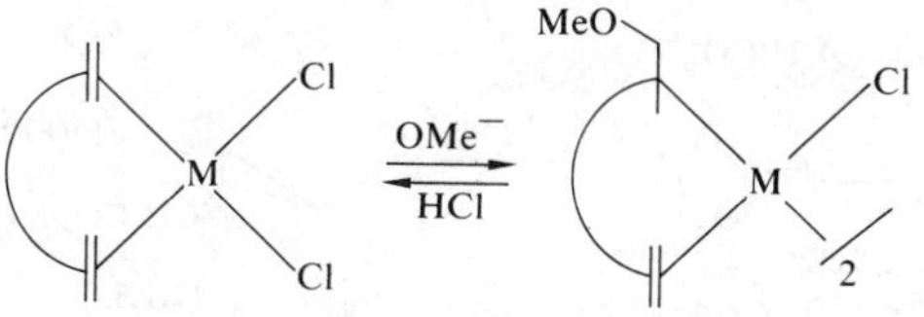

The stereochemistry of the alkoxy groups in these alkoxylated products has been shown to be *exo* by X-ray diffraction[99], n.m.r. spectroscopy and degradation studies[263]. Since the OMe group is *exo*, i.e. *trans*-attack has occurred, the reaction clearly does not involve coordination to the metal followed by a migration on to the olefinic double bond. Even if the metal were attacked first it is difficult to see how the Pt–OMe system could interact with the coordinated double bond since the two would be at right angles and for a four-centre addition to occur they would have to be in line[264].

Other nucleophiles (Y) will attack the coordinated diene system of MX_2(diolefin) in a similar way to give complexes of type [MX(diolefin-Y)]$_2$ [2.1.30] (M = Pd or Pt, X = halogen):

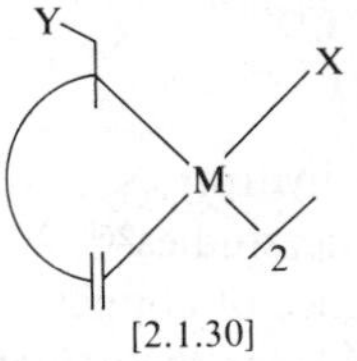

Secondary amines give compounds of this type in which Y = NR_2. Diethyl malonate, ethyl acetoacetate or β-diketones give complexes with Y = $CH(COOEt)_2$, $CH_3COCHCOOEt$ or $CR(COR)_2$ respectively. Thallous-acetylacetonate, -benzoylacetonate or -dibenzoylmethane react with MX_2(diolefin) to give complexes of the type [2.1.31]

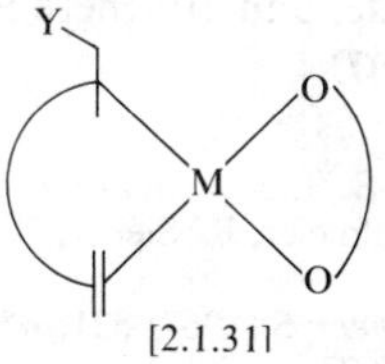

[263] J. K. Stille and R. A. Morgan, *J. Am. Chem. Soc.* **88** (1966) 5135.
[264] B. L. Shaw, *Chem. Commun.* (1968) 464.

where Y is a C-bonded group, e.g. $CH(COMe)_2$, and the bridging halogen has been replaced by a chelating acetonate system[13, 265]. Similar treatment of MX_2(diolefin) with silver acetate (1 mole per metal atom) gives [2.1.30], Y = OAc, X = halogen, but with 2 mole per metal atom the product has Y = OAc and X = OAc.

The bis(β-diketonate) derivatives [2.1.31] react with 1 mole of halogen acid (HX) to give compounds of type [2.1.30], Y = C-bonded β-diketonate, but an excess of halogen acid gives MX_2 (diolefin).

The di(acetylacetonate) complex of platinum [2.1.32] reacts with triphenylphosphine so that the chelating β-diketonate attacks the remaining double bond, giving [2.1.33].

There are a number of other reactions of these types of diolefin complexes with nucleophiles[13].

A very important catalytic industrial process is the oxidation of ethylene to acetaldehyde by an aqueous hydrochloric acid solution of palladium(II) and copper(II) salts in the presence of oxygen (air)[266-268]. Zeise's salt in a non-acidic aqueous solution is attacked by water to give acetaldehyde and platinum metal. Palladium(II)–ethylene complexes are attacked even more readily. In the catalytic process the cupric chloride reoxidizes palladium(0) to palladium(II). The process can be represented schematically by the following steps:

$$C_2H_4 + PdCl_2 + H_2O \rightarrow CH_3CHO + Pd^\circ + 2HCl \qquad \text{(oxidative hydrolysis of ethylene)}$$
$$Pd^\circ + 2CuCl_2 \rightarrow PdCl_2 + 2CuCl \qquad \text{(Pd oxidation)}$$
$$2CuCl + 2HCl + \tfrac{1}{2}O_2 \rightarrow CuCl_2 + H_2O \qquad \text{(Cu oxidation)}$$
$$C_2H_4 + \tfrac{1}{2}O_2 \rightarrow CH_3CHO \qquad \text{(overall reaction)}$$

The mechanism of the oxidative hydrolysis of olefins by palladium(II) salts has been worked out by some very fine kinetic studies[269-271]. These studies were done in aqueous solutions using $[PdCl_4]^{2-}$ in the absence of copper.

The following rate expression holds for the oxidation of ethylene, propylene, but-1-ene, and cis- and trans-but-2-ene[269, 270].

$$-\frac{d(\text{olefin})}{dt} = \frac{k[PdCl_4^-][\text{olefin}]}{[Cl^-]^2[H^+]}$$

It is also found (i) that when C_2H_4 is converted into acetaldehyde in D_2O no deuterium is incorporated into the acetaldehyde, and (ii) there is only a small isotope effect on using C_2D_4 in place of C_2H_4, $k_H/k_D = 1.07$.

[265] B. F. G. Johnson, J. Lewis and M. S. Subramanian, J. Chem. Soc. A (1968) 1993.
[266] J. Smidt, W. Hafner, R. Jira, J. Sedlmeier, R. Sieber, R. Rüttinger and H. Kojer, Angew. Chem. 71 (1959) 176.
[267] A. Aguilo, Advances Organomet. Chem. 5 (1967) 321.
[268] E. W. Stern, Catalysis Rev. 1 (1967) 73.
[269] P. M. Henry, J. Am. Chem. Soc. 86 (1964) 3246.
[270] P. M. Henry, J. Am. Chem. Soc. 88 (1966) 1595.
[271] M. N. Vargaftik, I. I. Moiseev and Y. K. Syrkin, Dokl. Akad. Nauk SSSR 147 (1962) 399.

The first step in the oxidation of ethylene is the rapid, reversible formation of a palladium–ethylene complex, with displacement of Cl^-

$$PdCl_4^{2-} + C_2H_4 \rightleftharpoons PdCl_3(C_2H_4)^- + Cl^- \quad (K_1)$$

This step causes one inverse dependence on $[Cl^-]$ in the expression for the overall reaction rate. Next is the replacement of another chloride by water,

$$PdCl_3(C_2H_4)^- + H_2O \rightleftharpoons PdCl_2(H_2O)(C_2H_4) + Cl^- \quad (K_2)$$

This equilibrium causes the second inverse dependence on $[Cl^-]$.
    Then follows the ionization of coordinated water,

$$PdCl_2(H_2O)(C_2H_4) + H_2O \rightleftharpoons [PdCl_2(OH)(C_2H_4)]^- + H_3O^+ \quad (K_3)$$

Coordinated water is much more acidic than uncoordinated water, e.g. $PtCl_2(H_2O)(C_2H_4)$ has a pK_a of *ca.* 5. Hence the concentration of $[PdCl_2(OH)(C_2H_4)]^-$ is quite appreciable even in acidic solutions. The conversion to $[PdCl_2(OH)(C_2H_4)]^-$ will show an inverse dependence on $[H^+]$. It is also found that when the reaction is carried out in D_2O there is a considerable isotope effect, $k_H/k_D = 4.05$.
    The rate-determining step is thought to be the *cis*-addition of the Pd–OH grouping across the coordinated ethylene. As discussed in section 2.1.4, n.m.r. studies show that coordinated mono-olefins can readily rotate around the $M—\overset{\displaystyle C}{\underset{\displaystyle C}{\|}}$ bond, so that a four-centre *cis*-addition is possible. Since it had been established that on using C_2D_4 as reactant all the deuteriums are retained in the acetaldehyde, a series of reversible additions (insertions) probably occurs, i.e. using C_2D_4 reacting in H_2O as the example

(for simplicity the chloride ligands are not shown).
    The rate of formation of acetaldehyde from ethylene will thus be proportional to the concentration of $[PdCl_2(OH)(C_2H_4)]^-$, i.e. to $[PdCl_4^{2-}]/[Cl^-]^2[H^+]$, as observed. The three equilibrium constants have been estimated to be $K_1 = 17.4$ mole l^{-1}, $K_2 = ca.$ 10^{-3} mole l^{-1} and $K_3 = ca.$ 10^{-6} mole l^{-1} (at 25°).
    The small isotope effect on the rate when C_2D_4 is used in place of C_2H_4 is supporting evidence for the above mechanism, i.e. it suggests that a C–H bond is not broken in the RDS.
    The rate constant for the rate-determining oxypalladation step, i.e. the addition of Pd–OH across the coordinated C=C, is not affected much by increasing alkyl substitution on the olefin. This suggests that there is little carbonium ion character in the transition state, i.e. the reaction involves a concerted non-polar four-centre addition. In contrast, the

oxymercuration or the oxythallation of olefins is greatly accelerated by alkyl substituents on the olefin, showing that there is some carbonium ion character in the transition state.

Olefin–metal complexes (particularly palladium complexes) are also attacked by nucleophiles in non-aqueous solvents[268]. In many cases vinyl compounds are among the products. Thus ethylene-palladous chloride reacts with alcohols to give acetals and vinyl ethers.

$$Pd_2Cl_4(C_2H_4)_2 + 4ROH \rightarrow 2CH_3CH(OR)_2 + Pd$$
$$Pd_2Cl_4(C_2H_4)_2 + 2ROH \rightarrow 2CH_2:CHOR + Pd$$

Ethylene is converted into vinyl acetate by palladium(II)/copper(II) systems in acetic acid containing acetate ion. The reaction has been developed into a commercial process, although details are not available[268]. There is a lot of published work on the conversion of olefins to vinyl(ic) acetate(s) by palladium(II) in copper-free acetate-containing media. Ethylene is readily absorbed by palladium chloride in acetic acid, but palladium metal is not precipitated, i.e. no oxidation occurs. However, on adding a soluble acetate such as sodium or lithium acetate, palladium metal forms and oxidation to vinyl acetate, ethylidene diacetate and acetaldehyde then takes place.

$$C_2H_4 + Pd^{2+} + 2CH_3COO^- \rightarrow CH_3COOCH:CH_2 + Pd + CH_3COOH$$
$$C_2H_4 + Pd^{2+} + 2CH_3COO^- \rightarrow (CH_3COO)_2CHCH_3 + Pd$$

When CH_3COOD is used as reactant there is practically no deuterium in the ethylidene diacetate. This shows that ethylidene diacetate is not formed by the addition of acetic acid to the vinyl acetate. Possible mechanisms for these reactions could be:

Vinyl acetate formation

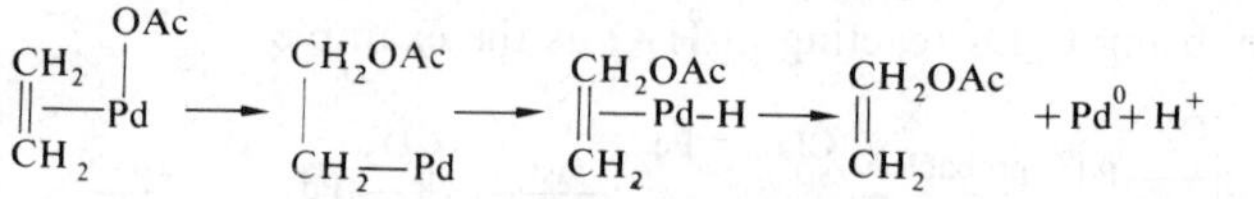

Ethylidine diacetate formation

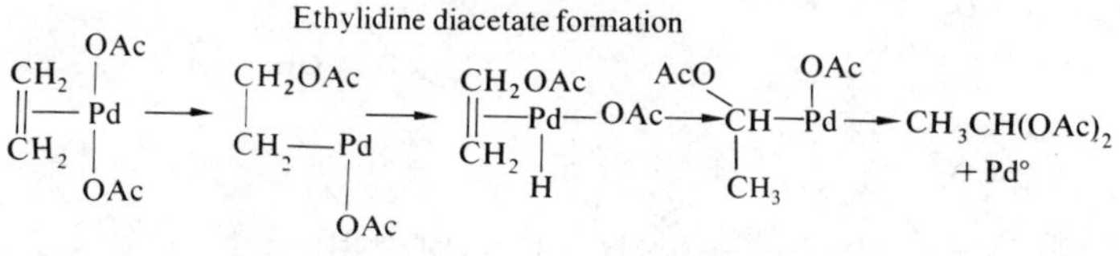

(for simplicity only some of the ligands and charges are shown). In the presence of copper(II) and large concentrations of chloride ion the major product is glycol diacetate (1,2-diacetoxyethane). This reaction producing glycol diacetate is first order in [Cu(II)], [Pd(II)] and [C_2H_4], and the rate of formation shows an inverse dependence on [Cl⁻] in the concentration range studied. It is thought that a chloride-bridged species containing both copper and palladium is involved[272].

Cyclohexene when treated with palladium chloride and sodium acetate in acetic acid gives no cyclohexenyl acetate; instead the products are

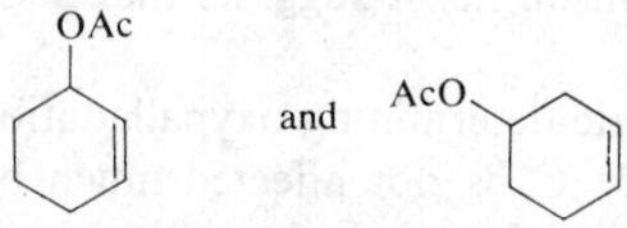

and

272 D. Clark, P. Hayden and R. D. Smith, Preprints Division of Petroleum Chemistry, American Chemical Society Meeting, April 1969, p. B 10

Clearly a cyclohexenyl acetate could not be formed by a mechanism analogous to that discussed above for ethylene because the ring system would prevent rotation about the single bond (for Pd–H to be eliminated the Pd and the H must presumably be *cis*).

Palladium(II) will also catalyse very efficient vinyl ester interchange, e.g.

$$CH_2:CHOAc + RCOOH \underset{\longleftarrow}{\overset{Pd(II)}{\rightleftharpoons}} CH_2:CHOOCR + AcOH$$

The Li_2PdCl_4 catalysed exchange of deuteriated vinyl acetate $CH_2:CHOOCD_3$ with acetic acid has been studied in detail. The reaction is first order in vinyl acetate. Using *cis*- or *trans*-1-acetoxyprop-1-ene and CD_3COOH it was found that vinyl exchange is coupled with isomerization *cis* → *trans* and vice versa. These results can be readily explained by an oxypalladation mechanism in which Pd–OAc adds *cis* to the coordinated double bond and after a rotation the elimination also occurs *cis*, e.g. from *cis*-$CH_3COOCH:CH\cdot CH_3$ and CD_3COOH the product is *trans*-$CD_3COOCH:CH\cdot CH_3$ (Fig. 7).

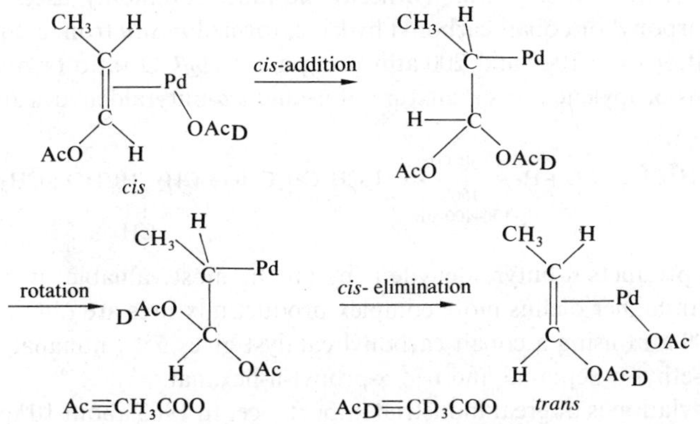

FIG. 7. Showing how palladium catalysed vinyl ester interchange of *cis*-$CH_3CH:CHOAc$ with CD_3COOH gives *trans*-$CH_3CH:CHOAc_D$.

The results could also be explained by a specific *trans*-addition and *trans*-elimination, but this seems less likely.

It is also found that 1-acetoxycyclopentene will not undergo vinyl ester exchange even after several hours in the presence of palladium(II). Clearly the ring system prevents rotation about the C–C single bond formed in the addition. This result shows that the addition and elimination of PdOOCR occurs stereospecifically in the same direction (probably *cis*).

2.1.5b. *Electrophilic Attack*

Examples of electrophilic attack on coordinated olefins are given in other sections and will not be discussed in detail here. Possibly the conversion of $Rh_2Cl_2(C_2H_4)_4$ by acid to an ethylrhodium complex involves direct protonation of coordinated ethylene. It may, however, involve formation of an intermediate rhodium hydride species followed by addition of Rh–H across the double bond.

[273] P. M. Henry, Preprints Division of Petroleum Chemistry, American Chemical Society Meeting, April 1969, p. B 15.

The cyclo-octa-1,5-diene-cobalt(I) complex Co(COD)Cp [2.1.34] when treated with [Ph₃C]⁺[BF₄]⁻ loses H⁻ to give the allylic complex [2.1.35].

In contrast, the corresponding rhodium or iridium complexes substitute on the C_5H_5 ring when treated with one mole of trityl cation but with two moles removal of H⁻ from the COD ring also occurs, giving [Co(C_8H_{11})(C_5H_4CPh₃)]⁺ [13,274].

2.1.5c. *Hydroformylation, Carbonylation and Carboxylation of Olefins*

(i) *Hydroformylation or the oxo reaction*[7b, 275-278]. In this reaction an olefin is converted into an aldehyde by the catalysed addition of the elements of hydrogen and carbon monoxide across the double bond. Hitherto the most commonly used catalyst has been dicobalt octacarbonyl or cobalt carbonyl hydride, formed *in situ* from a cobalt salt. Typical reaction conditions are 100° and 200 atm using a 1:1 H_2/CO ratio (synthesis gas). Under these conditions propylene gives a mixture of n- and *iso*-butyraldehydes and some propane, i.e.

$$CH_3CH:CH_2 + CO + H_2 \xrightarrow[\substack{150° \\ 100-400 \text{ atm}}]{Co_2(CO)_8} CH_3CH_2CH_2CHO + CH_3\underset{\underset{CH_3}{|}}{CH}CHO + CH_3CH_2CH_3$$

Of these three products n-butyraldehyde is by far the most valuable; it is formed in about 70% yield. With higher olefins more complex product mixtures are obtained, e.g. oct-1-ene at 100° and 200 atm using a cobalt carbonyl catalyst gives 65% nonanal, 22% α-methyl-n-octanal, 7% α-ethyl-n-heptanal and 6% α-propyl-n-hexanal.

Hydroformylation is of great industrial importance. In 1963 about 10^9 pounds of organic chemicals were produced by the hydroformylation of olefins. The largest usage is for the production of C_8-plasticizer alcohols from heptenes. Not surprisingly, the mechanism of the reaction has received a lot of attention. Figure 8 gives a simplified version of the most generally accepted mechanism.

As can be seen from Fig. 8, the reaction involves an olefin complex, olefin insertion, a metal alkyl complex, carbonyl insertion into the metal–carbon σ-bond, and oxidative addition and reductive elimination steps.

It is established that cobalt carbonyl hydride is the active catalytic species in the reaction. Increasing carbon monoxide pressure (concentration) has an inhibiting effect on hydroformylation. It is possible that this inhibition is caused by a reversal of step (1) (in Fig. 8). The addition of pent-1-ene to cobalt tetracarbonyl hydride to give pentylcobalt tetracarbonyls is inhibited by carbon monoxide at 0° [276, 277]. Alternatively, the carbon monoxide could compete with hydrogen for the cobalt in step (6), Fig. 8, and in this way inhibit the

[274] J. Lewis and A. W. Parkins, *J. Chem. Soc.* A (1967) 1150.
[275] C. W. Bird, *Transition Metal Intermediates in Organic Synthesis*, Logos Press, London; Academic Press, New York (1967).
[276] R. F. Heck, in *Organic Synthesis via Metal Carbonyls*, Vol. 1 (eds. I. Wender and P. Pino), Interscience, New York (1968), p. 373.
[277] R. F. Heck, *Advances Organomet. Chem.* 4 (1966) 243.
[278] A. J. Chalk and F. Harrod, *Advances Organomet. Chem.* 6 (1968) 119.

overall hydroformylation reaction. The rate of hydroformylation is decreased by increasing olefin substitution, e.g. terminal olefins react faster than straight-chain internal olefins, which in turn react faster than branched-chain internal olefins. This order is probably due to a decreasing ease of olefin complex formation in step (2), Fig. 8.

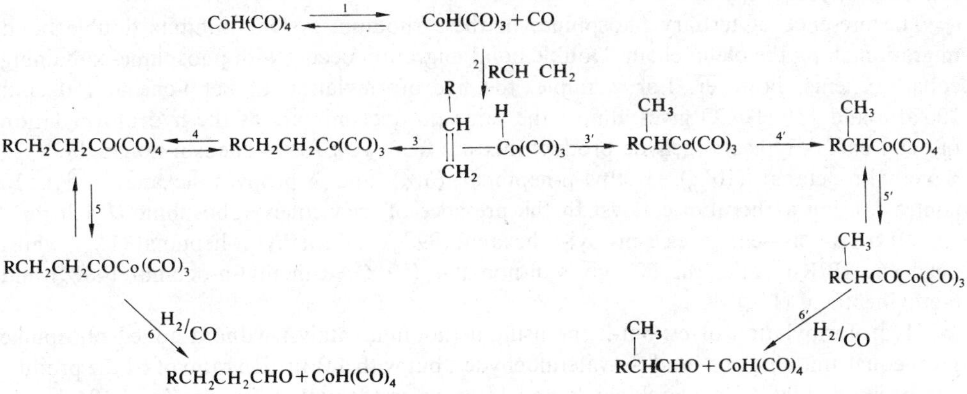

FIG. 8. Mechanism of the hydroformylation of olefins ($RCH:CH_2$) using a cobalt carbonyl hydride catalyst (formed *in situ*).

[1] Loss of a molecule of CO from $CoH(CO)_4$.	[4 4'] Addition of CO.
[2] Coordination of $RCH:CH_2$.	[5 5'] CO insertion.
[3] Addition of hydride to the β-carbon atom.	[6 6'] Addition of H_2, elimination of aldehyde.
[3'] Addition of hydride to the α-carbon atom.	

In the hydroformylation of propylene to butyraldehydes the ratio of n- to *iso*-butyraldehyde in the product increases as the partial pressure of the carbon monoxide is increased. With long-chain olefins the product mixture of saturated aldehydes is complex because under hydroformylation conditions the olefinic double bond can move along the carbon chain. With long-chain olefins the proportion of the more valuable terminally carbonylated product is increased by an increased partial pressure of carbon monoxide (which unfortunately also reduces the rate of hydroformylation).

An interesting and important development is the use of tertiary phosphine-containing catalysts[279]. Tertiary phosphine-substituted cobalt carbonyls such as $Co_2(CO)_6(PBu_3^n)_2$ are less active hydroformylation catalysts than dicobalt octacarbonyl under the usual conditions but more active at low pressures. What is important is that they prefer to hydroformylate the terminal position of alk-1-enes. Thus pent-1-ene at 150° and 30 atm gives 91% of the terminally carbonylated isomer. Under hydroformylation conditions $Co_2(CO)_6(PBu_3)_2$ is converted into the hydride $CoH(CO)_3PBu_3$, and when this adds to a terminal olefin the hydrogen adds to the 2-position (anti-Markownikoff addition), possibly because it is hydridic in character. The steric effect of the bulky tri-n-butylphosphine group might also promote anti-Markownikoff addition. When cobalt tetracarbonyl hydride adds to pent-1-ene at 0° there is about 50% terminal and 50% non-terminal addition (the hydrogen of cobalt tetracarbonyl hydride is acidic). An additional effect is that whereas cobalt carbonyl hydride as hydroformylation catalyst gives an aldehyde product, with phosphine ligands some reduction of the aldehyde to the corresponding alcohol also occurs.

Another important development in hydroformylation has been the discovery that

[279] L. H. Slaugh and R. D. Mullineaux, *J. Organomet. Chem.* 13 (1968) 469.

rhodium gives much more active catalysts than cobalt, being approximately 10^3 times as active under comparable conditions. Again the presence of tertiary phosphines improves the n-/iso-ratio of the hydroformylation product. In the absence of tertiary phosphines the ratio is smaller for rhodium than for cobalt but with large PR_3/Rh molar ratios the n-/iso-ratio can be large (> 10).

The presence of tertiary phosphines in these rhodium systems inhibits double bond migration along the olefin chain. Double bond migration occurs with phosphine-containing cobalt systems, however. For example, the hydroformylation of oct-4-ene at 150° and 200 atm and 1:1 H_2/CO gives almost the same product mixture as the hydroformylation of oct-1-ene. A typical aldehyde product mixture from trans-oct-4-ene is n-nonanal (78%), α-methyl-n-octanal (10%), α-ethyl-n-heptanal (6%) and α-propyl-n-hexanal (6%). In contrast, using a rhodium catalyst in the presence of tricyclohexylphosphine (P/Rh ratio ca. 60) trans-oct-4-ene gives α-propyl-n-hexanal (99%) and α-ethyl-n-heptanal (1%), whilst oct-1-ene (P/Rh ratio ca. 50) gives n-nonanal (59%), α-methyl-n-octanal (40%) and α-ethylheptanal (1%)[280].

Hydroformylation of cis-but-2-ene using a rhodium catalyst without added phosphine gives equal amounts of n- and iso-valeraldehydes, but with a PBu_3/Rh ratio of 62 the product composition is 99% isovaleraldehyde and 1% n-valeraldehyde. With a ratio of 124:1 only isovaleraldehyde could be detected[280].

Similarly, rhodium-catalysed hydroformylation of various methyl hexenoates without added phosphine gives all possible formylhexanoates. With a large amount of added phosphine the two isomers corresponding to the original position of the double bond only are formed, i.e. from methyl hex-3-enoate, only the three-formyl and the four-formyl derivatives.

Thus. the presence of tertiary phosphine ligands on the rhodium makes the rate of hydroformylation much faster than the rate of olefin isomerization. As shown in Fig. 9, during the reaction a hydrorhodium-olefin complex will be converted into an alkylrhodium complex [2.1.36]. This complex can undergo either carbon monoxide insertion to give [2.1.37], or hydride elimination, giving [2.1.38], in which the double bond of the original

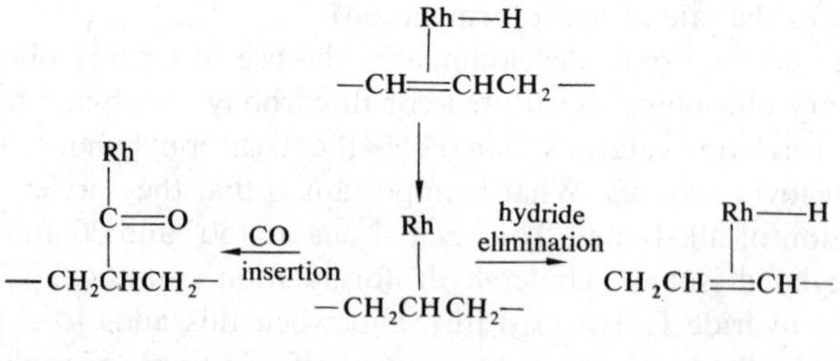

FIG. 9. Showing how, during hydroformylation, an intermediate alkylrhodium complex can undergo either carbon monoxide insertion or hydride elimination.

olefin has moved along the chain by one carbon atom. Without phosphine ligands on the rhodium these processes occur at a comparable rate, but with phosphine ligands the carbon monoxide insertion reaction becomes much faster than the hydride elimination reaction.

There has been an extensive study of rhodium(I) complexes such as trans-

280 B. Fell, W. Rupilius and F. Asinger, Tetrahedron Letters (1968) 3261.

$RhCl(CO)(PPh_3)_2$ and $RhH(CO)(PPh_3)_3$ as hydroformylation catalysts[281]. With *trans*-$RhCl(CO)(PPh_3)_2$ there is an induction period while the chloro-complex is being converted into a more active hydro-species. $RhH(CO)(PPh_3)_3$ will catalyse hydroformylation of alk-1-enes even at 25° and 1 atm of H_2/CO mixture. The rate of hydroformylation of alk-1-enes is at least 20 times that of alk-2-enes. With hex-1-ene at 25° and 1 atm, the ratio of straight-chain to branched-chain isomer in the product aldehyde is *ca*. 20, but at higher gas pressures the ratio falls, e.g. at 100 atm and 25° it is 3. An excess of triphenylphosphine inhibits hydro-formylation, presumably by blocking a coordination site and preventing attack by the olefin. With these triphenylphosphine–rhodium catalyst systems little olefin isomerization seems to occur. Thus hydroformylation of a *cis*/*trans*-pent-2-ene mixture gives an all-branched aldehyde product (i.e. there is no isomerization to pent-1-ene followed by terminal attack). A detailed mechanism for olefin hydroformylation using these rhodium–triphenyl-phosphine catalysts has been given[281]. It is similar to the Heck/Breslow mechanism for cobalt carbonyl catalysed hydroformylation, but the stereochemistry of the various catalytic intermediates is discussed.

The formation of other products of hydroformylation, such as formate esters, can be readily explained by postulating the addition of $CoH(CO)_4$ to aldehyde followed by inser-tion of carbon monoxide and hydrogenation[282].

$$RCHO + CoH(CO)_4 \longrightarrow RCH_2OCo(CO)_4 \xrightarrow{CO} RCH_2OCOCo(CO)_4 \xrightarrow{H_2} RCH_2OOCH + CoH(CO)_4.$$

Under very high pressures of H_2/CO, ethylene can also give diethyl ketone. This can be explained as follows:

$$CH_2:CH_2 \longrightarrow CH_3CH_2COCo(CO)_4 \xrightarrow{C_2H_4} CH_3CH_2COCH_2CH_2Co(CO)_4 \xrightarrow{H_2} CH_3CH_2COCH_2CH_3 + CoH(CO)_4$$

(ii) *Carboxylation and other carbonylation reactions.* Carboxylation[275] converts an olefin into a carboxylic acid (ester) by reacting it with carbon monoxide and water (alcohol), usually in the presence of mineral acid. A nickel catalyst such as the iodide is commonly used; under the vigorous conditions of the reaction (250° and 200 atm) this is converted into nickel carbonyl. Salts of other metals, e.g. palladium, platinum, cobalt, rhodium, iron, ruthenium, osmium and rhenium, have also been used. The stereochemistry of carboxylation of the double bond is not generally known, but for the carboxylation of bicyclo[2,2,1]hept-1-ene [2.1.39] it is a *cis*-addition since on treatment with $Ni(CO)_4$, D_2O and AcOD it gives the 3-*exo*-deuterio-2-*exo*-carboxylic acid [2.1.40] [283].

[2.1.39]   [2.1.40]

Propylene gives a mixture of n- and *iso*-butyric acids on carboxylation. The n-/*iso*- ratio depends on the catalyst and conditions. As with hydroformylation the ratio increases with increasing carbon monoxide pressure. The kinetics of carboxylation are complex. Thus with cyclohexene at pressures of carbon monoxide below 160 atm the rate is proportional to p_{CO} but at *ca*. 340 atm it is proportional to $1/p_{CO}$. The mechanism of carboxylation is not

[281] D. Evans, J. A. Osborn and G. Wilkinson, *J. Chem. Soc.* A (1968) 3133.
[282] R. F. Heck, *Mechanisms of Inorganic Reactions*, in Advances in Chemistry Series, No. 49, American Chemical Society (1964).
[283] C. W. Bird, R. C. Cookson, J. Hudec and R. O. Williams, *J. Chem. Soc.* (1963) 410.

known but it probably involves a hydrometal–olefin complex which isomerizes to an alkyl-metal complex; this in turn undergoes carbon monoxide insertion to give an acylmetal complex which on solvolysis gives the carboxylic acid or ester. Possibly the function of the mineral acid in the reaction mixture is to convert the metal complex into a hydride by protonation. Thus tetrakis(triphenylphosphine)palladium(0) will not catalyse carboxylation of ethylene or propylene in ethanol below 100°. However, in the presence of hydrogen chloride, esters are formed at 40–80° [284]. It has been shown that hydrogen chloride will convert tetrakis(triphenylphosphine)platinum(0) into a hydride, and it seems very likely that acid will similarly produce a palladium hydride species.

Olefin–palladium chloride complexes or mixtures of palladium chloride with olefin, react with carbon monoxide under pressure (40–100 atm) to give β-chloroacyl chlorides, e.g. ethylenepalladous chloride gives 2-chloropropionyl chloride. With alk-1-enes these carbonylations occur exclusively on the terminal carbon atom.

Carbonylation of non-conjugated diolefins frequently gives cyclic ketonic products. For example, hexa-1,5-diene reacts with carbon monoxide in aqueous acetone using dicobalt octacarbonyl as catalyst {at 100–200° and 70–340 atm of CO} as follows:

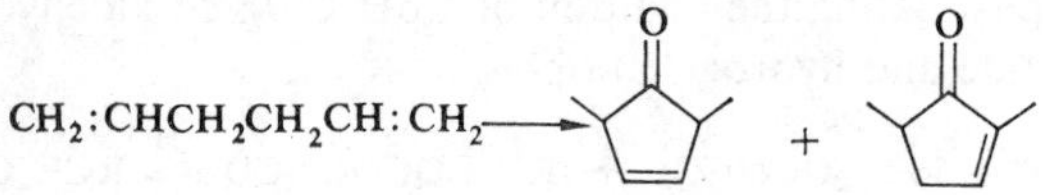

$$CH_2:CHCH_2CH_2CH:CH_2 \longrightarrow$$

Norbornadiene reacts with iron carbonyls to give a variety of cyclic ketonic products, e.g. [2.1.41] and [2.1.42].

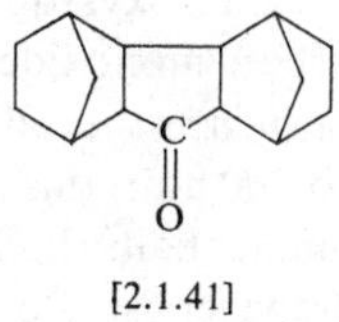

[2.1.41]

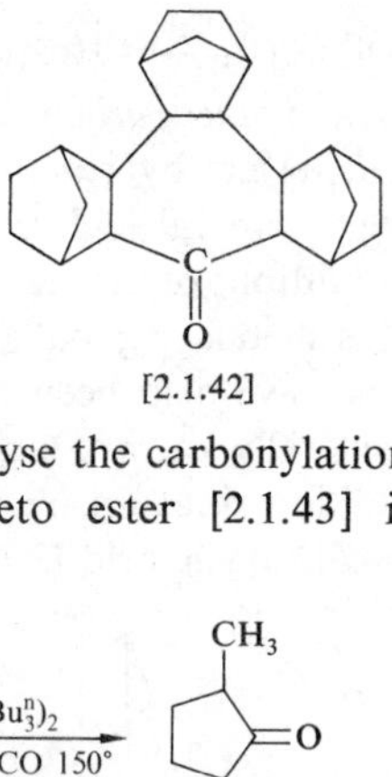

[2.1.42]

Tertiary phosphine–metal complexes can catalyse the carbonylation of non-conjugated diolefins, e.g. hexa-1,5-diene gives a cyclic keto ester [2.1.43] in methanol, using $PdI_2(PBu_3^n)_2$ as catalyst.

$$CH_2:CHCH_2CH_2CH:CH_2 \quad \xrightarrow[\text{100 atm CO 150°}]{Pd\,I_2(PBu_3^n)_2}$$

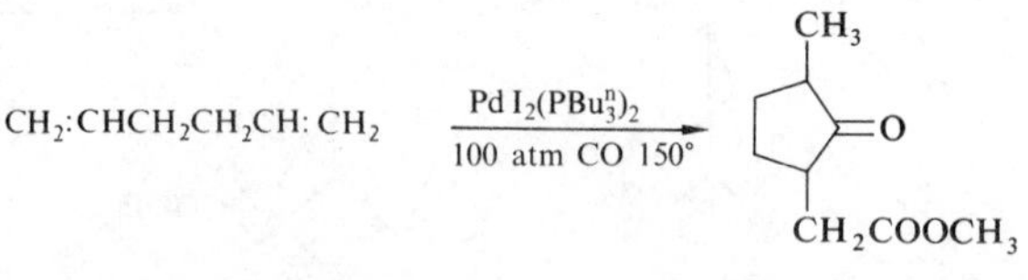

[2.1.43]

It is postulated that a palladium hydride intermediate adds to one of the double bonds of hexa-1,5-diene. Successive insertion reactions of carbon monoxide, the remaining olefinic bond and another molecule of carbon monoxide then follow to give an acyl–palladium complex which is then solvolysed to give the keto ester [2.1.43] [285].

Other diolefins (e.g. COD), give mono- and dicarboxylates when treated with carbon monoxide and methanol using $PdI_2(PBu_3)_2$ as catalyst.

[284] K. Bittler, N. V. Kutepow, D. Neubauer and H. Reis, *Angew. Chem. Int. Edn.* 7 (1968) 329.
[285] S. Brewis and P. R. Hughes, *Chem. Commun.* (1965) 489.

2.1.5d. *Oligomerization of Olefins*[275]

Many transition metal complexes will catalyse the dimerization or oligomerization of olefins. These catalysts include palladium chloride—a system generated from rhodium trichloride—and π-allylnickel halides in the presence of aluminium halides. The most thoroughly studied is the dimerization of ethylene by rhodium trichloride[286]. Ethylene is catalytically dimerized at 40° to linear butenes by a solution of rhodium trichloride in ethanol. The yields of butenes may be $>99\%$ on consumed ethylene. In the reaction there is an induction period during which the rhodium trichloride is converted into $Rh_2Cl_2(C_2H_4)_4$.

$$2RhCl_3 + 6C_2H_4 + 2H_2O \rightarrow Rh_2Cl_2(C_2H_4)_4 + 2CH_3CHO + 4HCl$$

However, $Rh_2Cl_2(C_2H_4)_4$ itself is not a catalyst and acid is an essential cocatalyst. A solution containing $Rh_2Cl_2(C_2H_4)_4$ and HCl is an effective catalyst with no induction period.

The mechanism of this rhodium catalysed dimerization of ethylene has been worked out in detail using a combination of n.m.r. spectroscopy with other techniques. The mechanism is outlined in Fig. 10. Some of the steps in the dimerization were studied by n.m.r.

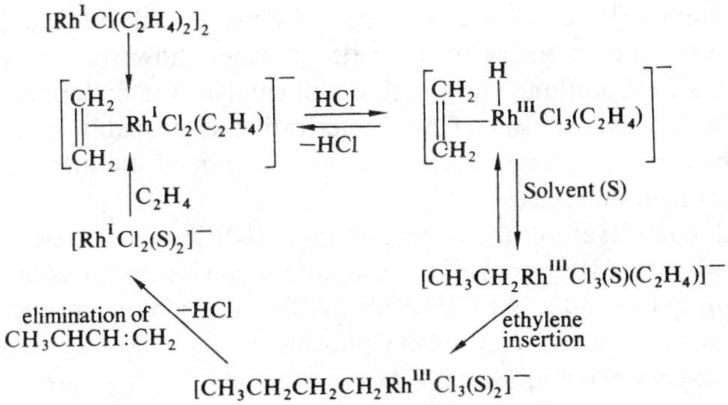

FIG. 10. The mechanism of ethylene dimerization by an $[RhCl(C_2H_4)]_2/HCl$ catalyst[286].

spectroscopy at low temperature. For example, the formation of an ethyl group by protonation of coordinated ethylene (this is thought not to be a direct attack but goes via a rhodium hydride intermediate as shown in Fig. 10). The insertion of coordinated ethylene into the rhodium–ethyl bond, which is the rate-determining step, was also studied by n.m.r. The activation parameters are $\Delta H^* = 16.6$ kcal mole^{-1}, $\Delta G^* = 22.7$ kcal mole^{-1} and $\Delta S^* = -20.1$ cal mole^{-1} deg^{-1}. The combination of $Rh_2Cl_2(C_2H_4)_4 + HCl$ is a good isomerization catalyst for olefins so that extensive isomerization of the product but-1-ene, to but-2-enes occurs[2]. (See section 2.1.5f [286].)

Other homogeneous systems which will catalyse dimerization or oligomerization of simple olefins are based on cobalt[287], palladium[155], nickel[288] and titanium[289].

[286] R. Cramer, *J. Am. Chem. Soc.* **87** (1965) 4717.

[287] L. S. Pu, A. Yamamoto and S. Ikeda, *J. Am. Chem. Soc.* **90** (1968) 7170.

[288] G. Wilke, B. Bogdanović, P. Hardt, P. Heimbach, W. Keim, M. Kröner, W. Oberkirch, K. Tanaka, E. Steinrücke, D. Walter and H. Zimmermann, *Angew. Chem. Int. Edn.* **5** (1966) 151.

[289] G. Henrici-Olivé and S. Olivé, *J. Organomet. Chem.* **16** (1969) 339.

The cobalt nitrogen complex $CoH(N_2)(PPh_3)_3$ will catalyse ethylene dimerization even at $0°$ [287]; small amounts of higher oligomers and ethane are also formed. The complex will also dimerize propylene, mainly to 2-methylpent-1-ene. The proposed mechanism is similar to that for the rhodium system described above and can be summarized as a series of reversible steps; these are outlined in Fig. 11. Several other steps and equilibria are possible in this system and the but-1-ene is further isomerized to but-2-enes by the catalyst.

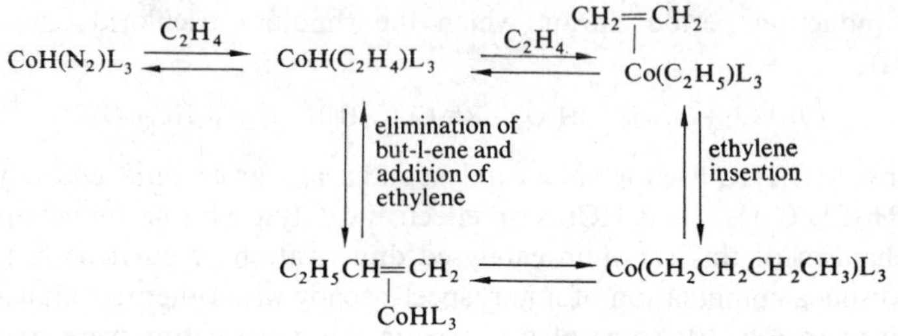

$$CoH(N_2)L_3 \;\xrightleftharpoons{}\; CoH(C_2H_4)L_3 \;\xrightleftharpoons[C_2H_4]{C_2H_4}\; \underset{|}{CH_2{=}CH_2} \;\; Co(C_2H_5)L_3$$

elimination of but-1-ene and addition of ethylene

ethylene insertion

$$\underset{CoHL_3}{\overset{|}{C_2H_5CH{=}CH_2}} \;\xrightleftharpoons{}\; Co(CH_2CH_2CH_2CH_3)L_3$$

FIG. 11. Mechanism for the dimerization of ethylene by $CoH(N_2)L_3$ as catalyst (L = PPh_3) [287].

When palladium chloride is reacted with ethylene in chloroform, $Pd_2Cl_4(C_2H_4)_2$ is formed. In the presence of traces of alcohols or water, however, the ethylene complex dissolves to give a red solution. The solution will catalyse the dimerization of ethylene to linear butenes[155]. The mechanism of the dimerization is probably similar to that of the rhodium or cobalt systems described above, the function of the alcohol (water) being to form a palladium hydride species.

π-Allylnickel complexes of the type $NiX(\pi\text{-}C_3H_5)PR_3$ (X = halogen, R = alkyl or aryl) when mixed with $AlBr_3$ or $AlCl_2Et$ promote dimerization or codimerization of olefins[288]. From propylene and $NiI(C_3H_5)PPh_3/AlBr_3$ the major products (*ca.* 70%) are 2-methylpentenes, but with tricyclohexylphosphine in place of triphenylphosphine, 2,3-dimethylbut-2-enes make up *ca.* 60% of the product.

There are many examples of Ziegler catalysts which will oligomerize olefins[289]. Thus pent-1-ene in the presence of a catalyst prepared from methyltitanium trichloride and methylaluminium dichloride at $-70°$ gives mainly dimers and some trimers. The catalyst is possibly $[TiCl_2Me]^+[AlCl_3Me]^-$. Higher polymers of ethylene are produced using soluble Ziegler catalysts formed from $TiCl_2Cp_2$ and $AlClEt_2$.

Ruthenium complexes, in the presence of hydrogen, will catalyse the conversion of acrylonitrile into 1,4-dicyanobut-1-ene. Other products such as propionitrile and adiponitrile are also formed. The ruthenium complexes include $RuCl_3$ or $RuCl_2(CH_2{:}CHCN)_4$. The formation of 1,4-dicyanobut-1-ene can be explained in terms of an initial addition of a ruthenium hydride to the double bond of acrylonitrile followed by insertion of more acrylonitrile and elimination of a β-hydrogen. However, the mechanism may be more complicated than this[290]. Such dimerizations of acrylonitrile could form the basis of a new route to polyamides of the Nylon type.

Palladium chloride at $70°$ will catalyse the copolymerization of ethylene and sulphur dioxide to the sulphone, $CH_3CH_2SO_2CH_2CH_2{:}CHCH_3$ (with a *trans* C$=$C double bond). A series of insertion reactions has been postulated to explain the product from this reaction[291].

[290] E. Billig, C. B. Strow and R. L. Pruett, *Chem. Commun.* (1968) 1307.
[291] H. S. Klein, *Chem. Commun.* (1968) 377.

The reactive diolefin NBD {bicyclo[2,2,1]hepta-2,5-diene} is converted catalytically into a variety of dimers, depending on the catalyst used[275]. Commonly occurring products are the various geometrical isomers of [2.1.44], i.e. *endo-trans-endo*, *exo-trans-exo* and *exo-trans-endo*. Other, more complex, dimers can be formed.

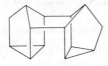

[2.1.44]

Catalysts include Ni(acrylonitrile)$_2$, Fe(NO)$_2$(CO)$_2$, Co$_2$(CO)$_6$(PPh$_3$)$_2$ or Fe$_2$(CO)$_9$. The unusual dimer "Binor-S" [2.1.45] is obtained with the catalyst Zn{Co(CO)$_4$}$_2$ [292]. These

[2.1.45]

dimerizations almost certainly do not involve hydride addition, insertion, etc.; instead, four-membered rings are probably formed from a bis(mono-olefin) complex by a similar mechanism to that which occurs during olefin dismutation[293] (section 2.1.5e).

Bis(acrylonitrile)nickel will catalyse the addition of acrylonitrile to the 2,6-position of NBD. The product [2.1.46] is formed in 95% yield.

CH$_2$———CHCN
[2.1.46]

2.1.5e. *Olefin Dismutation (Metathesis)*

This is one of the most interesting catalytic reactions to be discovered recently. Heterogeneous catalysts such as molybdenum or tungsten oxides on alumina or the corresponding hexacarbonyls on alumina will catalyse the disproportionation or dismutation of olefins. Thus propylene is broken down into its equilibrium mixture with ethylene and but-2-enes. Many such catalysts have been described, especially in the patent literature.

An important development was the discovery that homogeneous systems would also cause dismutation of olefins[294]. The catalytic activity can be very high; thus a solution of 0.1 mole of pent-2-ene in dry benzene (15 ml) when treated with 0.2 ml of an 0.5M solution of tungsten hexachloride and ethanol in benzene, followed by 0.2 ml of 0.2M Al$_2$Cl$_4$Et$_2$ in benzene, gives, after 2 min at room temperature, the equilibrium mixture of but-2-enes, pent-2-enes and hex-3-enes, i.e. in mole ratios of 1:2:1. Double-bond migration could not be detected. An equimolar mixture of but-2-ene and hex-3-ene after treatment with the same

292 G. N. Schrauzer, B. N. Bastian and G. A. Fosselius, *J. Am. Chem. Soc.* **88** (1966) 4890.
293 F. D. Mango and J. H. Schachtschneider, *J. Am. Chem. Soc.* **89** (1967) 2484.
294 N. Calderon, H. Y. Chen and K. W. Scott, *Tetrahedron Letters* (1967) 3327.

catalyst gives the same equilibrium mixture. Olefin dismutation reactions of this type can thus be represented as follows:

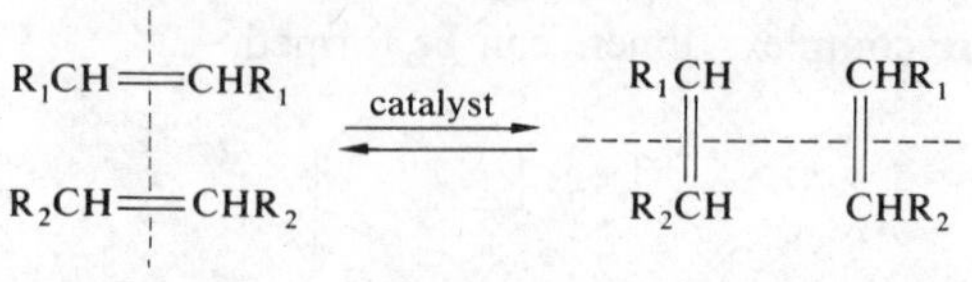

A catalyst system can also be prepared from tungsten hexachloride and n-butyl lithium[295]. This catalyst is not as active as the one described above. Using different ratios of LiBu/W the most active catalyst is obtained when the ratio is 2. This indicates that a tungsten(IV) species is probably the active catalyst, i.e.

$$WCl_6 \xrightarrow{} WCl_4Bu_2 \xrightarrow{\text{olefin}} WCl_4(\text{olefin})_2 + \text{decomposition products from the butyl groups}$$

It is suggested that two olefins coordinate to tungsten(IV) and that dismutation occurs via a cyclobutane complex, as shown in Fig. 12.

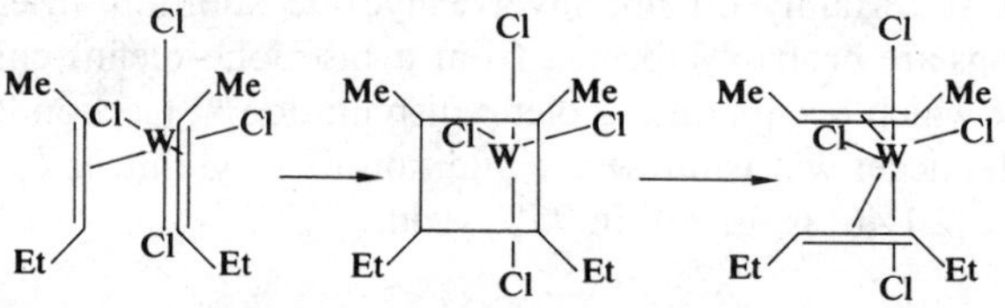

Fig. 12. Proposed mechanism for the dismutation of pent-2-ene to but-2-ene and hex-3-ene on $W^{IV}Cl_4$.

The fusion of two olefins to a cyclobutane ring is a symmetry-forbidden process. A theory has been developed which shows that cyclobutane formation can be allowed when the two olefins are coordinated to a transition metal[293]. The transition metal must have occupied and unoccupied orbitals of the correct symmetry, which can act as a source and a sink for electrons.

Other homogeneous systems which will catalyse dismutation are formed by treating $MCl_2(NO)_2L_2$, L = PPh_3, $(OPh)_3$, py, etc. (M = Mo or W), with alkylaluminium halides in chlorobenzene. The resultant brown homogeneous systems are very active catalysts. Treatment of pent-1-ene with such catalysts and allowing the ethylene formed to escape, gives initially oct-1-ene and then tetradec-7-ene. However, some of these systems also cause olefin isomerization, especially with ethylaluminium chloride, e.g. with hept-2-ene olefins from C_2 to C_{12} are produced.

An interesting observation on $MoCl_2(NO)_2L_2$/alkylaluminium halide catalysis is that there is a marked initial preference for cis-pent-2-ene to dismutate to cis-but-2-ene and cis-hex-3-ene. Similarly from trans-pent-2-ene the preferred initial products are trans-but-2-ene and trans-hex-3-ene[296]. Eventually the thermodynamically equilibrated mixture of cis- and trans-isomers is produced. This result for cis-pent-2-ene has been explained in terms of two olefins coordinating to molybdenum as shown in [2.1.47] (Fig. 13).

295 J. L. Wang and H. R. Menapace, J. Organic Chem. 33 (1968) 3794.
296 W. B. Hughes, Chem. Commun. (1969) 431.

In order for positive overlap of the olefin π-orbitals to occur during cyclobutane formation the olefins will rotate in a disrotatory manner; hence on dismutation *cis*-but-2-ene and *cis*-hex-3-ene will be formed (Fig. 13). The *trans*-pent-2-ene complex [2.1.48] will similarly give *trans*-but-2-ene and *trans*-hex-3-ene. Arrangements such as [2.1.49] would give *cis*-but-2-ene and *cis*-hex-3-ene from *trans*-pent-2-ene but are considered to be less likely because of steric repulsion between methyl or ethyl groups in close proximity, e.g. methyl in [2.1.49]. These less favourable arrangements would eventually give a thermodynamically equilibrated mixture of all the possible olefins, however.

There will obviously be many developments in this field of olefin dismutation and related reactions.

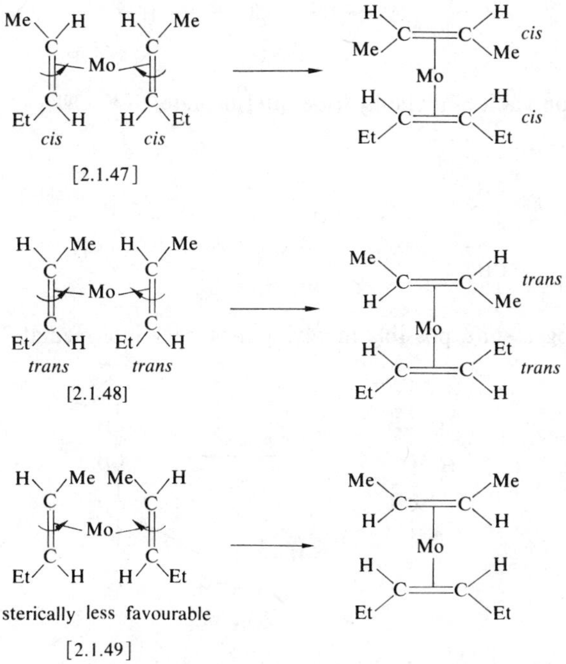

FIG. 13. The stereochemistry of the dismutation of *cis*- or *trans*-pent-2-enes to but-2-enes + hex-3-enes on a molybdenum catalyst (see text).

2.1.5f. *Olefin Isomerization and Other Hydrogen Transfer Reactions*[268, 275, 297, 298]

Hydrogen transfer reactions such as double-bond migration along carbon chains are catalysed homogeneously by transition metal complexes, e.g. $Fe(CO)_5$, $CoH(CO)_4$, $PdCl_2(PhCN)_2$ and $IrHCl_2(PEt_2Ph)_3$, etc. More frequently, however, a cocatalyst is required to activate the metal complex. The cocatalyst acts as a source of hydrogen (hydride); typical cocatalysts are acids, alcohols and molecular hydrogen. Active combinations of transition metal complexes with cocatalysts (shown in parentheses) include the following: $Ni\{P(OEt)_4\}_4$ (methanolic H_2SO_4); $RuCl_3$ (allyl alcohol); $Rh_2Cl_2(C_2H_4)_4$ (HCl or H_2), a

[297] G. C. Bond, *Ann. Rep. Chem. Soc.* **63** (1966) 27.
[298] M. Orchin, *Advances in Catalysis* **16** (1966) 2.

Rh^I–stannous chloride complex (H_2); Li_2PdCl_4 (CF_3COOH or H_2), $Pt_2Cl_4(C_2H_4)_2$ (ethanol or silanes); and the H_2PtCl_6–$SnCl_2$ complex (H_2). Usually without the cocatalyst the metal complex is inactive[299, 300].

Many of these catalysts are very active under mild conditions and will, for example, catalyse the isomerization of alk-1-ene, alk-2-enes, alk-3-enes, etc.

Three main mechanisms proposed for the transition metal complex catalysed migration of olefinic double bonds along carbon chains all involve the initial formation of a metal–olefin complex. The three mechanisms are:

(1) Hydride addition and elimination[297, 299, 300]:

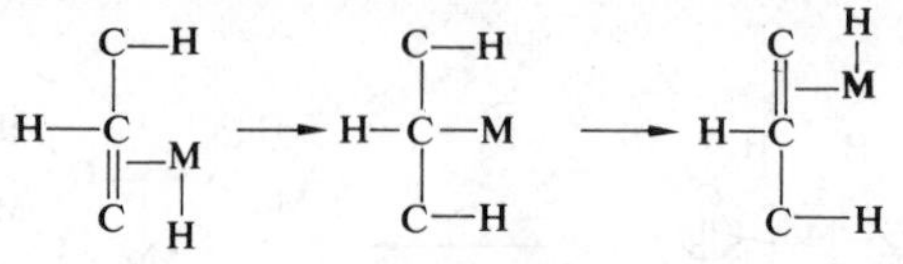

(2) Isomerization via a π-allylic hydride intermediate[297, 300, 301]:

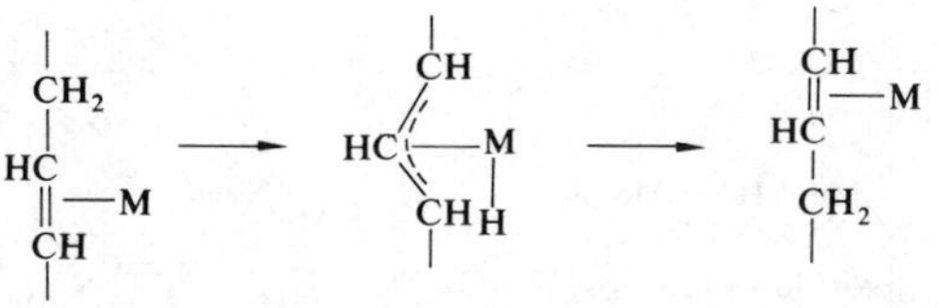

(3) A 1,2-hydrogen shift, possibly involving a carbene intermediate[302]:

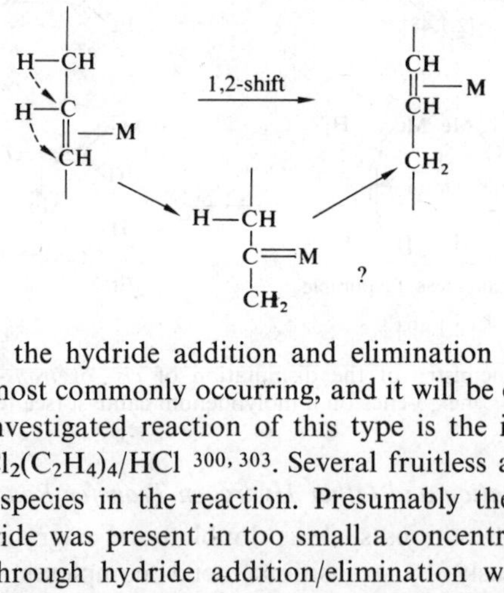

Of these three mechanisms the hydride addition and elimination mechanism is the one which is thought to be the most commonly occurring, and it will be discussed first.

The most thoroughly investigated reaction of this type is the isomerization of linear butenes, catalysed by $Rh_2Cl_2(C_2H_4)_4$/HCl [300, 303]. Several fruitless attempts were made to detect a rhodium hydride species in the reaction. Presumably these attempts were not successful because the hydride was present in too small a concentration. Strong evidence that the reaction occurs through hydride addition/elimination was obtained, however. Thus deuteration studies involving catalytic isomerization of but-1-ene in CH_3OD strongly suggest that H–D exchange and isomerization processes are merely different aspects of the same reaction since they respond equivalently to changes in reaction conditions. Thus (1)

299 R. Cramer and R. V. Lindsey, J. Am. Chem. Soc. 88 (1966) 3534.
300 J. F. Harrod and A. J. Chalk, J. Am. Chem. Soc. 88 (1966) 3491.
301 T. A. Manuel, J. Org. Chem. 27 (1962) 3941.
302 N. R. Davies, Australian J. Chem. 17 (1964) 212.
303 R. Cramer, J. Am. Chem. Soc. 88 (1966) 2272.

when ethylene is added to the reaction mixture both isomerization and deuteriation of but-1-ene are inhibited (ethylene displaces but-1-ene from the metal and is deuteriated instead). (2) At $-33°$ in CH_3OD, isomerization and deuteriation of but-1-ene simultaneously cease, whereas in a less basic medium than methanol (e.g. chloroform) but-1-ene may be isomerized at $-35°$, and under these conditions H–D exchange also occurs. (3) H–D exchange and isomerization under various conditions have about the same rate and, to a first approximation, one deuterium is introduced for each molecule of but-1-ene that is isomerized.

The above mentioned results (and others obtained by Cramer) can be accommodated by a mechanism involving hydride (deuteride) addition/elimination (Fig. 14).

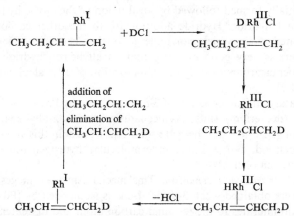

FIG. 14. Proposed hydride (deuteride) addition/elimination mechanism involved in the isomerization of but-1-ene to but-2-enes by rhodium(I) {i.e. $Rh_2Cl_2(C_2H_4)_4$} in the presence of acid[303]. For simplicity some of the ligands on the rhodium are not shown.

Deuterium analyses on the three butenes of the product mixture show that part of the but-1-ene is deuteriated whilst some of the but-2-ene contains no deuterium; clearly other exchange processes are also involved. Thus deuteriation of but-1-ene probably goes via a terminal alkyl–rhodium intermediate instead of the secondary alkyl intermediate involved in isomerization (Fig. 14). Replacement of the coordinated deuteriated but-1-ene by non-deuteriated but-1-ene could then give non-deuteriated but-2-ene after isomerization.

As described above, in the isomerization of but-1-ene in CH_3OD much deuteriated but-1-ene is formed initially. However, similar isomerization of cis-but-2-ene in CH_3OD gives a product mixture, composed almost entirely of cis- and trans-but-2-enes, in which about half of the trans-but-2-ene is deuteriated but very little of the cis-but-2-ene is deuteriated. These results strongly suggest that deuteriation involves a stereospecific addition/ elimination process, which is what one would expect of a reaction mechanism involving reversible rhodium hydride (deuteride) cis-addition. The formation of undeuteriated trans-but-2-ene from but-1-ene in CH_3OD involves a more complicated reaction sequence, with the production of deuteriated but-1-ene[303]. However, there is as yet no direct evidence in the above-mentioned reaction sequences to decide between direct protonation of a coordinated olefin and protonation of the metal to give a hydride, which then adds on to the carbon atom of a coordinated olefin. For references and a discussion of many other systems for which a hydride addition/elimination mechanism has been proposed, see the paper by Cramer and Lindsey[299].

An illustration of the importance of a cocatalyst occurs with the isomerization of but-1-ene by tetrakis(triethylphosphate)nickel(0) in the presence of acid. At 25° in 0.02M methanolic H_2SO_4, but-1-ene is 95% isomerized by this nickel complex in 5 min, but in the absence of acid less than 3% is isomerized after 4 days. However, in CH_3OD as solvent only a small amount of deuteriated butenes are formed. In order to explain this in terms of the hydride addition/elimination mechanism it has been suggested that olefin exchange on the metal occurs a lot faster than hydride (deuteride) formation or destruction.

π-Allylic hydride mechanism. This mechanism was proposed in 1962 [301] as a possible way of explaining the isomerization of pentenes by iron carbonyls. An olefin–iron tricarbonyl complex is initially formed followed by abstraction of the allylic hydrogen to form a π-allylic iron carbonyl hydride. Hydride addition on to the other carbon atom gives a rearranged coordinated olefin which can undergo further isomerization or be displaced by more olefin. There is very good evidence that an allyliciron hydride intermediate is formed in the iron pentacarbonyl catalysed isomerization of allyl alcohol to propionaldehyde (see section 3.1.5).

The palladium catalysed isomerization of $3\text{-}d_2\text{-hept-1-ene}$ by $PdCl_2(PhCN)_2$ seems to go via a 1,3-hydrogen (deuterium) shift. A mechanism with a π-allylicpalladium hydride intermediate has been suggested, but there are some features about this and similar reactions which are not understood. For example, intermolecular hydrogen (deuterium) transfer sometimes occurs between olefins[268, 300].

1,2-Hydrogen shift or carbene mechanism. This mechanism was suggested to account for the deuterioisomers obtained when $3\text{-}d_2\text{-oct-1-ene}$ is isomerized by $[Pd_2Cl_6]^{2-}$ in acetic acid[302]. No deuterium migrates to the terminal carbon atom and in deuterioacetic acid no deuterium is incorporated into the octenes from the solvent. It has since been pointed out that the results could be explained in terms of an abnormally large isotope effect operating in the isomerization of the partially deuteriated olefin (the $3\text{-}d_2\text{-oct-1-ene}$ used was about 70% isotopically pure)[268, 300]. The 1,2-hydrogen shift or carbene mechanism of olefin isomerization is now much less favoured than the hydride addition/elimination or π-allylic hydride mechanisms.

In addition to much work on the catalysed isomerization of alk-1-enes to the more thermodynamically stable alk-2-enes, alk-3-enes, etc., the migration of olefinic double bonds in other systems has been studied. These include

cyclo-octa-1,5-diene → cyclo-octa-1,4-diene → cyclo-octa-1,3-diene,

allylbenzene to propenylbenzene and allylic alcohols to aldehydes or ketones.

2.1.5g. *Catalytic Hydrogenation*[297]

Many transition metal complexes will catalyse the homogeneous hydrogenation of olefins. These complexes include $RhCl(PPh_3)_3$, $RuHCl(PPh_3)_3$, a $PtCl_2(PPh_3)_2\text{–}SnCl_2$ complex, $[Co(CN)_5]^{3-}$, $Co(CN)_2(bipy)$ and a complex formed by reducing $[RuCl_5(H_2O)]^{2-}$ with titanium(III) chloride. In all cases a hydrometal–olefin intermediate complex is formed. The transfer of hydrogen to the double bond occurs within the coordination sphere of the metal. Sometimes, as in hydrogenation by $RhCl(PPh_3)_3$, the hydrogen adds to the metal to give a dihydride and the two hydride ligands then add to a coordinated olefin either simultaneously (i.e. a concerted process) or stepwise. These processes can be represented by the following greatly simplified scheme. (M = metal complex with vacant coordination sites.)

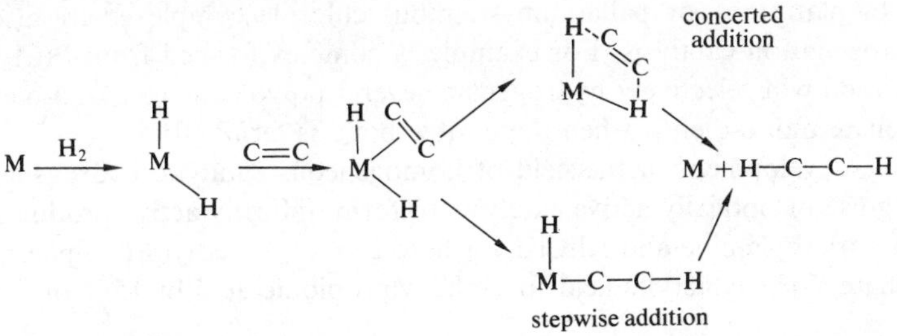

In other cases, e.g. with some ruthenium(II) chloride catalysed hydrogenations the reaction with hydrogen is heterolytic (H$^-$···H$^+$) and the metal hydride adds to the olefin to give an alkyl complex which is then split off as alkane by attack of a proton (from the solvent), viz. (M = metal complex, as before)

$$M \xrightarrow[-H^+]{H_2} M\text{---}H \xrightarrow{C=C} M\text{---}H \to M\text{---}C\text{---}C\text{---}H \xrightarrow{H^+} M + H\text{---}C\text{---}C\text{---}H$$

One of the most active and most thoroughly investigated catalysts is RhCl(PPh$_3$)$_3$ [304]. A 10^{-3}M benzene solution of this catalyst will cause such rapid hydrogenation of cyclohexene at 1 atm and 25° that the heat evolved raises the solution almost to its boiling point. The mechanism of catalytic hydrogenation of olefins by RhCl(PPh$_3$)$_3$ is summarized in Fig. 15.

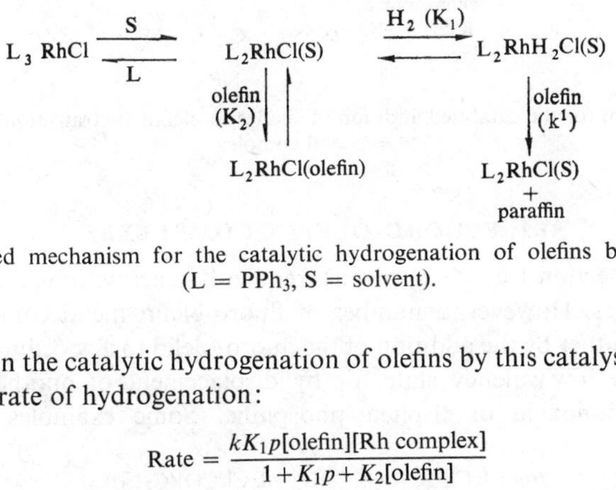

FIG. 15. Proposed mechanism for the catalytic hydrogenation of olefins by RhCl(PPh$_3$)$_3$ [305] (L = PPh$_3$, S = solvent).

Kinetic studies on the catalytic hydrogenation of olefins by this catalyst give the following expression for the rate of hydrogenation:

$$\text{Rate} = \frac{kK_1 p[\text{olefin}][\text{Rh complex}]}{1 + K_1 p + K_2[\text{olefin}]}$$

K_1, K_2 are respectively the equilibrium constants for interaction of the solvated species RhCl(solvent)(PPh$_3$)$_2$ with H$_2$ and olefin; k is the rate constant for the RDS, i.e. for the reaction of the dihydro species RhH$_2$Cl(solvent)(PPh$_3$)$_2$ with olefin and p is the concentration of H$_2$ in solution (Fig. 15).

[304] S. Montelatici, A. van der Ent, J. A. Osborn and G. Wilkinson, J. Chem. Soc. A (1968) 1054.
[305] F. H. Jardine, J. A. Osborn and G. Wilkinson, J. Chem. Soc. A (1967) 1574.

The ruthenium(II) hydride complex $RuHCl(PPh_3)_3$ is an excellent selective hydrogenation catalyst for alk-1-enes[193]. Non-terminal alkenes are hydrogenated at a very much slower rate (usually at least 10^3 times as slow).

A series of platinum– or palladium–stannous chloride complexes are also very good selective hydrogenation catalysts. For example, a complex formed from $PtCl_2(PPh_3)_2$ and stannous chloride will selectively hydrogenate several polyolefins to mono-olefins but will only hydrogenate mono-olefins when the double bond is terminal[306].

A probable development in the field of homogeneous catalytic hydrogenation will be the increasing use of optically active catalysts to form optically active products. A catalyst prepared from triethylamine and $RhCl_3L_3$, where $L = (-)$ methylpropylphenylphosphine, will hydrogenate α-phenylacrylic acid to α-phenylpropionic acid of 15% optical purity[307].

2.1.5h. *Hydrosilation of Olefins Catalysed by Transition Metal Complexes*

Some Group VIII metal complexes are extremely active catalysts for the addition of silanes to olefins. The catalysts include chloroplatinic acid, $Pt_2Cl_4(C_2H_4)_2$, $Rh_2Cl_2(C_2H_4)_2$ [308] and dicobalt octacarbonyl[278]. The suggested mechanism for this catalysis is similar to that proposed for catalytic hydrogenation. It is outlined in Fig. 16 and involves oxidative addition of the silane to the metal–olefin complex followed by an addition of M–H across the coordinated olefin. The alkylated silane is then liberated by a reductive elimination step involving attack on the metal by more olefin.

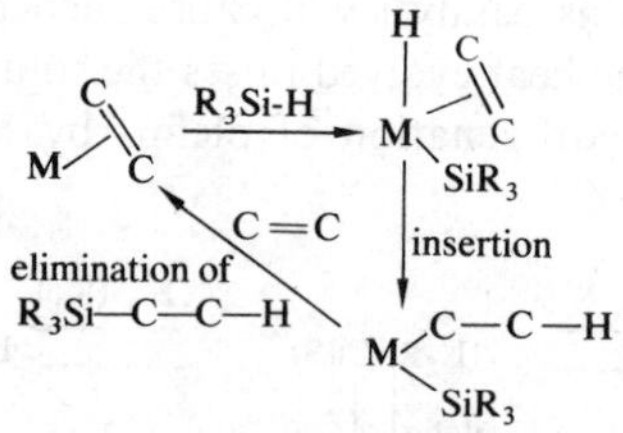

FIG. 16. Mechanism for the catalysed addition of silanes to olefins by transition metal complexes. M = metal complex.

2.2 FLUORO-OLEFIN COMPLEXES

As described in section 1.6, fluoro-olefins frequently react with metal complexes to give perfluoroalkyl species. However, a number of fluoro-olefin–metal complexes are known. They are prepared, either by the addition of the fluoro-olefin to a coordinatively unsaturated metal complex in a low valency state, or by displacement of another ligand, such as ethylene, carbon monoxide or triphenylphosphine. Some examples are given below ($L = PPh_3$).

$$\textit{trans-}IrCl(CO)L_2 + C_2F_4 \rightarrow IrCl(CO)(C_2F_4)L_2$$

With displacement of ethylene:

$$Rh(acac)(C_2H_4)_2 + C_2F_4 \rightarrow Rh(acac)(C_2H_4)(C_2F_4) \text{ [199]}$$
$$Ni(C_2H_4)L_2 + C_2F_4 \rightarrow Ni(C_2F_4)L_2$$

[306] R. W. Adams, G. E. Batley and J. C. Bailar, *J. Am. Chem. Soc.* **90** (1968) 6051.
[307] W. S. Knowles and M. J. Sabacky. *Chem. Commun.* (1968) 1445.
[308] A. J. Chalk and J. F. Harrod, *J. Am. Chem. Soc.* **87** (1965) 16.

With displacement of triphenylphosphine (L):

$$ML_4 + \text{fluoro-olefin} \rightarrow M(\text{fluoro-olefin})L_2 + 2L$$

where M = Pd or Pt; fluoro-olefin = C_2F_4, C_3F_6, C_2F_3Cl, perfluorobut-2-ene, Dewar hexafluorobenzene;

$$RhClL_3 + \text{fluoro-olefin} \rightarrow RhCl(\text{fluoro-olefin})L_2$$

where fluoro-olefin = C_2F_4, C_2F_3Cl, C_2F_3H.

With displacement of carbon monoxide:

$$\textit{trans-}Ru(CO)_3Q_2 + F_2C\text{:}C(CN)_2 \rightarrow Ru(CO)_2\{F_2C\text{:}C(CN)_2\}Q_2 \quad \text{(two isomers)}$$

where Q = $P(OCH_2)_3CEt$. The C_2F_4 complex is also known.

$$Fe(CO)_5 + \text{fluoro-olefin} \xrightarrow{h\nu} Fe(CO)_4(\text{fluoro-olefin}) \text{ [310]}$$

where fluoro-olefin = $C_2F_2Cl_2$, C_2F_3Cl, C_3F_6, perfluorocyclopentene, perfluorocyclohexene.

The structure of $RhCl(C_2F_4)(PPh_3)_2$ has been determined by X-ray diffraction[311]. The geometry and some bond lengths are shown in [2.2.1].

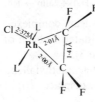

[2.2.1]

The two carbon atoms are 0.44 Å from the plane of the four fluorine atoms. The coordination around the rhodium (taking the C_2F_4 as monodentate) is not planar but distorted towards tetrahedral.

It is likely that in these perfluoro-olefin–metal complexes there is extensive back donation from the metal to the antibonding π^*-perfluoro-olefin ligand orbitals. Thus the bonding is similar to that in tetracyanoethylene complexes (section 2.1.3) and other complexes formed from olefins with electro-negative substituents. With very extensive back donation the perfluoro-olefin–metal bonding could be represented as a metallacyclopropane ring, as in [2.1.21].

As mentioned above, in structure [2.2.1] the fluorines are bent away from the metal, suggesting a change in hybridization of the carbons from sp^3 to sp^2, and in the ^{19}F n.m.r. spectrum of the complex $Fe(C_2ClF_3)(CO)_4$ the coupling constants $J(F–F)$ are very different from those of the free ligand and also suggest a change in hybridization from sp^2 to sp^3 on complexation. The ^{19}F n.m.r. spectrum of $Pt(C_3F_6)(PPh_3)_2$ similarly suggests that the perfluoropropane is σ-bonded to the metal, as in [2.1.21] [309].

Not many reactions of coordinated fluoro-olefins have been reported, but an example of a thermal rearrangement and examples of protonation reactions are given below. The complex $Pt(CF_2\text{:}CFCl)(PPh_3)_2$ on heating to just below its melting point (206–208°) rearranges to the vinyl complex $\textit{trans-}Pt^{II}Cl(CF\text{:}CF_2)(PPh_3)_2$. Fluoro-olefin complexes of platinum or rhodium are readily protonated to give alkyl–metal complexes[309], i.e.

$$Pt(C_2F_4)(PPh_3)_2 + CF_3COOH \rightarrow Pt(OOCCF_3)(CF_2CF_2H)(PPh_3)_2$$
$$RhCl(C_2F_3X)(PPh_3)_2 + HCl \rightarrow RhCl_2(CFXCF_2H)(PPh_3)_2$$

where X = F, Cl or H.

309 D. M. Barlex, R. D. W. Kemmitt and G. W. Littlecott, *Chem. Commun.* (1969) 613.

310 R. Fields, M. M. Germain, R. N. Haszeldine and P. W. Wiggans, *Chem. Commun.* (1967) 243.

311 P. B. Hitchcock, M. McPartlin and R. Mason, *Chem. Commun.* (1969) 1367.

2.3. ALLENE COMPLEXES

The complexes which allenes form with transition metals are analogous to complexes formed by olefins and are prepared in a similar way, e.g. displacement of a ligand by allene, reductive allenation, etc. In the methods described below, "alln" is an abbreviation for allene or a substituted allene.

2.3.1. Preparation

Allenes will displace ethylene from ethyleneplatinous chloride to give the corresponding allene complex

$$Pt_2Cl_4(C_2H_4)_2 + 2alln \rightarrow Pt_2Cl_4(alln)_2 + 2C_2H_4$$

where alln = allene, 1,1-dimethylallene, tetramethylallene.

Allenes similarly displace ethylene from $Rh(acac)(C_2H_4)_2$:

$$Rh(acac)(C_2H_4)_2 + 2alln \rightarrow Rh(acac)(alln)_2$$

With $Rh(acac)(CO)(C_2H_4)$ binuclear complexes of type $\{Rh(acac)(CO)\}_2(alln)$ are formed in which a rhodium is coordinated to each of the allenic double bonds, i.e. giving an allene bridged structure (see Table 3).

Iron enneacarbonyl reacts with tetramethylallene to give the complex $Fe(CO)_4$(tetra-methylallene)[312]. Allenes will also displace triphenylphosphine from rhodium(I) or platinum(0) complexes:

$$RhCl(PPh_3)_3 + allene \rightarrow RhCl(PPh_3)_2(alln)$$
$$Pt(PPh_3)_4 + allene \rightarrow Pt(PPh_3)_2(alln)$$

Platinum–alln complexes of the type $Pt(PPh_3)_2(alln)$ can also be prepared by the hydrazine reduction of cis-$PtCl_2(PPh_3)_2$ in the presence of alln (e.g. with alln or 1,3-diphenyl-allene).

$[Ir(diphos)_2]^+$ will take up allene reversibly to give $[Ir(diphos)_2(alln)]^+$.

2.3.2. Structures and Bonding

The structures of four rhodium–allene complexes have been determined by X-ray diffraction (see Table 11). The structures of $Rh(acac)(1,1$-dimethylallene$)_2$ [2.3.1] and $Rh_2(acac)_2(CO)_2(alln)$ [2.3.2] are shown.

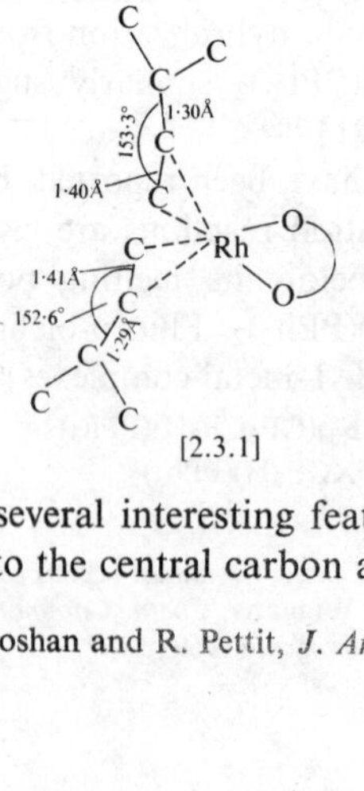

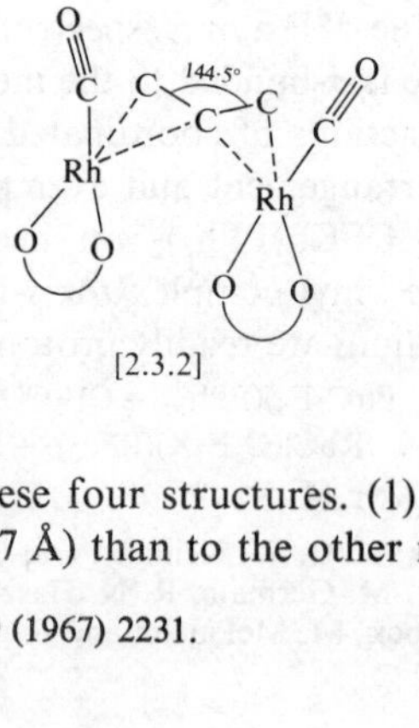

There are several interesting features about these four structures. (1) The rhodium is always closer to the central carbon atom (2.03–2.07 Å) than to the other rhodium-bonded

---

[312] R. Ben-Shoshan and R. Pettit, J. Am. Chem. Soc. 89 (1967) 2231.

TABLE 11. SOME STRUCTURAL DATA FOR RHODIUM–ALLENE COMPLEXES
(Bond lengths in Å)

	Coordinated C=C	Uncoordinated C=C	C–C–C angle (°)	Rh–C distances	References
Rh(acac)(1,1-dimethylallene)₂	1.40 1.41	1.30 1.29	153.3 152.6	2.07, 2.13 2.06, 2.13	P. Racanelli, G. Pantini, A. Immirzi, G. Allegra and L. Porri, *Chem. Commun.* (1969) 361
Rh(acac)(1,1,3-tetramethylallene)₂	1.37 1.38 1.37	1.33 1.32	147.2 148.9 144.5	2.03, 2.18 2.03, 2.18 2.05, 2.12 2.06, 2.14	T. G. Hewitt, K. Anzenhofer and J. J. De Boer, *Chem. Commun.* (1969) 312
Rh₂(acac)₂(CO)₂(allene)					P. Racanelli, G. Pantini, A. Immirzi, G. Allegra and L. Porri, *Chem. Commun.* (1969) 361
RhI(allene)(PPh₃)₂	1.35	1.34	158	2.04, 2.17	T. Kashiwagi, N. Yasuoka, N. Kasai and M. Kukudo, *Chem. Commun.* (1969) 317

allene carbon atom (2.12–2.18 Å). (2) Although the errors are rather large, the C–C separation of the rhodium-bonded carbon atoms is greater than that of the other allene C–C separation. This is what one would expect, since back donation from the d-orbitals of the rhodium into the antibonding π^*-orbitals of the two rhodium-bonded allene carbon atoms would weaken the C–C bonding (as in olefin complexes, section 2.1.3a). (3) The non-bonded allene carbon atom is considerably bent away from the rhodium. The C–C–C angles of 144.5–158° (Table 11) can be interpreted in terms of a rehybridization of the allene central carbon atom from sp towards sp^2. (4) In the 1,1-dimethyl-allene complex [2.3.1] the unsubstituted allenic double bond is coordinated to the rhodium. Similarly, n.m.r. shows that in the platinum complex $Pt(PPh_3)_2(1,1\text{-dimethylallene})$ the unsubstituted double bond is coordinated to the platinum[313].

2.3.3. Physical Methods

2.3.3a. *Nuclear Magnetic Resonance Spectroscopy*

Some allene complexes show variable temperature n.m.r. spectra. At room temperature the ^1H n.m.r. spectrum of $Fe(CO)_4(\text{tetramethylallene})$ consists of a singlet, but at $-60°$ it consists of three singlets with intensity ratios of $1:1:2$. The effect is reversible and shows that a rapid valence tautomerism is occurring at room temperature,

$$Me_2C{=}C{=}CMe_2 \rightleftarrows Me_2C{=}C{=}CMe_2$$
$$\hspace{1.5em}|\hspace{9em}|$$
$$\hspace{1.5em}Fe(CO)_4\hspace{5em}Fe(CO)_4$$

but becomes slow (or stops) at $-60°$ [312].

The platinum(II) complex $Pt_2Cl_4(\text{tetramethylallene})_2$ also shows a variable temperature ^1H n.m.r. spectrum, over the range $-70°$ to $+50°$. During these changes ^{195}Pt–H coupling is preserved, showing that the rate process is intramolecular[313]. A rapid valence tautomerism such as is found with the iron complex (above) is probably occurring. Complexes of the type $Pt(PPh_3)_2(\text{alln})$ or $RhCl(PPh_3)_2(\text{alln})$ show no evidence of valence tautomerism, however. Nuclear magnetic resonance data on complexes of the type $Pt(PPh_3)_2(\text{alln})$ have been interpreted in terms of a four-coordinated planar structure, i.e. as a platinacyclopropane.

2.3.3b. *Infrared Spectroscopy*

So far few data have been published, but $Pt(PPh_3)_2(\text{alln})$ shows an infrared absorption band at 1680 cm⁻¹, assigned to $\nu(C{=}C{=}C)$, shifted *ca.* 250 cm⁻¹ from the value in free allene. The partially decomposed $[Ir(\text{diphos})_2(\text{alln})]^+$ shows a band at 1640 cm⁻¹, attributed to complexed allene.

2.3.4. Reactions

Few reactions of coordinated allenes have been reported. Allene is displaced by carbon monoxide from $RhCl(PPh_3)_2(\text{alln})$ and by oxygen (O_2) from $[Ir(\text{diphos})_2(\text{alln})]^+$. The formation of allylic complexes from allenes, and some related reactions, are discussed in section 3. An interesting pentamerization of allene occurs when it reacts with $Rh_2Cl_2(C_2H_4)_4$. One of the products, $RhCl(C_{15}H_{20})$, has been formulated as chloro-(1,2,5,6,8-pentamethylenecyclododecane)rhodium(I)[314].

[313] K. Vrieze, H. C. Volger, M. Gronert and A. P. Praat, *J. Organomet. Chem.* **16** (1969) 19.
[314] S. Otsuka, K. Tani and A. Nakamura, *J. Chem. Soc.* A (1969) 1404.

2.4. ACETYLENE COMPLEXES[149, 151, 315, 316]

Acetylene complexes are prepared by reacting an acetylene with a metal complex. It is convenient to discuss the methods of preparation in terms of the types of ligands being displaced.

2.4.1a. *Preparation by Displacement of Halide, a Solvent Molecule or by Addition to a Coordinatively Unsaturated Species*

Some acetylenes (ac) form complexes, completely analogous to ethylene-platinous chloride or Zeise's salt. Thus di-tert-butyl-acetylene reacts with Na_2PtCl_4, as shown, and the product undergoes bridge splitting reactions with KCl or amines (am)[317].

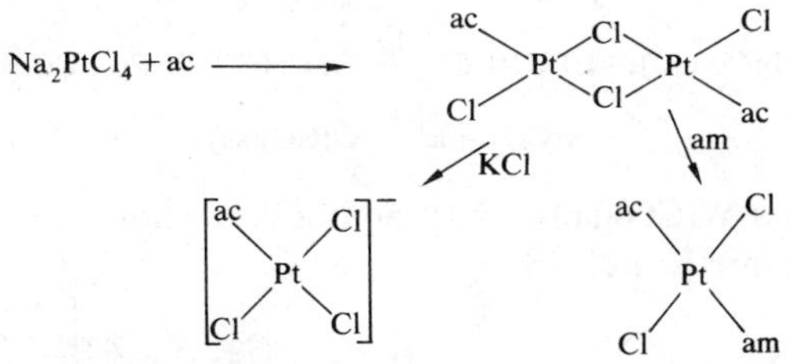

Other disubstituted acetylenes RC⋮CR react similarly, provided one of the Rs is t-butyl; otherwise decomposition occurs. Complexes of type M[PtCl₃(ac)] (M = Na or K) are formed from M_2PtCl_4 and acetylenic diols or their methyl ethers. For diols of the type RMe(OH)C⋮C(OH)MeR the stability of the complexes decreases in the order of R, Me > Ph > H. The diol ethers give less stable complexes[151].

Acetylenes also form complexes with copper(I) or silver(I), e.g. acetylene reacts with copper(I) chloride in dilute acid to give colourless crystalline complexes of type $(CuCl)_3(C_2H_2)$ or $(CuCl)_2(C_2H_2)$. Disubstituted acetylenes also react, e.g. but-2-yne gives CuCl(but-2-yne). Hex-3-yne reacts with silver nitrate slowly to give a labile adduct $AgNO_3 \cdot C_6H_{10}$ [318] and distribution studies show that complexes of types [Ag(ac)]⁺ and [Ag₂(ac)]²⁺ can be formed in aqueous solution.

Some coordinatively unsaturated compounds in low valency states form complexes with acetylenes, especially if the acetylene carries electro-negative substituents. Thus the complex *trans*-IrCl(CO)(PPh₃)₂ forms adducts IrCl(CO)(PPh₃)₂(ac), probably with the stereochemistry shown in [2.4.1] [319].

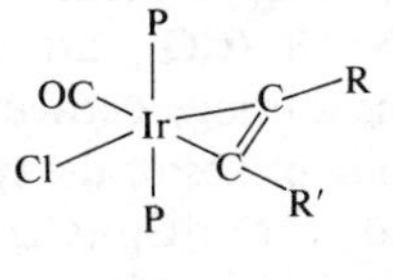

[2.4.1]

315 F. L. Bowden and A. B. P. Lever, "Transition metal chemistry of acetylenes", in *Organomet. Chem. Rev.* **3** (1968) 227.

316 W. Hübel, *Organometallic Derivatives from Metal Carbonyls and Acetylenic Compounds in Organic Syntheses via Metal Carbonyls*, Vol. 1, Interscience, New York (1968).

317 J. Chatt, R. G. Guy and L. A. Duncanson, *J. Chem. Soc.* (1961) 827.

318 A. E. Comyns and H. J. Lucas, *J. Am. Chem. Soc.* **79** (1957) 4341.

319 J. P. Collman and J. W. Kang, *J. Am. Chem. Soc.* **89** (1967) 844.

(ac) = $ROOC \cdot C : C \cdot COOR$ (R = H, Me or Et) or $PhC : C \cdot COOEt$. Other low valence metal complexes, which react similarly, are $RhCl(CO)(PPh_3)_2$ to give $RhCl(CO)(PPh_3)_2(ac)$ and VCp_2 to give $VCp_2(MeOOC \cdot C : C \cdot COOMe)$, which is paramagnetic[315].

2.4.1b. *By Displacement of Carbon Monoxide*

When acetylenes react with metal carbonyls, C–C bond formation may occur to give complex products (see section 2.4.4). However, there are some examples of simple displacement of carbon monoxide by the acetylene. Thus $VCp(CO)_4$ reacts with pent-1-yne or hex-1-yne to give $VCp(CO)_2(ac)$ and $VCp(CO)_3PPh_3$ reacts with tolan ($PhC : CPh$) to give $VCp(CO)(PPh_3)(PhC : CPh)$. Ultraviolet light may promote displacement of carbon monoxide, e.g.

$$MnCp(CO)_3 + CF_3C : CCF_3 \xrightarrow{h\nu} MnCp(CO)_2(CF_3C : CCF_3)$$

$$W(CO)_6 + ac \xrightarrow{h\nu} W(CO)(ac)_3$$

The tungsten complexes $W(CO)(ac)_3$, where ac is a disubstituted acetylene, are thought to have the unusual structure [2.4.2]

[2.4.2]

Acetylenes will displace the two bridging carbon monoxides from $Co_2(CO)_8$, in a particularly smooth reaction:

$$Co_2(CO)_8 + ac \rightarrow Co_2(CO)_6(ac) + 2CO$$

Kinetic measurements show the intermediate formation of $Co_2(CO)_7(ac)$. Many complexes of type $Co_2(CO)_6(ac)$ have been made and the more electro-negative the acetylene substituents the more stable is the complex (as shown by displacement studies)[320]. Acetylenes will also displace two bridging carbon monoxides from $Co_4(CO)_{12}$ to give $Co_4(CO)_{10}(ac)$.

The bridging carbonyls from $Ni_2Cp_2(CO)_2$ are similarly displaced by acetylenes to give $Ni_2Cp_2(ac)$. The acetylene forms a bridge between the two metal atoms, just as in the cobalt complexes[321]. Molybdenum complexes of the type $MoCpR(CO)_3$ (R = H, Me or Et) react with tolan to give a low yield of $Mo_2Cp_2(CO)_4(PhC : CPh)$. An acetylene bridging structure has also been suggested for this complex.

Although acetylenes usually react with iron carbonyls to give complex ligands there are a few examples where the acetylene remains intact. Thus tolan reacts with $Fe_2(CO)_9$ to give a low yield of $Fe_3(CO)_9(PhC : CPh)$, with the structure [2.4.3] [322], as shown by X-ray diffrac-

[320] G. Cetini, O. Gambino, R. Rossetti and E. Sappa, *J. Organomet. Chem.* **8** (1967) 149.
[321] O. S. Mills and B. W. Shaw, *Acta Cryst.* **18** (1965) 562.
[322] J. F. Blount, L. F. Dahl, C. Hoogzand and W. Hübel, *J. Am. Chem. Soc.* **88** (1966) 292.

tion. Further reaction of $Fe_3(CO)_9(PhC\!:\!CPh)$ with tolan under mild conditions gives $Fe_3(CO)_8(PhC\!:\!CPh)_2$, a violet complex with the structure [2.4.4][323].

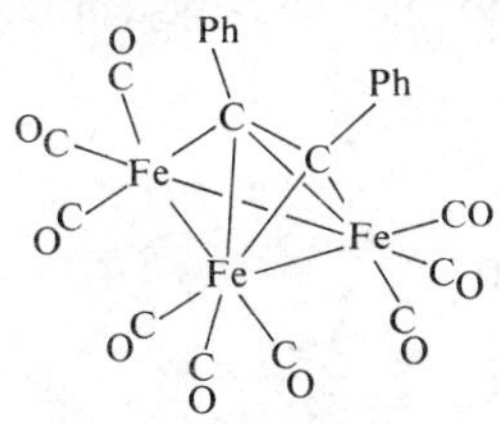

[2.4.3] 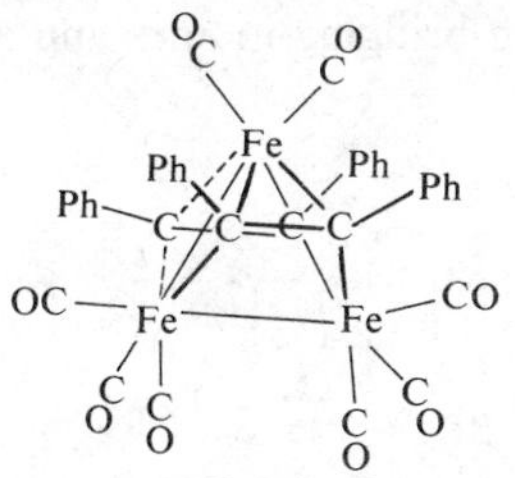   [2.4.4]

Reaction mixtures generated from iron carbonyls and acetylenes with bulky substituents give complexes of the type $Fe(CO)_4(ac)$ and $Fe_2(CO)_7(ac)$. The proposed structures[316] for these two complexes are shown in [2.4.5] and [2.4.6].

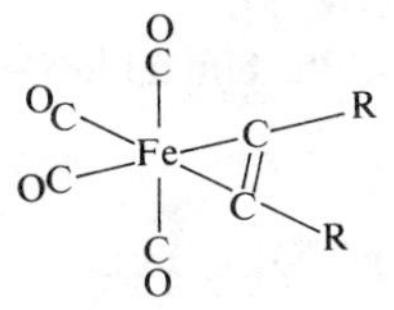

[2.4.5]

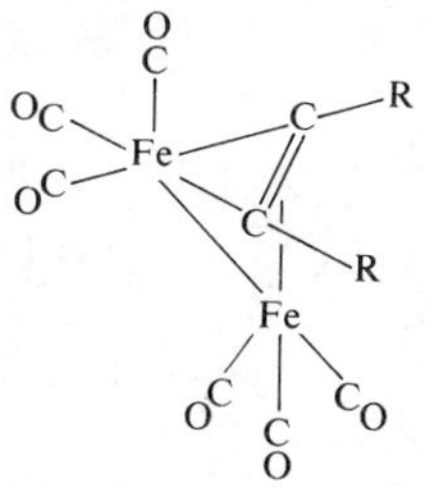

[2.4.6]

2.4.1c. *Displacement of Other Ligands*

The nitrogen of the iridium–N_2 complex *trans*-$IrCl(N_2)(PPh_3)_2$ is displaced by acetylenes (ac) with electro-negative substituents, e.g. $MeOOC\cdot C\!:\!C\cdot COOMe$, to give acetylene complexes $IrCl(ac)(PPh_3)_2$ [16].

Acetylenes will displace acetonitrile from $W(acetonitrile)_3(CO)_3$ to give the complexes of type $W(CO)(ac)_3$, discussed in section 2.4.1b. They will also displace ethylene from ethyleneplatinous chloride (e.g. $Bu^tC\!:\!CBu^t$) and olefins from $M(PPh_3)_2(olefin)$ to give $M(PPh_3)_2(ac)$ (M = Ni or Pt).

Zerovalent triphenylphosphine–platinum or –palladium complexes react with acetylenes as follows:

$$M(PPh_3)_4 + ac \rightarrow M(PPh_3)_2(ac) \text{ [324]}$$

where M = Pd or Pt. The more electro-negative the acetylene substituents the more stable the complex.

2.4.1d. *Reductive Acetylenation and Miscellaneous Methods*

Reduction of $PtCl_2(PPh_3)_2$ with hydrazine in the presence of acetylenes gives complexes of the type $Pt(PPh_3)_2(ac)$ [325]. This method is reported not to work for palladium.

[323] R. P. Dodge and V. Schomaker, *J. Organomet. Chem.* 3 (1965) 274.
[324] E. O. Greaves, C. J. L. Lock and P. M. Maitlis, *Can. J. Chem.* 46 (1968) 3879.
[325] J. Chatt, G. A. Rowe and A. A. Williams, *Proc. Chem. Soc.* (1957) 208.

When nickel carbonyl is treated with o-di-iodobenzene, black crystals are formed of what was originally formulated as a nickel–benzyne complex[326]. More recent work has shown the complex to be the polymeric catena-[μ-(o-phthaloyl-C,C'O)]-μ-iodo-iodo-nickel(IV) with bridging iodines and a bridging phtholoyl–nickel system [2.4.7][326a].

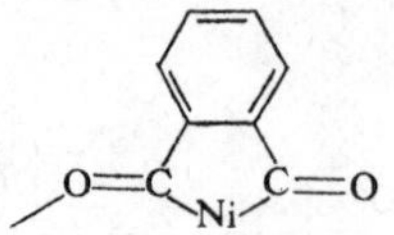

[2.4.7]

2.4.2. Some Examples of Transition Metal–Acetylene Complexes

A list of the more important types of metal–acetylene complexes is now given, metal by metal. As before, the majority of these compounds are discussed further in section 2.4.

Vanadium

VCp$_2$(ac)   ac = MeOOCC⫶CCOOMe [327].
VCp(CO)$_2$(ac)   ac = PrC⫶CH, BuC⫶CH [19].
VCp(CO)(ac)PPh$_3$ ac = PhC⫶CPh. Unstable[19].

Niobium

NbCp(CO)$_2$(ac)   From NbCp(CO)$_4$+PhC⫶CPh [329].

Chromium

Cr(CO)$_2$Ar(ac)   ac = HC⫶CH, PhC⫶CPh acetylene dicarboxylic
                             ester. Ar = hexamethylbenzene (most stable) to
                             benzene (less stable), etc.

Molybdenum

Mo(CO)(PhC⫶CPh)$_3$ From Mo(CO)$_6$+PhC⫶CPh [331].

Tungsten

W(CO)(ac)$_3$ ac = disubstituted acetylene[331, 332].

Manganese

MnCp(CO)$_2$(ac)   ac = PhC⫶CPh or CF$_3$C⫶CCF$_3$ [333, 334].

[326] E. W. Gowling, S. F. A. Kettle and G. M. Sharples, *Chem. Commun.* (1968) 21.
[326a] N. A. Bailey, S. E. Hall, R. W. Jotham and S. F. A. Kettle, *Chem. Commun.* (1971) 282.
[327] R. Tsumura and N. Hagihara, *Bull. Chem. Soc. Japan* **38** (1965) 861.
[328] R. Tsumura and N. Hagihara, *Bull. Chem. Soc. Japan* **38** (1965) 1901.
[329] A. N. Nesmeyanov, K. N. Anisimov, N. E. Kolobova and A. A. Pasynskii, *Bull. Soc. Acad. Sci. USSR, Chem. Sect.* **4** (1966) 746.
[330] W. Strohmeier and H. Hellmann, *Chem. Ber.* **98** (1965) 1598.
[331] W. Strohmeier and D. von Hobe, *Z. Naturforsch.* **19b** (1964) 959.
[332] D. P. Tate, J. M. Augl, W. M. Ritchey, B. L. Ross and J. G. Grasselli, *J. Am. Chem. Soc.* **86** (1964) 3261.
[333] J. L. Boston, S. O. Grim and G. Wilkinson, *J. Chem. Soc.* (1963) 3468.
[334] W. Strohmeier and D. von Hobe, *Z. Naturforsch.* **16b** (1961) 402.

Rhenium

ReCl(ac)$_2$ — From rhenium trichloride and PhC⋮CH; also with
ac = 3-methylbut-1-yne-3-ol [335].

Iron

Fe(CO)$_4$(ac) — ac = Bu^{t}C⋮CBu^t, PhC⋮CSiMe$_3$ [316].
Fe$_2$(CO)$_7$(ac) — ac = Me$_3$SiC⋮CSiMe$_3$ or Bu^{t}C⋮CBu^t [316].
Fe$_3$(CO)$_9$(PhC⋮CR) — R = Ph or Me [322].
Fe$_3$(CO)$_8$(PhC⋮CPh)$_2$ [323]

Cobalt

Co$_2$(CO)$_6$(ac) — Deeply coloured. Many examples. See text[336].
Co$_4$(CO)$_{10}$(ac) — ac = PhC⋮CPh, EtC⋮CEt, etc.[337].
Co$_2$(CO)$_6$PhC⋮C·C⋮CPhCo$_2$(CO)$_6$ — From Co$_2$(CO)$_8$ + PhC⋮C·C⋮CPh [338].
Co$_2$(CO)$_6$·R$_2$M(C⋮CR')$_2$·Co$_2$(CO)$_6$ — R, R', alkyl or aryl. M = Si, Ge, Sn [339].
[Co(CH$_3$NC)$_4$MeOOC·C$_2$COOMe](NO$_3$)$_2$ [340]

Rhodium

RhCl(ac)L$_2$ — ac = PhC⋮CPh. L = PPh$_3$, AsPh$_3$ or SbPh$_3$ [201].
ac = CF$_3$C⋮CCF$_3$ [341], 2,7-dimethyloct-3-en-5-yn-2,7-diol [342].

RhCl(CO)(PPh$_3$)$_2$(ac) — ac = HOOC·C⋮C·COOH, HOOC·C⋮CH [319].

Iridium

IrCl(CO)(PPh$_3$)$_2$(ac) — ac = ROOC·C⋮C·COOR. R = H, Me, Et.
ac = PhC⋮C·COOEt [319] with ac = CF$_3$C⋮CCF$_3$
the addition is reversible in the solid state[343].

IrCl(PPh$_3$)$_2$(ac) — ac = PhC⋮CPh, MeOOC·C⋮C·COOMe [319].

Nickel

Ni(ac)(PPh$_3$)$_2$ — ac = PhC⋮CPh, MeC⋮CMe [344],
MeOOC·C⋮C·COOMe, PhC⋮CMe [324].

Ni$_2$Cp$_2$(ac) — dark green, crystalline[345].
Ni$_4$(CO)$_3$(ac)$_3$ — ac = CF$_3$C⋮CCF$_3$. [346].

335 R. Colton, R. Levitus and G. Wilkinson, *Nature* **186** (1960) 233.
336 H. W. Sternberg, H. Greenfield, R. A. Friedel, J. H. Wotiz, J. Markby and I. Wender, *J. Am. Chem. Soc.* **81** (1959) 1629.
337 L. F. Dahl and D. L. Smith, *J. Am. Chem. Soc.* **84** (1962) 2450.
338 W. Hübel and R. Merényi, *Chem. Ber.* **96** (1963) 930.
339 S. D. Ibekwe and M. J. Newlands, *J. Chem. Soc. A* (1967) 1783.
340 W. C. Kaska and M. E. Kimball, *J. Inorg. Nucl. Chem. Letters*, **4** (1968) 719.
341 M. J. Mays and G. Wilkinson, *J. Chem. Soc.* (1965) 6629.
342 H. Singer and G. Wilkinson, *J. Chem. Soc. A* (1968) 849.
343 G. W. Parshall and F. N. Jones, *J. Am. Chem. Soc.* **87** (1965) 5356.
344 G. Wilke and G. Herrmann, *Angew. Chem.* **74** (1962) 693.
345 J. F. Tilney-Bassett, *J. Chem. Soc.* (1963) 4784.
346 R. B. King, M. I. Bruce, J. R. Phillips and F. G. A. Stone, *Inorg. Chem.* **5** (1966) 684.

Palladium

Pd(PR₃)₂(ac) PR_3 = PBu_3, PMe_2Ph, PPh_3. ac =
                                             $MeOOC \cdot C \vdots C \cdot COOMe$, $CF_3C \vdots CCF_3$ [324].

Platinum

Pt(PR₃)₂(ac)   R = Ph. ac = variety of substituted tolans[325],
                                           cyclo-octyne[347]. PR_3 = PMe_2Ph. ac =
                                           $CF_3C \vdots CCF_3$ [324].

Pt(ac)₂   ac = acetylenic diol[347a].

K[PtCl₃(ac)], Pt₂Cl₄(ac)₂   ac = various t-butylacetylenes[317], various
cis- and *trans*-PtCl₂(ac)amine   acetylenic diols or their methyl ethers[348].

(Et₃NH)[PtCl₃PhC $\vdots$ CPh] [349]

Copper

CuCl(MeC $\vdots$ CMe) [350]
Cu₃Cl₃·HC $\vdots$ CH, Cu₂Cl₂·HC $\vdots$ CH [351]

Silver

Ag(ac)⁺, Ag₂(ac)²⁺   ac = EtC $\vdots$ CEt, PrC $\vdots$ CMe; existence demonstrated
                                            in solution by distribution studies[352].

2.4.3. Structures, Bonding and Stability

2.4.3a. *Structures*

X-ray crystallography has shown that an acetylene ligand can be bonded to just one metal atom, as in *trans*-PtCl₂(BuᵗC $\vdots$ CBuᵗ)*p*-toluidine; or form a bridge between two metal atoms, as in Co₂(CO)₆(PhC $\vdots$ CPh); or be bonded to several metal atoms, e.g. four in Co₄(CO)₁₀(EtC $\vdots$ CEt). Some transition metal–acetylene complexes whose structures have been determined by X-ray diffraction, are listed in Table 12.

In the complex *trans*-PtCl₂(BuᵗC $\vdots$ CBuᵗ)(*p*-toluidine) the two acetylenic carbon atoms are equidistant from the platinum and lie on a line perpendicular to the coordination plane {as in PtCl₂(C₂H₄)NHMe₂}. In contrast, the line of the two acetylenic carbon atoms in Pt(PPh₃)₂(PhC $\vdots$ CPh) is at an angle of only 14° from the P–Pt–P plane. Some other angles and bond lengths for this compound are shown in [2.4.8].

[2.4.8]

---

[347] T. L. Gilchrist, F. J. Graveling and C. W. Rees, *Chem. Commun.* (1968) 821.
[347a] F. D. Rochon and T. Theophanides, *Can. J. Chem.* **46** (1968) 2973.
[348] A. O. Sheveleva and S. V. Bukhovets, *Russ. J. Inorg. Chem.* **12** (1967) 508.
[349] S. V. Bukhovets and N. K. Pukhov, *Zhur. Neorg. Khim.* **3** (1958) 1714; *Chem. Abs.* **53** (1959) 21338 i.
[350] F. L. Carter and E. W. Hughes, *Acta Cryst.* **10** (1957) 801.
[351] R. Vestin, *Acta chem. scand.* **8** (1954) 533.
[352] W. S. Dorsey and H. J. Lucas, *J. Am. Chem. Soc.* **78** (1956) 1665.

TABLE 12. SOME TRANSITION METAL–ACETYLENE COMPLEXES WHOSE STRUCTURES HAVE BEEN DETERMINED BY X-RAY CRYSTALLOGRAPHY

	C–C (Å)	C—C:C angle (°)	References
trans-PtCl₂(BuᵗC:CBuᵗ)p-toluidine	1.27	165	G. R. Davies, W. Hewertson, R. H. B. Mais and P. G. Owston, Chem. Commun. (1967) 423
Pt(PPh₃)₂(PhC:CPh)	1.32	140	J. O Glanville, J. M. Stewart and S. O. Grim, J. Organomet. Chem. 7 (1967) 9
Co₂(CO)₆(PhC:CPh)	1.46	137, 139	W. Sly, J. Am. Chem. Soc. 81 (1959) 18
Ni₂(Cp)₂(PhC:CPh)			O. S. Mills and B. W. Shaw, Acta Cryst. 18 (1965) 562
Co₂(CO)₆(perfluorocyclohex-3-en-1-yne)	1.29	118, 123 [a]	N. A. Bailey, M. R. Churchill, R. Hunt, R. Mason and G. Wilkinson, Proc. Chem. Soc. (1964) 401
Co₄(CO)₁₀(EtC:CEt)	1.44	—	L. F. Dahl and D. L. Smith, J. Am. Chem. Soc. 84 (1962) 2451
Fe₃(CO)₉(PhC:CPh)	1.41	118, 131	J. F. Blount, L. F. Dahl, C. Hoogzand and W. Hübel, J. Am. Chem. Soc. 88 (1966) 292
Fe₃(CO)₈(PhC:CPh)₂	1.40, 1.38	122, 128 123, 125	R. P. Dodge and V. Schomaker, J. Organomet. Chem. 3 (1965) 274

[a] In a ring system.

In both these complexes the substituents on the acetylene are bent back, suggesting substantial back coordination from the platinum, i.e. a rehybridization of the acetylenic carbon atoms occurs on complexation.

In $Co_2(CO)_6(PhC\!:\!CPh)$ the acetylenic carbons of the diphenylacetylene lie in a line almost at right angles to the Co–Co bond, as shown in [2.4.9].

Ph

Ph   C

C

C   Co(CO)₃

(CO)₃Co

[2.4.9]

The C–C distance of 1.46 Å is 0.27 Å greater than in the free acetylene, indicative of extensive back coordination from the cobalts. The Co–Co distance is 2.47 Å compared with 2.52 Å in $Co_2(CO)_8$. In the nickel complex $Ni(Cp)_2(PhC\!:\!CPh)$ the acetylene is also at right angles to the Ni–Ni bond (which is very short at 2.33 Å). In $Co_2(CO)_6$(perfluorocyclohex-3-ene-1-yne) the acetylenic carbon atoms form a bridge between the two cobalts, as shown in [2.4.10].

F   CF₂

C   CF₂

FC   C

C   Co(CO)₃

(CO)₃Co

[2.4.10]

In $Co_4(CO)_{10}(EtC\!:\!CEt)$ the acetylene is bonded to all four cobalts, forming σ-bonds to two of them and μ-bonds to the remaining two, [2.4.11]

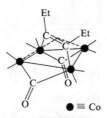

[2.4.11]

Each Co has two terminal COs, which are not shown.

The structure of $Fe_3(CO)_9(PhC\!:\!CPh)$ is complex, with a triangle of iron atoms bonded to the acetylene in a delocalized manner. In $Fe_3(CO)_8(PhC\!:\!CPh)_2$ the two acetylenes lie on either side of a triangle of iron atoms; the Fe–Fe bond lengths are 2.46, 2.47 and 2.59 Å. Again the bonding is delocalized.

2.4.3b. *Bonding*

The bonding of $Bu^tC\!:\!CBu^t$ to the platinum in *trans*-$PtCl_2(Bu^tC\!:\!CBu^t)(p\text{-toluidine})$ or $Pt_2Cl_4(Bu^tC\!:\!CBu^t)_2$ is clearly analogous to that of ethylene in $PtCl_2(C_2H_4)NHMe_2$ or $Pt_2Cl_4(C_2H_4)_2$, i.e. there is a σ-donor bond from the acetylenic π-orbitals into a vacant metal orbital and back donation from filled metal d-orbitals into π^*-orbitals of the acetylene. In complexes of these types $v(C\!:\!C)$ is lowered by *ca.* 200 cm^{-1} on coordination.

In complexes of type $Pt(PPh_3)_2(ac)$, however, $v(C:C)$ is lowered by much more, e.g. for ac = $CF_3C:CCF_3$ by 425 cm^{-1}. The value for $v(C:C)$ approaches that of a double bond and it has been suggested that a good representation of the bonding would be in terms of a platinacyclopropene structure [2.4.12], with each carbon bonded to the platinum by a bent bond.

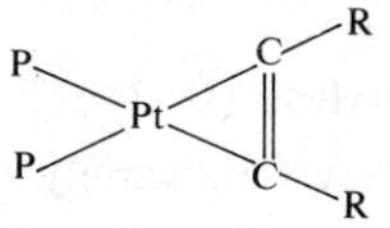

[2.4.12]

The almost planar structure for $Pt(PPh_3)_2(PhC:CPh)$ [2.4.8] supports this formulation.

Since acetylenes have two sets of π-orbitals at right angles they can form a bivalent bridging group, as in $Co_2(CO)_6PhC:CPh$. We have seen in section 2.4.3a that they can also participate in more complex bonding arrangements, e.g. to three or four metal atoms. In these complexes the acetylenic carbons are farther apart than in the free acetylene (see Table 12).

2.4.3c. *Stability*

A complex formed between an acetylene and a metal in a low valency state is undoubtedly stabilized by electronegative substituents on the acetylene. For complexes of type $Pt(PPh_3)_2(ac)$ the order of affinities of acetylenes for the platinum is

$4-NO_2C_6H_4C:CC_6H_4-4-NO_2 > PhC:CPh > PhC:CH \sim AlkC:CAlk > AlkC:CH > HC:CH$ [317].

With binuclear complexes of type $Co_2(CO)_6(ac)$ the order is

$CF_3C:CCF_3 > MeOOCC:CCOOMe > PhC:CPh > MeC:CPh > MeC:CMe > PhC:CH \geqslant$

$MeC:CH > HC:CH > Et_2NCH_2C:CCH_2NEt_2$ [320].

Thus back donation of electrons from the metal(s) to the π^*-orbitals of the acetylene is more important than donation of electrons from the acetylene to the metal.

The stable acetylene complexes of platinum(II), however, do not have acetylenes with electro-negative substituents. The remarkable stabilizing effect of the t-butyl group has not been explained; it is possibly not electronic in origin and may be steric, i.e. by preventing attack leading to decomposition. Perhaps the absence of hydrogens on the α-carbon atom is also an important factor.

TABLE 13. C:C STRETCHING FREQUENCIES (cm^{-1}) FOR SOME ACETYLENE COMPLEXES [315, 324]

	$CF_3C:CCF_3$	MeOOCC:CCOOMe	PhC:CPh
VCp$_2$(ac)	1800	1821	
MnCp(CO)$_2$(ac)	1919		*ca*.1900
RhCl(PPh$_3$)$_2$(ac)	1917		1919
IrCl(CO)(PPh$_3$)$_2$(ac)	1773		
Ni(PPh$_3$)$_2$(ac)	1790		
Pd(PPh$_3$)$_2$(ac)	1811, 1838	1830, 1845	
Pt(PPh$_3$)$_2$(ac)	1775	1782	1768, 1740

2.4.4. Infrared Spectroscopy

As mentioned in section 2.4.3, the stretching frequency of the coordinated triple bond decreases on complexation. In complexes such as $Pt_2Cl_4(Bu^tC\!:\!CBu^t)_2$ the decrease is only *ca.* 200 cm^{-1}, but in complexes where extensive back donation occurs it is as much as 425 cm^{-1}. Table 13 gives $\nu(C\!:\!C)$ frequencies for some acetylene complexes.

2.4.5. Reactions of Coordinated Acetylenes[111a, 268, 277, 282, 275, 315]

Coordinated acetylenes show a very wide variety of reactions. These range from simple insertions, through oligomerization to complex reactions with metal carbonyls, giving products with very unusual structures.

2.4.5a. *Insertion Reactions*

In an insertion reaction a coordinated acetylene, RC:CR, interacts with a metal–ligand bond, M–X, to give an "insertion" product,

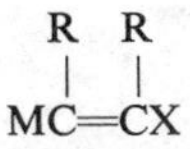

$$\begin{array}{cc} R & R \\ | & | \\ \end{array}$$
$$MC\!=\!CX$$

Although the mechanism is not known in detail, it seems likely that a four-centre concerted *cis*-addition of M–X across the coordinated triple bond is involved. Examples of "insertion" of acetylenes into metal–hydrogen, metal–carbon, metal–hydroxyl, metal–chlorine and metal–metal bonds are known.

(1) *Metal–hydrogen*, An example of this occurs when $MnH(CO)_5$ is treated with $CF_3C\!:\!CCF_3$, giving $Mn\{C(CF_3)\!:\!CHCF_3\}(CO)_5$. Carboxylation reactions, e.g. the acrylic ester synthesis, probably involve insertion into a metal–hydride bond and are discussed in section 2.4.5b.

(2) *Metal–carbon.* Alkyl– or acyl–cobalt carbonyls undergo insertion reactions with acetylenes, e.g. hex-3-yne and acetylcobalt tetracarbonyl give a good yield of the π-allylic complex 2,3-diethyl-π-(2,4)-penteno-4-lactonylcobalt tricarbonyl [2.4.13]. The following mechanism has been suggested[277, 282]:

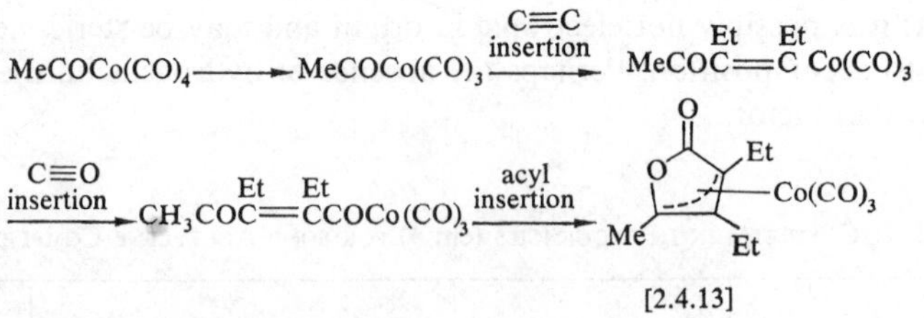

[2.4.13]

Acetylene insertion reactions into a metal–carbon bond occur in the synthesis of hexa-2,5-dienoic esters (section 3.5).

(3) *Metal–hydroxyl.* The hydration of acetylenes to ketones by ruthenium(III) chloride probably involves insertion into an Ru–OH bond[353].

(4) *Metal–chlorine.* An example is the formation of $Pd_2Cl_2(CR\!:\!CClCR'R''NMe_2)_2$ from palladium chloride and $RC\!:\!C\!\cdot\!CR'R''NMe_2$ (see section 1.1.2).

[353] J. Halpern, B. R. James and A. L. W. Kemp, *J. Am. Chem. Soc.* **83** (1961) 4097.

(5) *Metal–metal*. Acetylene insertion into a metal–metal bond seems to occur when potassium pentacyanocobaltate, which may have a Co–Co bond, reacts with acetylene to give $[(CN)_5CoCH:CHCo(CN)_5]^{6-}$ [282].

2.4.5b. *Hydrocarboxylation and Carbonylation of Acetylenes*[275, 282]

Acetylene reacts with nickel tetracarbonyl at about 40° under aqueous acidic conditions to give acrylic acid. The reaction is catalytic at 150° and 30 atm. Many solvents have been used, but water must be present; with aqueous alcohols acrylic esters are produced. The reaction involves overall *cis*-addition of H–COOH to the triple bond. Monosubstituted acetylenes give the 2-substituted acrylic acid, e.g. phenylacetylene gives atropic acid, $CH_2:C(Ph)COOH$; the addition is therefore Markownikoff. Originally a mechanism with a cyclopropeneone intermediate was suggested, but a much more likely mechanism is the following[282]:

$$Ni(CO)_4 + HX \rightleftharpoons NiHX(CO)_2 + 2CO$$

$$NiHX(CO)_2 \xrightarrow{HC:CR} \begin{array}{c} HC\equiv CR \\ | \\ HNiX(CO)_2 \end{array}$$

elimination of $CH_2:CRCOOR'$ ↑ $R'OH$ $\downarrow$ $\begin{array}{c} C\equiv C \\ insertion \end{array}$

$$CH_2:CRCONiX(CO)_2 \xleftarrow[insertion]{CO} CH_2:CRNiX(CO)_2$$

$\downarrow$ CO

$$CH_2:CRCOX + Ni(CO)_4$$

Many different acetylenes have been hydrocarboxylated. Other metal carbonyls such as iron pentacarbonyl are less effective than nickel tetracarbonyl. Dicobalt octacarbonyl in methanol converts acetylene into dimethyl succinate as the major product. Palladium chloride catalyses the carboxylation of acetylenes: usually a mixture of products is obtained. A feature is the occurrence of di- and polycarboxylation rather than monocarboxylation[354].

Carboxylation of propargylic halides gives more complex products. Thus 3-chlorobut-1-yne and nickel tetracarbonyl give the allenic acid, penta-2,3-dienoic acid, in 10% yield[355]. A possible mechanism is:

$$MeCHClC:CH \rightarrow \underset{\underset{Ni(CO)_3}{|}}{MeCHClC:CH} \rightarrow MeCH=C=CHNi(CO)_3 \xrightarrow{CO} MeCH=CH=CHCONi(CO)_3 \xrightarrow{H_2O} MeCH=C=CH\cdot COOH$$

Similar treatment of 1-chlorobut-2-yne gives 2-methylbuta-2,3-dienoic acid, but carboxylation of propargylic iodides gives cyclic keto-carboxylic acids[275].

2.4.5c. *Oligomerization of Acetylenes*[356, 357]

The most extensively studied processes are cyclic oligomerization of acetylene catalysed by nickel complexes. This work was initiated by the cyclo-octatetraene (COT) synthesis of Reppe, discovered in 1940,

$$C_2H_2 \xrightarrow[THF]{Ni(CN)_2} C_8H_8 + benzene$$

[354] J. Tsuji, *Accounts Chem. Res.* **2** (1969) 144.
[355] E. R. H. Jones, G. H. Whitham and M. C. Whiting, *J. Chem. Soc.* (1957) 4628.
[356] G. N. Schrauzer, *Angew. Chem. Int. Edn.* **3** (1964) 185.
[357] V. O. Reikhsfeld and K. L. Makovetskii, *Russian Chem. Rev.* **35** (1966) 510.

Many nickel complexes have been studied as catalysts for COT formation, the best being octahedral derivatives of nickel(II) with ligands such as acac, salicylaldehydato or ethyl acetoacetate[356]. Solvents used include dioxan (the best), THF or benzene. Water or other strongly donating ligands kill the catalyst, probably by competing with the acetylene for coordination sites. The principal products are COT and benzene, but small amounts of styrene, phenylbutadiene and vinylcyclo-octatetraene are formed, and even traces of naphthalene and azulene. Usually substituted acetylenes cannot be tetramerized, an exception being methyl propiolate which gives 1,2,4,6-tetracarbomethoxycyclo-octatetraene. Co-polymerization of acetylene with a substituted acetylene is possible, e.g. but-2-yne and acetylene give a 19% yield of 1,2-dimethylcyclo-octatetraene using a nickel acetylacetonate catalyst.

The mechanism of cyclo-octatetraene formation has attracted a lot of attention. In 1956 Longuett-Higgins and Orgel suggested that a cyclobutadienenickel(II) complex could be an intermediate in the polymerization of acetylene to COT. Since then, however, several mechanistic studies have shown that it is probably not an intermediate. Substituted cyclobutadienes may be intermediates in some cyclotrimerizations of acetylenes to arenes, however.

The best catalysts for cyclotetramerization of acetylene to COT are labile octahedral derivatives of nickel(II) with ligands of weak field strength, i.e. paramagnetic complexes, such as [Ni(acac)$_2$]$_3$ and Ni(salicylaldehydato)$_2$. Complexes such as Ni(dimethylglyoximate)$_2$ are inactive. A mechanism for COT formation has been suggested which involves the coordination of four acetylenes to the nickel, as shown in Fig. 17. The coordinated acetylenes are in the right conformation for a concerted or electrocyclic reaction to give COT. In this mechanism solvent molecules (S) such as THF or dioxan occupy two coordination sites.

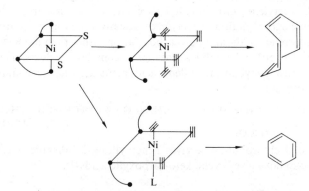

Fig. 17. Proposed mechanism for the cyclotetramerization of acetylene to cyclo-octatetraene and cyclotrimerization to give benzene (see text).

A good donor ligand such as triphenylphosphine prevents COT formation; instead benzene is formed. One mole of triphenylphosphine is sufficient, but larger amounts do not seem to inhibit benzene formation. A mechanism for cyclotrimerization has been proposed with one ligand position blocked by L (triphenylphosphine, etc., see Fig. 17). Bidentate ligands such as α,α'-dipyridyl or 1,10-phenanthroline destroy the catalytic activity. Many acetylenes have been cyclotrimerized to arenes using nickel complex catalysts. An interesting example is cyclo-octyne, which is trimerized by nickel cyanide but gives unusual products with other nickel complexes[358] (Fig. 18).

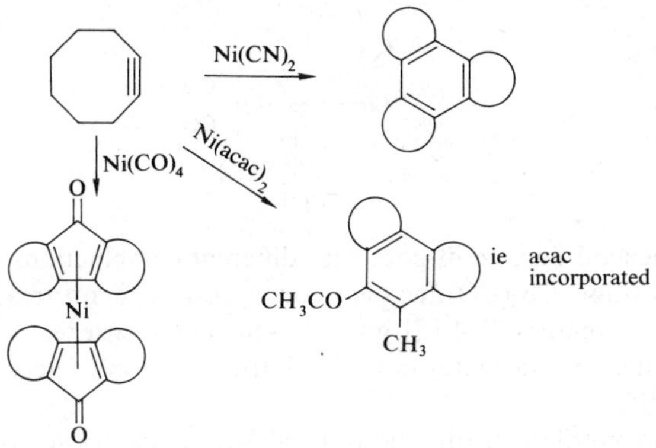

FIG. 18. Some cyclization reactions of cyclo-octyne with nickel complexes. The loops represent —$(CH_2)_6^-$ groups[358].

One of the first nickel catalysts to be used for trimerizing acetylenes to arenes was $Ni(CO)_2(PPh_3)_2$. With monosubstituted acetylenes the 1,2,4-trisubstituted benzene is usually the main cyclic product, e.g. phenylacetylene gives 1,2,4-triphenylbenzene (20%) and 1,3,5-triphenylbenzene (1%). A very detailed study of the cyclic trimerization and linear polymerization of ethyl propiolate by nickel complexes of type $Ni(CO)_2L_2$ has defined optimum conditions, etc., and determined how the nature of the ligand L affects the rate and product distribution of the reaction[359]. The main disadvantage of these nickel(0) catalysts is their tendency to give linear oligomers. Some other metal carbonyls or their derivatives are more specific at promoting cyclic trimerization, although vigorous reaction conditions may be necessary, e.g. 250–280°. Bis(tetracarbonylcobalt)mercury has been used as a catalyst for trimerization of a large number of acetylenes. There are also many examples of the acetylenes being copolymerized to mixed arenes[360]. One remarkable result is the formation of 1,4-(trimethylsilyl)-2,3,5,6-hexaphenylbenzene [2.4.14] from $Co_2(CO)_6(PhC:CPh)$ and bis(trimethylsilyl)acetylene[361]. The formation of [2.4.14] could be explained if a cyclo-butadiene intermediate were involved.

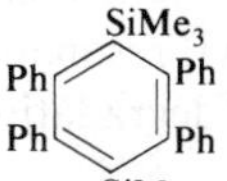

[2.4.14]

358 G. Wittig and P. Fritze, *Justus Liebigs Ann. Chem.* **712** (1968) 79.
359 L. S. Meriwether, M. F. Leto, E. C. Colthup and G. W. Kennerly, *J. Org. Chem.* **27** (1962) 3930.
360 W. Hübel and C. Hoogzand, *Chem. Ber.* **93** (1960) 103.
361 U. Krüerke and W. Hübel, *Chem. Ber.* **94** (1961) 2829.

When $Co_2(CO)_6(HC\vdots CH)$ is treated with t-butylacetylene a complex of stoichiometry $Co_2(CO)_4(HC\vdots CH)(Bu^tC\vdots CH)_2$ is obtained. X-ray analysis[362] shows this compound to have the "flyover" structure [2.4.15]

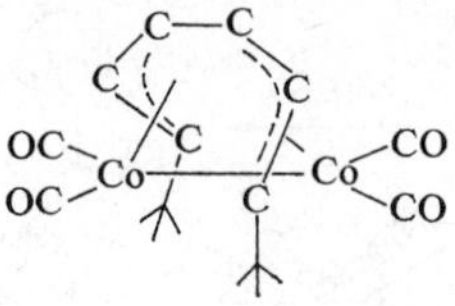

[2.4.15]

The reaction is a general one, i.e. it goes with different combinations of acetylenes. These products of composition $Co_2(CO)_4(ac)(ac')_2$ on pyrolysis or on treatment with bromine give arenes, e.g. the complex [2.4.15] gives 1,2-di-t-butylbenzene. Analogues of [2.4.15] are therefore probably intermediates in the cyclotrimerization of acetylenes by derivatives of cobalt carbonyl[316].

Dichlorobis(benzonitrile)palladium also catalyses arene formation from acetylenes. Low temperature n.m.r. and other studies suggest that the species shown in Fig. 19 are involved in the trimerization of but-2-yne to hexamethylbenzene.

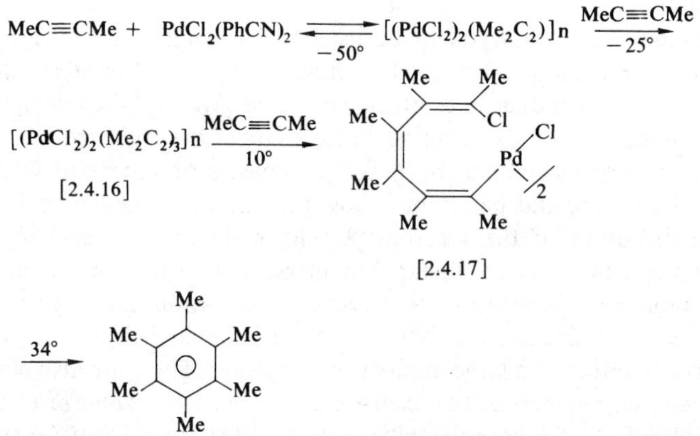

Fig. 19. Proposed intermediates in the cyclotrimerization of but-2-yne to hexamethylbenzene, catalysed by $PdCl_2(PhCN)_2$.

The complex [2.4.17] was isolated as yellow crystals in 45–50% yield. Both [2.4.16] and [2.4.17] show six separate n.m.r. (methyl) resonances. The structure [2.4.17] is based on the n.m.r. spectrum and on the products of degradation; the structure of the complex [2.4.16] is unknown[363].

The tetraphenyl(ethoxy)cyclobutenyl complexes formed from diphenylacetylene and sodium chloropalladate, reacting in ethanol, are discussed in section 3. Dichlorobis(benzonitrile)palladium reacts with diphenylacetylene to give a series of very unstable compounds of composition $(PdCl_2)_n(C_4Ph_4)_2$ ($n = 3, 5$ or 6). These could be tetraphenylcyclobutadiene

362 O. S. Mills and G. Robinson, *Proc. Chem. Soc.* (1964) 187.
363 P. M. Maitlis, H. Reinheimer, H. Dietl and J. Moffat, Reprints of the Division of Petroleum Chemistry Inc. American Chemical Society Meeting, April 1969.

complexes but they give octaphenylcyclo-octatetraene when treated with triphenylphosphine and therefore could have some other structure. 1-Phenylbut-1-yne reacts with palladium chloride in methanol to give two complexes, $[PdCl(PhC_2Et)_2]_2$ and $[PdCl(PhC_2Et)_4]_2$. The structures of these are not known.

Some remarkable products of unknown structure are formed from acetylenic alcohols and palladium chloride, e.g. 4-methylbut-1-yn-3-ol(ac) with palladium chloride in acetic acid solution gives a monomeric product of composition $PdCl(ac)_4$. The nature of this product is not known, but a band at 1630–1680 cm^{-1} in its infrared absorption spectrum suggests the presence of an organic carbonyl group; there may also be coordinated C=C bonds[364]. When methyl(phenyl)acetylene is trimerized with dichlorobis(benzonitrile)-palladium as catalyst some 1,2,3-trimethyl-4,5,6-triphenylbenzene is formed, i.e. C–C bond fission must occur at some stage. This result is possibly evidence for a cyclobutadiene intermediate[365].

Some interesting studies on iridium–acetylene complexes show that cyclotrimerization of acetylenes can go via a metallacyclopentadiene[16]. Treatment of *trans*-IrCl(N$_2$)L$_2$ (L = PPh$_3$) with dimethyl acetylene dicarboxylate (RC⋮CR) gives an acetylene complex [2.4.18] with displacement of N$_2$.

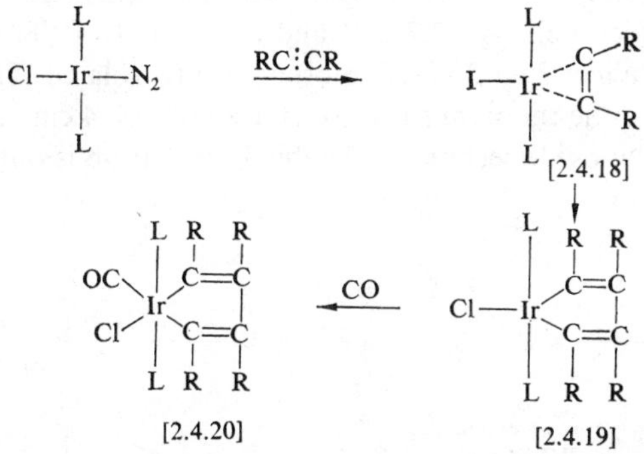

The acetylene complex [2.4.18] reacts further with more acetylene to give the iridiacyclo-pentadiene complex [2.4.19]. This complex is coordinatively unsaturated and will take up carbon monoxide to give [2.4.20]. Above 100° the iridiacyclopentadiene complex [2.4.19] catalyses trimerization of the dimethyl acetylene dicarboxylate to hexacarbomethoxyben-zene. When the iridiacyclopentadiene complex, prepared from $CD_3OOC·C⋮C·COOCD_3$, is treated with $CH_3OOC·C⋮C·COOCH_3$ the following reaction occurs:

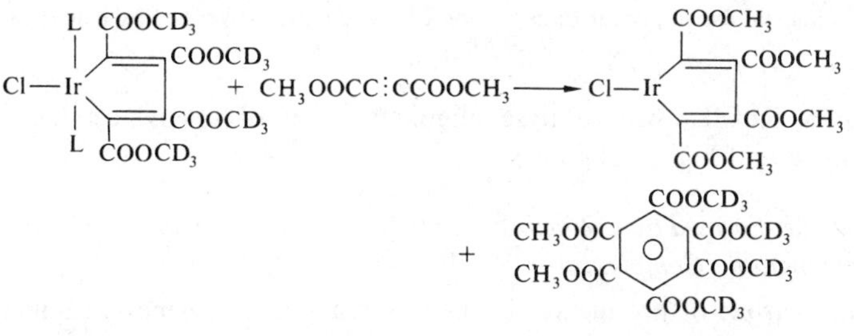

364 Y. Y. Kharitonov, T. I. Beresneva, G. Y. Mazo and A. V. Babaeva, *Russ. J. Inorg. Chem.* 13 (1968) 1129.
365 H. Dietl and P. M. Maitlis, *Chem. Commun.* (1968) 481.

This work shows that trimerization of acetylenes can go through metallocyclopentadienes; such a mechanism had been suggested previously for cobalt-catalysed trimerizations[359]. A similar series of steps occurs in the conversion of tolan to hexaphenylbenzene using a cyclopentadienyl cobalt complex (see Fig. 23, p. 891).

Arylchromium complexes convert acetylenes into arenes, although frequently an aryl group from the chromium is incorporated into the product. Thus after an induction period triphenylchromium tetrahydrofuranate reacts exothermally with but-2-yne to give various products including 1,2,3,4-tetramethylnaphthalene, hexamethylbenzene and π-arene-chromium complexes. Tri-1-naphthylchromium similarly gives 1,2,3,4-tetramethylphenan-threne. Alkylchromium complexes react similarly, e.g. trimethylchromium tetrahydro-furanate and diphenylacetylene gives cis- and trans-stilbenes, α-methylstilbene, α-benzyl-styrene and bibenzyl; the methyl groups or the solvent (THF) are the source of hydrogen[366]. Trimerization of $CD_3C:CCH_3$ by triphenylchromium tetrahydrofuranate gives no 1,2,3-trideuteriomethyl-4,5,6-trimethylbenzene in the product mixture, showing that a cyclobutadiene intermediate is not involved[367].

Metal complexes will also catalyse linear oligomerization of terminal acetylenes to enynes, dienynes, etc. The mechanism of dimerization has been worked out in some detail with $RhCl(PPh_3)_3$ as catalyst[342]. This complex catalyses dimerization of 3-methylbut-1-yn-3-ol to 2,7-dimethyloct-3-en-5-yn-2,7-diol and at the end of the reaction the complex $RhCl(PPh_3)_2(2,7\text{-dimethyloct-3-en-5-yn-2,7-diol})$ is readily isolated. The diol is coordinated via its acetylenic bond to the rhodium since $v(C:C)$ is only 1914 cm^{-1}, 205 cm^{-1} lower than for the free diol. The proposed mechanism for the dimerization is outlined in Fig. 20.

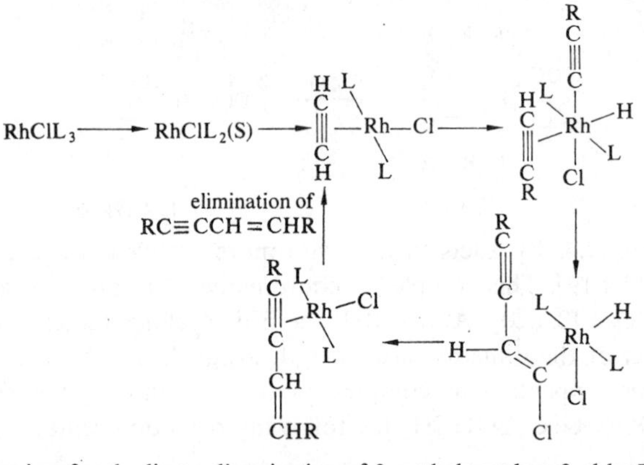

FIG. 20. Mechanism for the linear dimerization of 3-methylpent-1-yn-3-ol by $RhClL_3$ (L = PPh$_3$, R = Me$_2$C(OH)—)[342].

The complex $RhClL_3$ will catalyse oligomerization of phenylacetylene to trans-1,4-diphenylbutenyne and higher oligomers.

2.4.5d. Compounds Formed by Condensation of Acetylenes with Carbon Monoxide, and Related Organometallic Complexes[111a, 315, 316]

An amazing variety of products is formed by reacting acetylenes with metal carbonyls. The acetylenes can condense with themselves and/or carbon monoxide to give complexes

366 M. Michman and H. H. Zeiss, J. Organomet. Chem. 15 (1968) 139.
367 G. M. Whitesides and W. J. Ehmann, J. Am. Chem. Soc. 90 (1968) 804.

with unusual ligands. Table 14 gives the compositions of products formed by reacting acetylenes with iron carbonyls; more than one structure is possible with the same stoichiometry.

TABLE 14. COMPOSITIONS OF COMPLEXES FORMED FROM IRON CARBONYLS AND ACETYLENES (ac) [368]

	1 Alkyne	2 Alkynes	3 Alkynes	4 Alkynes	5 Alkynes
1Fe	$Fe(CO)_4ac$	$Fe(CO)_3(ac)_2$ $Fe(CO)_4(ac)_2$	$Fe(CO)_2(ac)_3$	$Fe(CO)_3(ac)_4$	$Fe(CO)_4(ac)_5$
2Fe	$Fe_2(CO)_6ac$ $Fe_2(CO)_7ac$	$Fe_2(CO)_6(ac)_2$ $Fe_2(CO)_7(ac)_2$	$Fe_2(CO)_6(ac)_3$ $Fe_2(CO)_7(ac)_3$		$Fe_2(CO)_6(ac)_5$
3Fe	$Fe_3(CO)_{10}ac$	$Fe_3(CO)_8(ac)_2$	$Fe_3(CO)_8(ac)_3$		$Fe_3(CO)_{10}(ac)_5$

Some of the ligands which can be formed are summarized in Fig. 21.

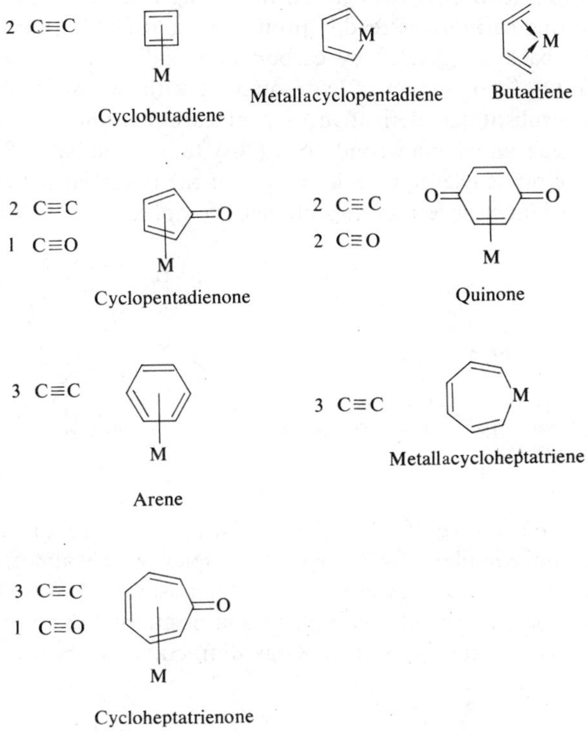

FIG. 21. Ligands formed by reacting acetylenes with metal carbonyls.

It is convenient to discuss first the nature of the products formed from acetylenes and iron carbonyls. The products from other metals will then be discussed in the order titanium to nickel, group by group.

[368] E. Weiss, R. G. Merényi and W. Hübel, *Chem. Ber.* **95** (1965) 1155.

(1) *Iron.* Reppe and Vester found that pentacarbonyliron and acetylene react in aqueous ethanol to give ethyl acrylate, hydroquinone and a yellow complex $C_{22}H_{14}Fe_2O_{10}$. This with acid gives hydroquinone and $C_8H_4FeO_4$, which in turn is easily oxidized to $C_{14}H_8Fe_2O_6$. $C_8H_4FeO_4$ has since been identified as cyclopentadienoneiron tricarbonyl [2.4.21] and $C_{14}H_8Fe_2O_6$ as [2.4.22] (by X-ray crystallography).

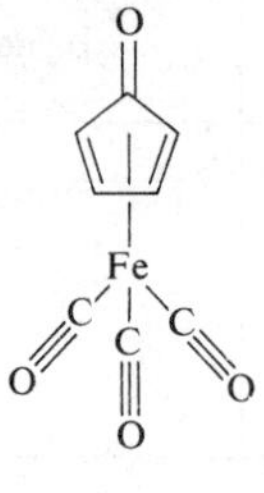

[2.4.21] 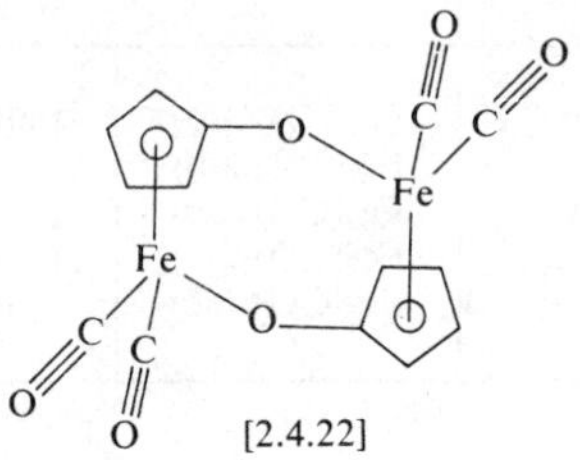   [2.4.22]

The complex [2.4.21] has a high dipole moment (4.45 D) and a very low infrared stretching frequency for the ring carbonyl (1634 cm^{-1}), indicating a lot of back coordination from metal to ring. The ring carbonyl is readily protonated and [2.4.21] forms salts with acids. [2.4.22] is reduced back to [2.4.21] by carbon monoxide under pressure. The original complex $C_{22}H_{14}Fe_2O_{10}$ is an adduct of hydroquinone with two molecules of [2.4.21].

The labile monosubstituted derivative of pentacarbonyliron, i.e. Fe(CO)$_4$(ac), reacts with diphenylacetylene under mild conditions (20°) to give initially Fe(CO)$_4$(PhC:CPh) and then, with more diphenylacetylene, a mixture of the binuclear ketonic "flyover" complex [2.4.23] and the binuclear ferracyclopentadiene complex [2.4.24]. R is Ph in both cases.

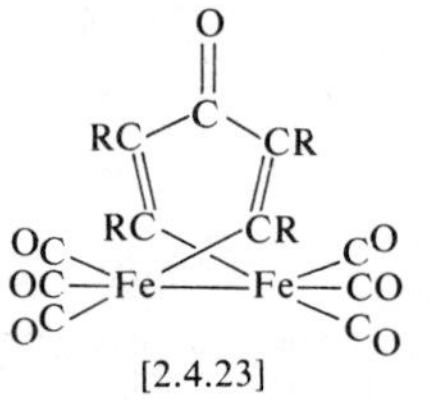

[2.4.23] 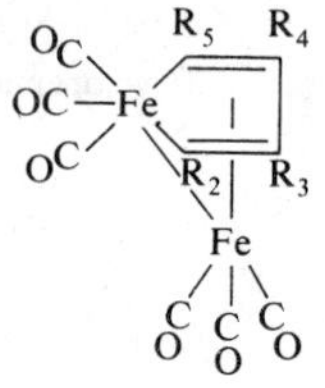   [2.4.24]

With hex-3-yne, Fe(CO)$_4$(Bu^{t}C:CBu^t) reacts to give a trace of [2.4.23], R = Et, a tetraethylquinone-iron complex [2.4.25] and a complex which almost certainly has the remarkable structure [2.4.26], with ethyl groups at positions $1 \rightarrow 6$. The corresponding complex made from phenylacetylene with phenyls at positions 1, 3 and 5 and hydrogens at 2, 4 and 6 has had its structure proved by X-ray diffraction; see below.

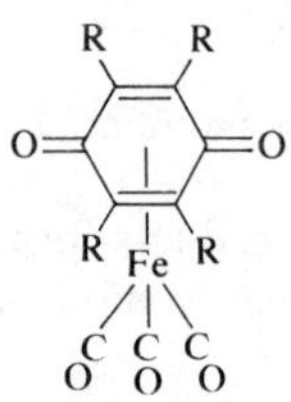

[2.4.25] 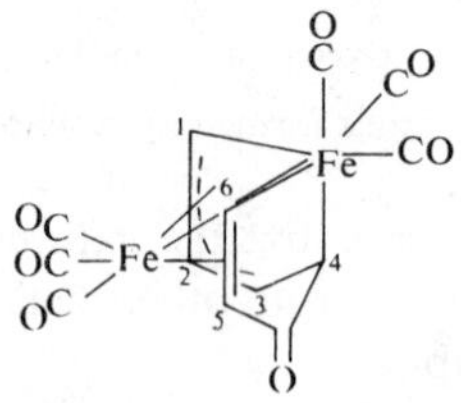   [2.4.26]

The complexes of type [2.4.26] have an allylic–iron bond, an olefin–iron bond, three σ-C–Fe bonds and one Fe–Fe bond.

The compounds of types [2.4.23] and [2.4.24], formed from Fe(CO)$_4$(ac) and acetylenes, have also been made directly from Fe(CO)$_5$, Fe$_2$(CO)$_9$ or Fe$_3$(CO)$_{12}$ and acetylenes, but usually under more vigorous conditions. Other products formed by reacting unsubstituted iron carbonyls with acetylenes include tricarbonyl(tropone)iron complexes. When phenyl-acetylene reacts with dodecacarbonyltri-iron below 60°, the complex [2.4.26] with phenyls at positions 1, 3 and 5 and hydrogens at 2, 4 and 6 is obtained. This complex decomposes on alumina to give a mixture of the two isomeric tricarbonyl(triphenyltropone)iron complexes [2.4.27] and [2.4.28].

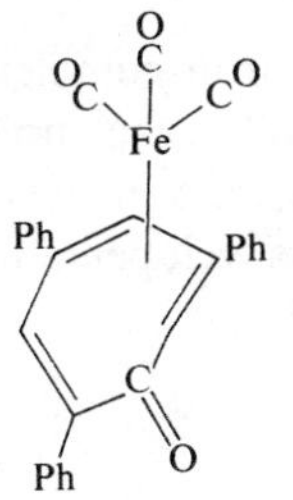

[2.4.27] 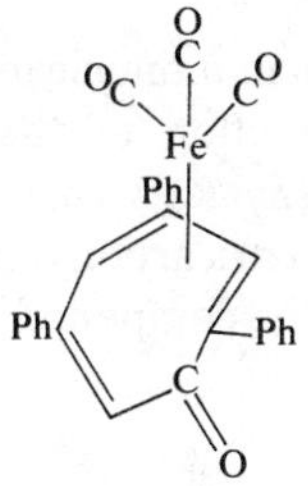　　　　　　　　　　[2.4.28]

The X-ray structure of the isomer [2.4.27] shows the ring to be folded. A mixture of both isomers is formed by reacting 2,4,6-triphenyltropone with dodecacarbonyltri-iron so, although the structure of the second isomer has not been determined by X-ray diffraction, [2.4.28] seems very likely.

Many ferracyclopentadienyliron complexes of type [2.4.24] have been made. One of the first examples was the product C$_{10}$H$_4$Fe$_2$O$_8$, formed from acetylene and pentacarbonyliron reacting in an alkaline medium. This complex has structure [2.4.24] with R$_2$ = R$_5$ = OH and R$_3$ = R$_4$ = H. An unusual example is the complex of composition Fe$_2$(CO)$_6$(PhC$_2$Ph) formed from dodecacarbonyltri-iron and diphenylacetylene. It has the phenyl(benzferracyclopentadiene) structure [2.4.29] with an Fe–Fe distance of 2.64 Å [369].

Treatment of Fe$_3$(CO)$_{12}$ with acetylene gives two complexes of composition Fe$_2$(CO)$_6$(C$_2$H$_2$)$_2$. The low melting orange isomer has the structure [2.4.30] whilst the red isomer has the cyclopentadienyl structure [2.4.31]. These two structures have been determined by X-ray diffraction. Pyrolysis of the orange isomer [2.4.30] gives a compound of composition Fe$_2$(CO)$_4$(C$_2$H$_2$)$_6$ for which a biscyclopentadienyl bridged structure has been proposed but not confirmed.

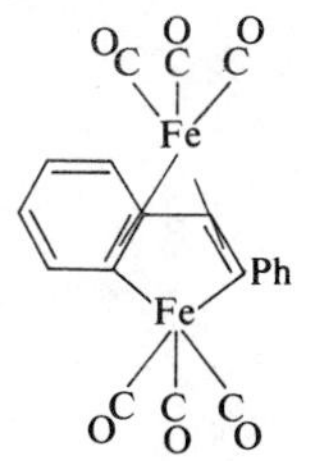

[2.4.29] 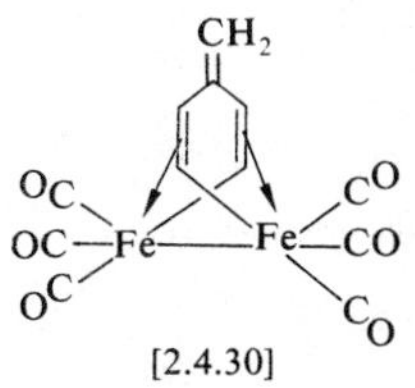　　　　　　[2.4.30] 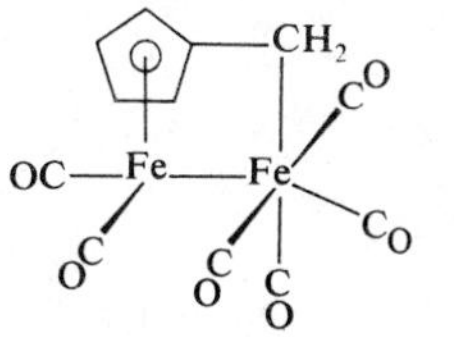　　　　　　[2.4.31]

[369] Y. Degrève, J. Meunier-Piret, M. van Meersche and P. Piret, Acta Cryst. 23 (1967) 119.

Tetraphenylcyclobutadieneiron tricarbonyl [2.4.32] is one of the products formed from $Fe(CO)_3(COT)$ or $Fe(CO)_5$ and diphenylacetylene. Its structure was shown by X-ray crystallography[370].

[2.4.32]

[2.4.33]

It is possible that a mononuclear ferracyclopentadienyliron complex [2.4.33] is an intermediate in the formation of the tetraphenylcyclobutadiene complex [2.4.32]. Another product from diphenylacetylene and $Fe_3(CO)_{12}$ is the tetraphenyl derivative [2.4.34] in Fig. 22, i.e. 1,1,1-tricarbonyl-2,3,4,5-tetraphenylferracyclopentadieneiron tricarbonyl. Some reactions of this compound are outlined in Fig. 22.

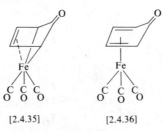

FIG. 22. Some reactions of 1,1,1-tricarbonyl-2,3,4,5-tetraphenylferracyclopentadieneiron tricarbonyl[316].

Perfluorobut-2-yne reacts with iron pentacarbonyl to give only one product, tetrakis-(trifluoromethyl)cyclopentadieneiron tricarbonyl. The structure of this compound has been determined by X-ray diffraction[371], and the angles and bond lengths can perhaps be described in terms of the σ- and π-bonded structure [2.4.35] rather than the π-bonded structure [2.4.36]. The ring is folded as shown.

[2.4.35]

[2.4.36]

---

[370] R. P. Dodge and V. Schomaker, *Acta Cryst.* **18** (1963) 614.
[371] N. A. Bailey and R. Mason, *Acta Cryst.* **21** (1966) 652.

Methylphenylpropiolate reacts with enneacarbonyldi-iron to give various products including the unexpected cyclopentadienyl derivative [2.4.37].

[2.4.37]

(2) *Titanium*. Diphenylacetylene will displace both carbon monoxides from $Ti(CO)_2Cp_2$ to give an air-sensitive complex $Ti(C_2Ph_2)_2Cp_2$. Two possible structures for this compound are dicyclopentadienyltitano(tetraphenyl)cyclopentadiene or dicyclopentadienyl(tetraphenylcyclobutadiene)titanium[315].

(3) *Molybdenum*. Several complexes have been made from molybdenum carbonyls and acetylenes[315, 316]. The structures have been assigned on infrared and chemical evidence but not yet confirmed by X-ray crystallography. By reacting $Mo(CO)_6$ with $PhC\colon CPh$ at 160–170°, complexes are produced with the formulations $Mo(CO)_2(Ph_4C_4)_2$, $Mo(CO)_2(Ph_4C_4)(Ph_4C_5O)$, $Mo_2(CO)_4(Ph_4C_4)(Ph_2C_2)$ and $Mo(Ph_5C_5)_2$. On infrared and degradative evidence the first complex has been formulated as dicarbonyl(tetraphenylcyclobutadiene)molybdenum, the second as dicarbonyl(tetraphenylcyclobutadiene)(tetraphenylcyclopentadienone)molybdenum and the third as two carbonyl(tetraphenylcyclobutadiene)molybdenum groups bridged by diphenylacetylene. The last compound, $Mo(Ph_5C_5)_2$, is paramagnetic ($\mu = 2.75$–3.17 BM) and has been formulated as bis(pentaphenylcyclopentadienyl)molybdenum(II). It is oxidized to a molybdenum(III) species $(Ph_5C_5)_2Mo^+Br_3^-$ with a magnetic moment of 3.47–3.5 BM. On being heated in methylnaphthalene $Mo(Ph_5C_5)_2$ gives a 42% yield of pentaphenylcyclopentadiene. Thus the evidence for formulating the original complex as bis(pentaphenylcyclopentadienyl)molybdenum(II) is good, but it is difficult to see how such a ligand could be formed, and the reaction is worth re-investigation.

Ethyl(cyclopentadienyl)molybdenum tricarbonyl reacts with diphenylacetylene to give a complex which has been formulated as 1,2,3,4-tetraphenylbenzene(cyclopentadienyl)-molybdenum.

(4) *Manganese*. By reacting decacarbonyldimanganese with acetylene the π-dihydropentalenyl complex [2.4.38] was obtained.

[2.4.38]

The olefinic double bond can be hydrogenated and the hydrogenation product has been synthesized by an independent route, thus confirming the structure [2.4.38] [315].

(5) *Ruthenium*. Dodecacarbonyltriruthenium has been reacted with various acetylenes. Hex-3-yne gives a tetraethylcyclopentadienone complex analogous to [2.4.21].

372 L. F. Dahl, R. J. Doedens, W. Hübel and J. Nielson, *J. Am. Chem. Soc.* **88** (1966) 446.

Diphenylacetylene at 200° gives a 1,1,1-tricarbonyl-2,3,4,5-tetraphenylcyclopentadiene-ruthenium tricarbonyl complex analogous to [2.4.24] [373].

(6) *Cobalt* [98, 315, 316]. Several unusual products are obtained by reacting cobalt carbonyls with acetylenes. Complexes of composition $Co_2(CO)_4(ac)_3$ with the "flyover" structure [2.4.15] have already been mentioned. When acetylene complexes of type $Co_2(CO)_6(RC:CR)$ are treated with carbon monoxide, 3 moles are absorbed. The products $Co_2(CO)_9(RC_2R)$ have the bridging lactone structure [2.4.39].

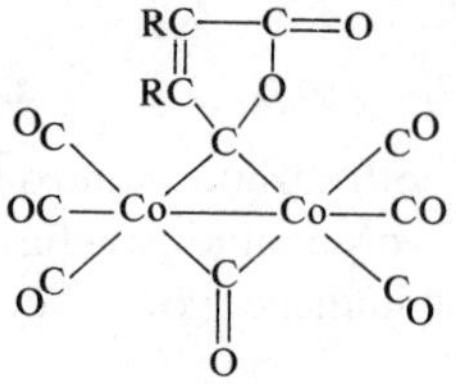

[2.4.39] 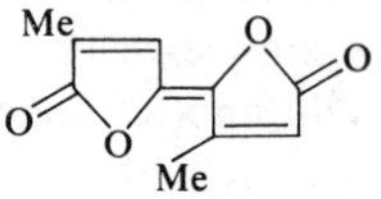   [2.4.40]

The structure of the complex with R = H has been determined by X-ray diffraction. An analogue of [2.4.39] is probably involved in the synthesis of diolides [2.4.40] from propyne, $Co_2(CO)_8$ and acetic anhydride.

Cyclopentadienyl(dicarbonyl)cobalt reacts with but-2-yne or hexafluorobut-2-yne in ultraviolet light to give complexes of the composition $CoCp(CO)(ac)_2$. These complexes have folded cyclopentadienone ligands, as found for the cyclopentadienoneiron tricarbonyl complexes. The bonding has beeni nterpreted in terms of the $\pi - \sigma$-structure [2.4.41] rather than the π-structure [2.4.42] [98].

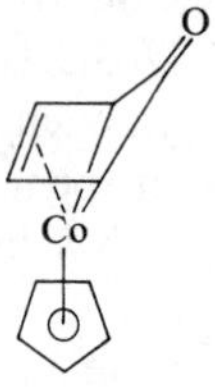

[2.4.41]

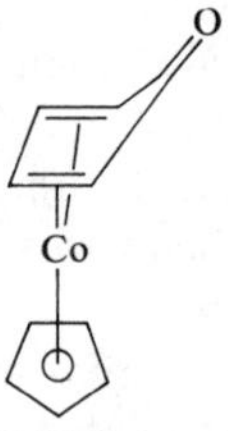

[2.4.42]

Some binuclear cyclopentadienone complexes with the probable structure [2.4.43] are formed by heating complexes of type $Co_2(CO)_6(ac)$ with alk-1-ynes, $RC:CH$, where R is a bulky substituent such as t-butyl.

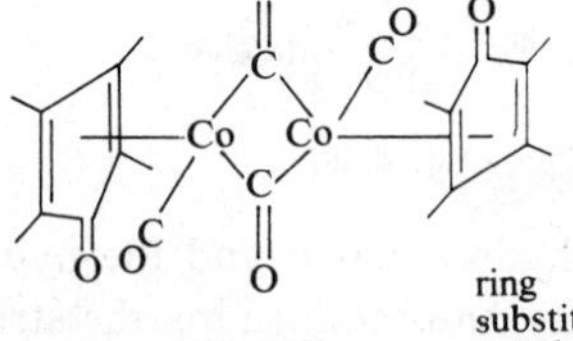

[2.4.43]   ring
              substituents
              not shown

Some of the "flyover" complex [2.4.15] is also formed in this reaction.

[373] M. I. Bruce and J. R. Knight, *J. Organomet. Chem.* **12** (1968) 411.

There is some interesting chemistry associated with products formed by treating $CoI_2Cp(PPh_3)$ with isopropylmagnesium bromide in the presence of diphenylacetylene (tolan)[374]. A labile di(isopropyl)–cobalt(III) species is presumably formed, but this reacts with the diphenylacetylene to give a green crystalline acetylene complex [2.4.44] (Fig. 23). This reacts with more tolan to give the red-brown air-stable cobalticyclopentadiene complex [2.4.45]. This can be converted into a tetraphenylcyclobutadienecobalt complex by heating; a tetraphenylcyclopentadienone complex when treated with carbon monoxide; or give hexaphenylbenzene with more tolan.

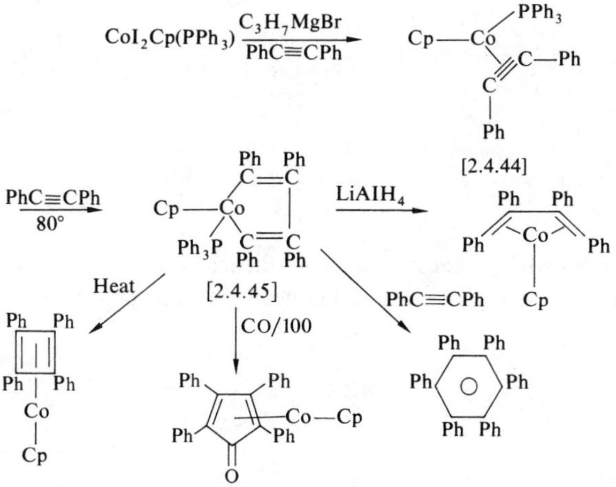

FIG. 23. Some reactions of the tetraphenylcobalticyclopentadiene complex $CoCp(C_4Ph_4)PPh_3$ [374].

(7) *Rhodium.* Treatment of $RhCp(CO)_2$ with hexafluorobut-2-yne give two complexes, $RhCp(C_9F_{12}O)$ and $RhCp(C_{12}F_{18})$. The first complex is a tetrakis(trifluoromethyl)cyclopentadienone(cyclopentadienyl)rhodium and the second is a hexakis(trifluoromethyl)-benzenecyclopentadienylrhodium, which has the structure [2.4.46]. The bonding of the arene ring is considered to be closer to the $\pi + \sigma$-structure [2.4.46] rather than a purely π-bonded structure[98].

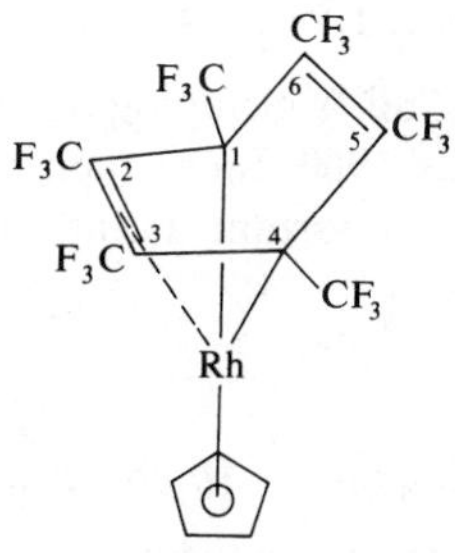

[2.4.46]

The dihedral angle between $C_1-C_2-C_3-C_4$ and $C_1-C_6-C_5-C_4$ is 48°. The distance C_2-C_3 is 1.42 Å, whilst C_1-C_2 and C_3-C_4 are 1.48 Å and 1.53 Å respectively. C_5-C_6 is 1.31 Å. The distance C_1-C_4 of 2.56 Å is 0.24 Å less than in benzene.

374 H. Yamazaki and N. Hagihara, *J. Organomet. Chem.* **7** (1967) P. 22.

A number of products are formed from $Rh_2Cl_2(CO)_4$ and diphenylacetylene. Two of them, $Rh_2Cl_2(C_4Ph_4CO)_2$ and $[RhCl(CO)(C_4Ph_4CO)]_n$, are thought to be tetraphenyl-cyclopentadienonerhodium complexes[375].

(8) *Nickel*[315, 316]. Tetracarbonylnickel reacts with diphenylacetylene to give $Ni(tetra-cyclone)_2$. The complex has a very low value for $\nu(CO)$ (1597 cm^{-1}) compared with 1715 cm^{-1} in free tetracyclone. This indicates a lot of back donation from the nickel to the tetracyclone rings, and perhaps the compound could be represented as a dicyclopentadienyl complex [2.4.47].

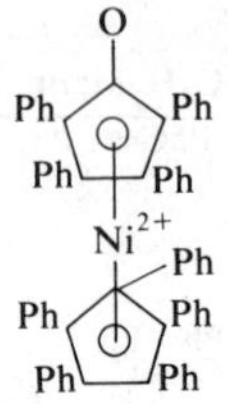

[2.4.47]

Nickelocene reacts with dimethyl acetylene dicarboxylate to give an adduct. The X-ray structure shows that the acetylene has added across a cyclopentadienyl ring [2.4.48] [98].

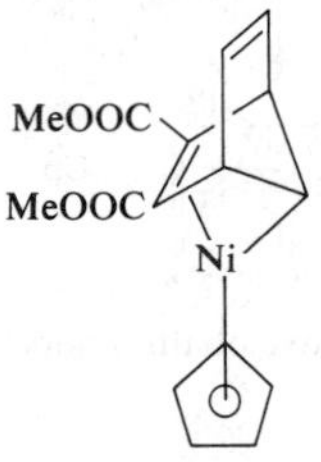

[2.4.48]

Hexafluorobut-2-yne adds on similarly.

3. π-ALLYLIC AND RELATED COMPLEXES

3.1. π-ALLYLIC COMPLEXES

The discovery that an allylic ("all") grouping could be π- or sandwich-bonded to a transition metal came as recently as 1960. Since then the field has developed rapidly, and allylic complexes are known to be important as intermediates in homogeneous catalysis. There are several reviews[151, 214, 288, 376-378].

3.1.1. Preparation

3.1.1a. *By Treating a Metal Halide with an Allylic Grignard Reagent*

An example is the preparation of diallylnickel (as a mixture of *cis*- and *trans*-isomers):

$$NiBr_2 + CH_2:CH \cdot CH_2MgBr \rightarrow Ni(C_3H_5)_2$$

375 S. McVey and P. M. Maitlis, *Can. J. Chem.* **44** (1966) 2429.
376 M. L. H. Green and P. L. I. Nagy, *Advances Organomet. Chem.* **2** (1964) 325.
377 M. I. Lorbach, B. D. Babitskii and V. A. Kormer, *Russian Chem. Rev.* **36** (1967) 476.
378 E. O. Fischer and H. Werner, *Z. Chemie* **2** (1962) 174.

Many other allylic complexes have been prepared by this method, e.g. of types $M(C_3H_5)_2$, Pd, Pt; $M(C_3H_5)_3$, V, Cr, Fe, Co, Rh, Ir; $M(C_3H_5)_4$, Zr, Th, Nb?, Ta?, Mo, W; $M_2(C_3H_5)_4$, Cr, Mo [288].

Halide complexes may also be converted into π-allylic complexes, e.g. $TiClCp_2$ gives $TiC_3H_5Cp_2$ [379] and $[RuCl_2(COD)]_x$ gives $Ru(C_3H_5)_2(COD)$ [380]. Treatment of complexes of the type mer-$IrCl_3L_3$ (L = tertiary phosphine) with allMgCl gives compounds of type $IrCl_2allL_2$ or $[Ir(all)_2L_2]^+$ [381].

3.1.1b. Formation of π-Allylic from σ-Allylic Complexes

In this method a monodentate σ-allylic ligand displaces another ligand and gives a π-allylic complex [382]:

$$[Mn(CO)_5]^- \xrightarrow{C_3H_5Cl} CH_2{:}CHCH_2Mn(CO)_5 \xrightarrow[\text{or } 80°]{\text{U.V. Light}} HC{\Big\langle}{\genfrac{}{}{0pt}{}{CH_2}{CH_2}}{-}Mn(CO)_4$$

$Mo(\sigma\text{-}C_3H_5)Cp(CO)_3$ [383] and $Fe(\sigma\text{-crotyl})Cp(CO)_2$ [122] are similarly converted into π-allylic species with loss of one CO molecule, and the σ-benzyl complex $Mo(CH_2Ph)Cp(CO)_3$ gives a π-benzyl complex $Mo(\pi\text{-}CH_2Ph)Cp(CO)_2$. The $[\sigma\text{-allylCo(CN)}_5]^{3-}$ and $[\pi\text{-allylCo(CN)}_4]^{2-}$ ions equilibrate in solution; the position of equilibrium depends on the concentration of cyanide ion [384]. Some σ- and π-allyliciridium(III) species can also be interconverted in solution [18].

Rapid σ-allylic/π-allylic interconversions are responsible for many of the temperature dependent n.m.r. spectra associated with allylic systems (see section 3.1.4a).

3.1.1c. From Allylic Halides or Alcohols

Palladium forms more allylic complexes than any other metal, and the best method of synthesizing the simple ones of type $Pd_2Cl_2(all)_2$ is to treat an aqueous methanolic solution of Na_2PdCl_4 with the allylic chloride and carbon monoxide, e.g.

$$PdCl_4^{2-} + C_4H_7Cl + CO + H_2O \rightarrow \tfrac{1}{2}[C_4H_7PdCl]_2 + CO_2 + 2HCl$$
$$\text{2-methylallyl chloride} \qquad\qquad \text{100\% yield}$$

An allylic alcohol may also be used. A closely related method for synthesizing allylic–rhodium(III) complexes, $Rh_2Cl_2(all)_4$, is by treating $Rh_2Cl_2(CO)_4$ with the allylic chloride in aqueous methanol, preferably maintaining the pH within the range 5–7 by adding a base [385].

There are many examples of the preparation of π-allylic complexes by the oxidative addition of allylic halides to metal complexes in low valence states, e.g. from metal carbonyls:

$$Fe(CO)_5 + C_3H_5I \rightarrow FeI(\pi\text{-}C_3H_5)(CO)_3 \text{ [386]}$$
$$[Et_4N][MoCl(CO)_5] + C_3H_5Br \rightarrow [Et_4N][Mo_2Br_3(\pi\text{-}C_3H_5)_2(CO)_4] \text{ [387]}$$
$$Co(CO)_2Cp + C_3H_5I \rightarrow CoI(\pi\text{-}C_3H_5)Cp \text{ [388]}$$

[379] H. A. Martin and F. Jellinek, J. Organomet. Chem. 12 (1968) 149.

[380] J. Powell and B. L. Shaw, J. Chem. Soc. A (1968) 159.

[381] J. Powell and B. L. Shaw, J. Chem. Soc. A (1968) 780.

[382] W. R. McClellan, H. H. Hoehn, H. N. Cripps, E. L. Muetterties and B. W. Howk, J. Am. Chem. Soc. 83 (1961) 1601.

[383] M. Cousins and M. L. H. Green, J. Chem. Soc. (1963) 889.

[384] J. Kwiatek and J. K. Seyler, J. Organomet. Chem. 3 (1965) 421.

[385] J. Powell and B. L. Shaw, J. Chem. Soc. A (1968) 583.

[386] R. F. Heck and C. R. Boss, J. Am. Chem. Soc. 86 (1964) 2580.

[387] H. D. Murdoch, J. Organomet. Chem. 4 (1965) 119.

[388] R. F. Heck, J. Org. Chem. 28 (1963) 604.

Olefin complexes may also be used, e.g. addition of allylic chlorides to $Rh_2Cl_2(C_2H_4)_4$ followed by butadiene gives compounds of the type $Rh_2Cl_4(all)_2C_4H_6$ [389] and the cyclo-octene complex of iridium(I), $[IrCl(CO)(C_8H_{14})_2]_2$, reacts rapidly with 2-methylallyl chloride to give the π-2-methylallylic complex $IrCl_2(C_4H_7)(CO)(C_8H_{14})$. Oxidative addition has been used to make a mixed Ni/Pd allylic complex:

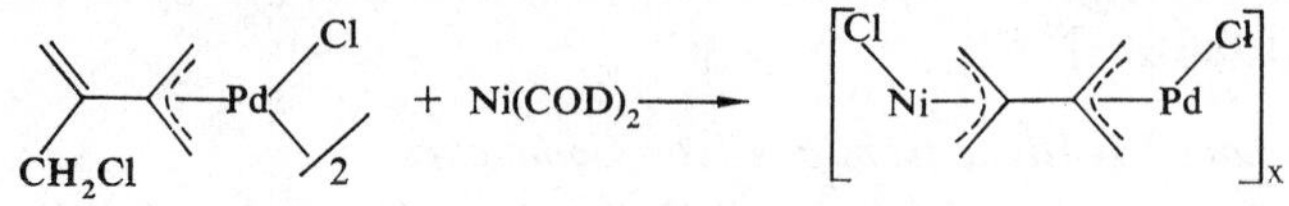

Tertiary phosphines may also be displaced in the oxidative addition. Addition of allylic halides to $Pt(PPh_3)_4$ gives cationic π-allylic species $[Ptall(PPh_3)_2]^+$ and $RhCl(PPh_3)_3$ with allyl chloride gives $RhCl_2(\pi\text{-}C_3H_5)(PPh_3)_2$ [69, 390].

π-Allylpalladium bromide can be made by the action of allyl bromide on palladium metal, but allyl chloride or iodide do not give the corresponding complexes $[Pd_2X_2(C_3H_5)_2]$ (X = Cl or I).

3.1.1d. *From Mono-olefins*

This important method has been used very extensively for making palladium complexes of the type $Pd_2Cl_2all_2$ [391], e.g.

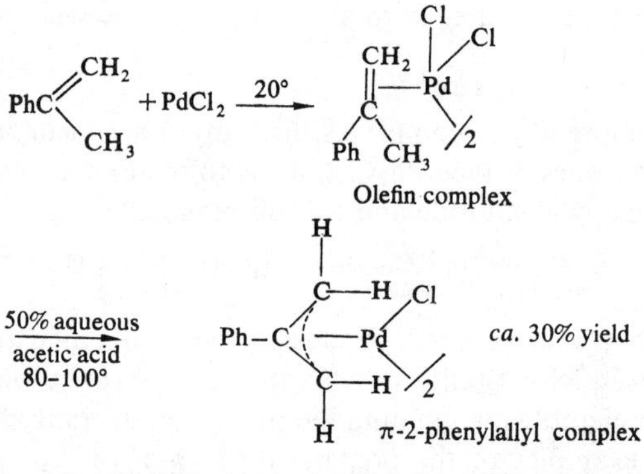

Olefin complex

π-2-phenylallyl complex

When non-cyclic olefins are used there must be a substituent other than hydrogen on the 2-position of the resultant allylic–palladium complex; otherwise palladium metal forms and the olefin is oxidized. With cyclic olefins the ring system prevents elimination of hydride (i.e. oxidation of the olefin does not occur easily) and hydrogen may be the substituent on the central (2-position) of the allylic system. Cyclododecene gives a mixture of all three possible isomers of 1,3-nonamethylene-π-allylpalladium chloride, viz. the syn-, syn-, the syn-, anti- and the anti-, anti-. (See note to Table 15 for an explanation of syn- and anti-.) Sometimes mixtures are obtained, e.g. 2,3-dimethylpent-2-ene gives 1,1,2,3-tetramethylallyl-, 1-ethyl-1,2-dimethylallyl- and 1,1-dimethyl-2-ethylallylpalladium chloride (chlorine bridged dimer).

[389] J. Powell and B. L. Shaw, *J. Chem. Soc.* A (1968) 597.
[390] H. C. Volger and K. Vrieze, *J. Organomet. Chem.* **9** (1967) 527.
[391] R. Hüttel and H. Dietl, *Chem. Ber.* **98** (1965) 1753.

The method is greatly improved by reacting sodium chloropalladate(II) with the olefin in acetic acid at 85° in the presence of sodium acetate, when high yields of π-allylicpalladium chlorides are obtained (Table 15)[392]. Using CH_3COOD as solvent partial deuteriation of the allylic complexes occurs.

TABLE 15. PREPARATION OF π-ALLYLICPALLADIUM COMPLEXES OF THE TYPE $Pd_2Cl_2(all)_2$ BY REACTING Na_2PdCl_4 WITH AN OLEFIN IN HOAc/NaOAc AT 85° [392]

Olefin	π-Allyl substituent[a]	Yield (%)
Isobutene	2-methyl	98
2,4,4-Trimethylpent-1-ene	2-neopentyl	95
2,4,4-Trimethylpent-2-ene	anti-1-tertbutyl-2-methyl	94
Ethyl 3-methylbut-2-enoate	syn- and anti-1-carboethoxy-2-methyl	89

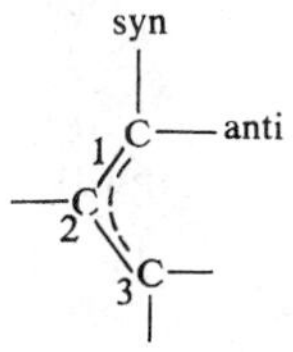

αβ-Unsaturated-carboxylic acids, -esters, -amides or -ketones react especially easily with palladium chloride or the chloropalladate ion to give keto- or carboxylato-allylic–palladium chloride complexes, i.e. loss of a proton to give the allylic system is facilitated by the activating effect of the carboxyl or keto group[393].

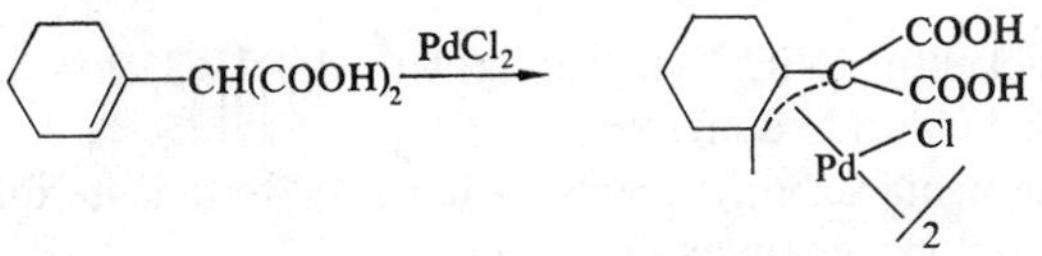

α,β-Unsaturated esters also react readily with palladium chloride, giving carbethoxyallylic complexes. Acetoacetic ester gives syn-1-carbethoxy-2-hydroxyallylpalladium chloride, the structure of which has been determined by X-ray diffraction.

Triphenylmethyl fluorborate will remove hydride ion from an olefin complex to give an allylic complex[13]:

$$CoCp(COD) + Ph_3C^+BF_4^- \rightarrow [CoCp(C_8H_{11})]^+$$

where C_8H_{11} = cyclo-octa-2,5-dienyl.

3.1.1e. From Conjugated Diolefins

π-Allylic ligands can be formed from conjugated diolefins in two ways: either (1) through attack on the diene system by a nucleophile or an electrophile (proton), or (2) by oligomerization of the diene.

392 H. C. Volger, Rec. trav. chim. 88 (1969) 225.
393 R. Hüttel and H. Schmid, Chem. Ber. 101 (1968) 252.

(1) *Formation by attack on the diene.* Groups or atoms which can attack the terminal carbon atom of the diene giving an allylic system include H, RCO, Cl, OR and OAc.

Cobalt tetracarbonyl hydride reacts with butadiene to give a mixture of the *syn-* and *anti-*isomers of π-1-methylallylcobalt tricarbonyl[394]. Similarly, the hydrido pentacyano-cobaltate ion $[CoH(CN)_5]^{3-}$ reacts with butadiene to give the π-1-methylallyltetracyano-cobaltate ion, $[Co(C_4H_7)(CN)_4]^{2-}$ [384].

Treatment of $TiCl_2Cp_2$ with isopropylmagnesium bromide in the presence of a 1,3-diene gives allylic–titanium(III) complexes $Ti(C_nH_{2n-1})Cp_2$; thus penta-1,3-diene gives the 1,3-dimethylallyl complex. The reaction probably goes via $TiHCp_2$, which adds to the diene system[379]. The rhodium hydride complex $RhH(PPh_3)_4$ reacts with 1,3-dienes to give π-allylic complexes of the type $Rh(\pi\text{-all})(PPh_3)_2$ [395]. Reduction of cobalt chloride in ethanol in the presence of butadiene gives $Co(C_8H_{13})C_4H_6$, m.p. 35°. The C_8H_{13} ligand is probably formed by addition of a hydrogen to butadiene followed by insertion of a second molecule of butadiene to give a branched-chain ligand [3.1.1].

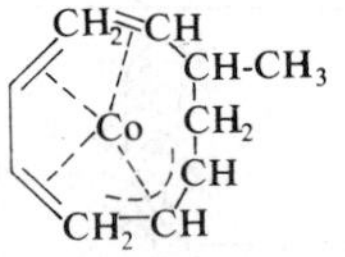

[3.1.1.]

Coordinated 1,3-dienes may be protonated to give π-allylic complexes[13, 396], e.g.

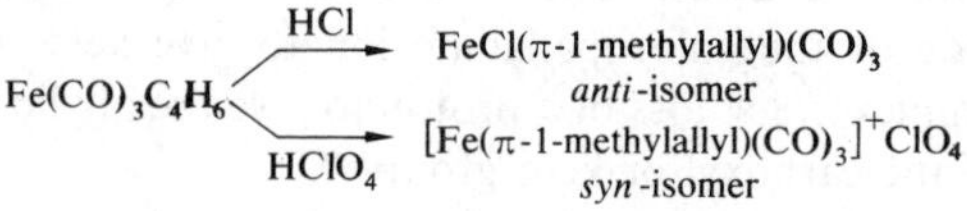

Cyclo-octatetraenenickel(0), $(NiC_8H_8)_x$, reacts with dry HCl at $-80°$ to give the red π-cyclo-octatrienylnickel chloride dimer[288].

Phenylmanganese pentacarbonyl reacts with 1,1,4,4-tetradeuteriobutadiene to give the benzoyl-substituted π-allylic complex [3.1.2] [397].

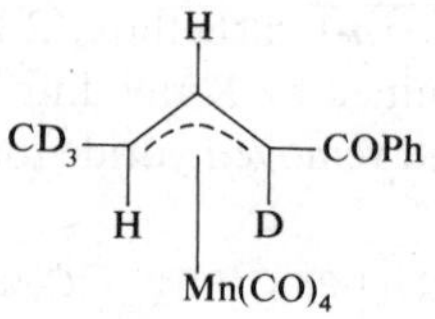

[3.1.2.]

Reaction mixtures of butadiene, $NaCo(CO)_4$ and alkyl or acyl halides, give acyl-substituted π-allyliccobalt tricarbonyl complexes[394].

394 R. F. Heck, *Accounts Chem. Res.* **2** (1969) 10.
395 C. A. Reilly and H. Thyret, *J. Am. Chem. Soc.* **20** (1967) 5144.
396 G. F. Emerson and R. Pettit, *J. Am. Chem. Soc.* **84** (1962) 4591.
397 M. Green and R. I. Hancock, *J. Chem. Soc.* A (1968) 109.

Many acyclic and cyclic 1,3-dienes react rapidly with sodium chloropalladate in alcohols to give alkoxy-π-allylicpalladium chloride complexes[398]. With unsymmetrical dienes the alkoxy group attacks the most heavily substituted carbon atom, e.g.

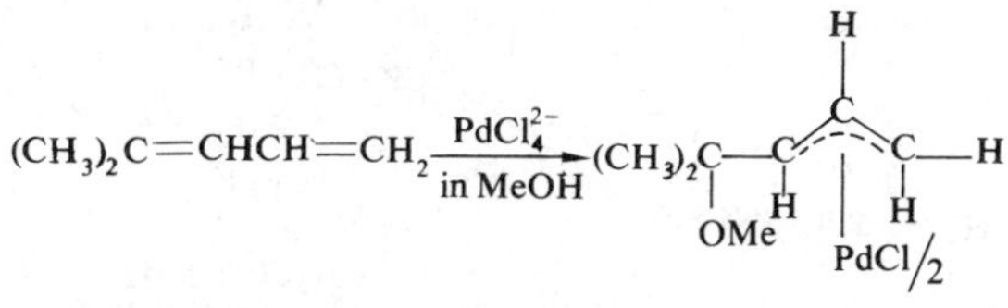

$$(CH_3)_2C=CHCH=CH_2 \xrightarrow[\text{in MeOH}]{PdCl_4^{2-}} (CH_3)_2C-C=C-H$$

Similarly, 1,3-dienes and palladium chloride in acetic acid give acetoxyallylicpalladium complexes.

Butadiene reacts with palladium chloride at $-40°$ to give the olefin complex $[C_4H_6PdCl_2]_2$, but above $-20°$ this rearranges to give the 4-chlorobut-2-enyl complex $[C_4H_6ClPdCl]_2$, i.e. an allylic complex[399].

(2) *By the oligomerization of conjugated dienes (mainly butadiene).* Many examples of this type of synthesis are known. Some are discussed in section 3.1.5 and we shall mention only a few here.

Reduction of nickel acetylacetonate in the presence of *trans, trans, trans*-cyclododeca-1,5,9-triene gives the red nickel(0) complex $NiC_{12}H_{18}$ [3.1.3]. This reacts with butadiene at $-40°$ to give cyclododecatriene and a new nickel(II) complex $NiC_{12}H_{18}$ with the α,ω-bisallylic structure [3.1.4] [288].

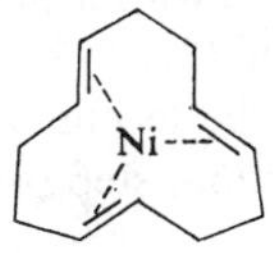

[3.1.3] 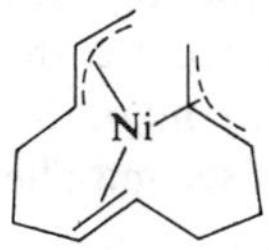   [3.1.4]

The complex [3.1.4] is too unstable for its structure to be determined by X-rays, but a complex $RuCl_2(C_{12}H_{18})$ with a completely analogous α,ω-bisallylic ligand is formed from ruthenium trichloride and butadiene (see section 3.1.3).

Reduction of nickel acetylacetonate in the presence of butadiene and a donor ligand L such as a phosphite, causes dimerization of the butadiene to give an octa-2,6-dien-1,8-diyl complex [3.1.5]. Ruthenium trichloride reacts with isoprene to give the bridged chlorine dimer $[RuCl_2(C_{10}H_{16})]_2$. This has the α,ω-bisallylic ligand 2,7-dimethylocta-2,6-diene-1,8-diyl, formed by tail-to-tail linear dimerization of isoprene[400].

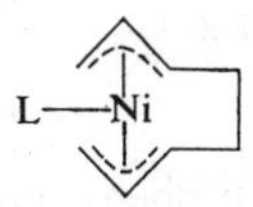

[3.1.5]

[398] S. D. Robinson and B. L. Shaw, *J. Organomet. Chem.* **3** (1965) 367.
[399] M. Donati and F. Conti, *Tetrahedron Letters* (1966) 1219.
[400] L. Porri, M. C. Gallazzi, A. Colombo and G. Allegra, *Tetrahedron Letters* (1965) 4187.

3.1.1f. *Miscellaneous Methods*

Treatment of palladium(II) salts with allenes gives π-allylic complexes, the nature of which depends on the reaction conditions.

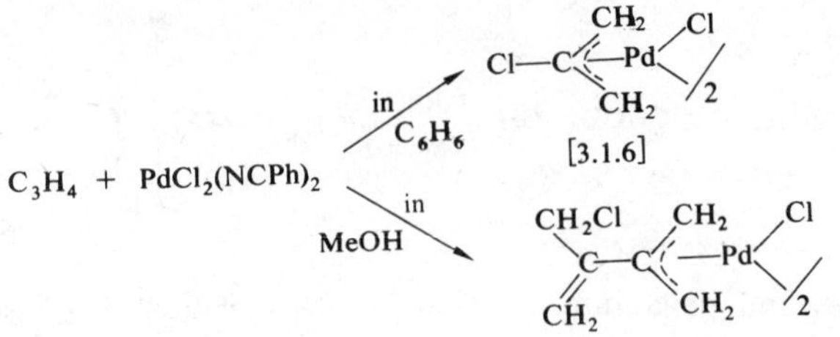

[3.1.6]

[3.1.7]

The formation of the 2-chloroallyl complex [3.1.6] probably involves the migration of coordinated chlorine on to coordinated allene (i.e. an insertion reaction as in step (1) below).

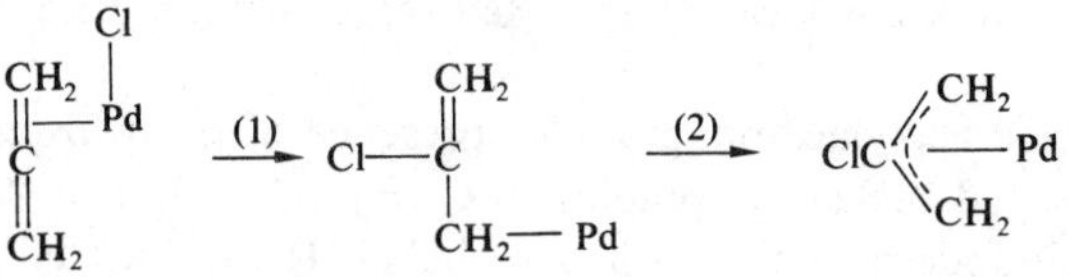

(Other ligands omitted for clarity.)

On treating $Pd_2Cl_2(\pi$-2-chloroallyl$)_2$, i.e. [3.1.6], with triphenylphosphine, allene is eliminated and $PdCl_2(PPh_3)_2$ is formed, i.e. steps (1) and (2) can be reversed. The complex [3.1.7] is probably formed by chlorine migration on to the terminal carbon atom of coordinated allene, step (3), followed by a second allene insertion, step (4) [401].

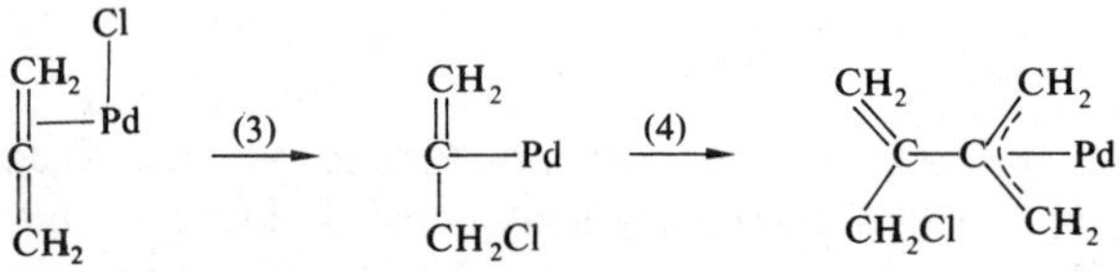

Allenes react with enneacarbonyldi-iron to give bridging bi-π-allylic complexes of the type [3.1.8].

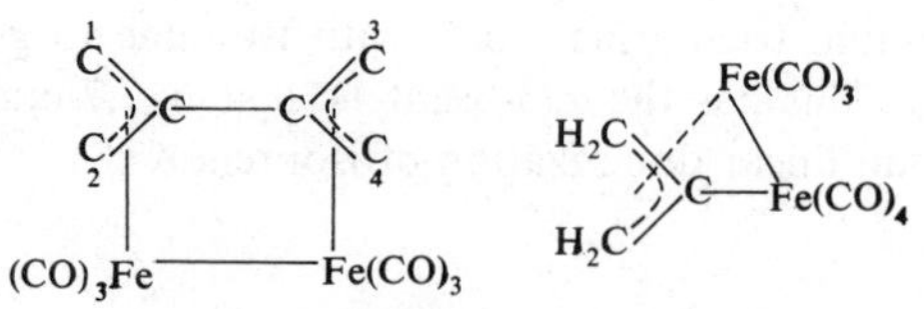

[3.1.8]   [3.1.9]

Phenylallene gives two isomers, one with phenyl groups on positions 1 and 3 and the other with phenyl groups on positions 1 and 4. Allene reacts to give a compound of type [3.1.8] and one of type [3.1.9] [402].

[401] M. S. Lupin, J. Powell and B. L. Shaw, *J. Chem. Soc.* A (1966) 1687.
[402] R. Ben-Shoshan and R. Pettit, *Chem. Commun.* (1968) 247.

From acenaphthylene and pentacarbonyliron a complex $Fe_2(CO)_5C_{12}H_{18}$ is formed. This has the π-allylic-π-cyclopentadienyl structure [3.1.10] with an Fe–Fe bond length of 2.77 Å [403].

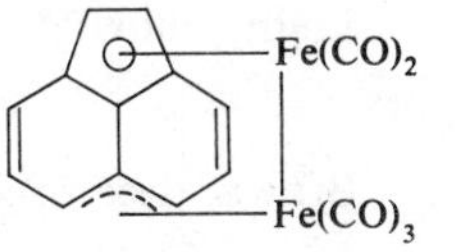

[3.1.10]

Treatment of sodium chloropalladate with diphenylacetylene in ethanol gives the *endo*-ethoxycyclobutenyl complex $[Pd_2Cl_2\{C_4(OEt)Ph_4\}_2]$ [3.1.11]. This with hydrogen chloride gives the tetraphenylcyclobutadiene complex [3.1.12] which in turn with ethanol gives the *exo*-ethoxycyclobutenyl complex [3.1.13] [404].

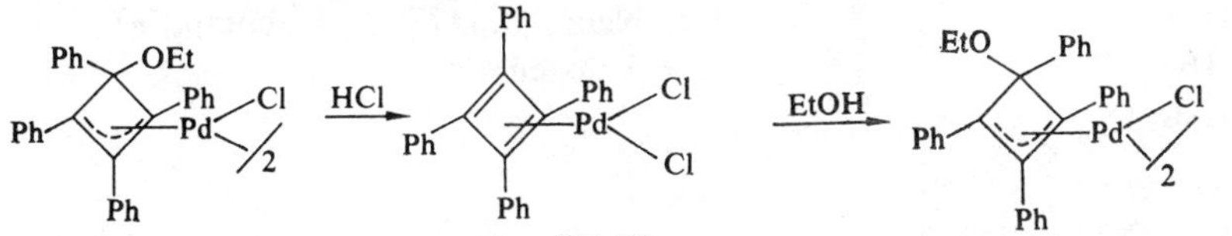

[3.1.11] [3.1.12] [3.1.13]

Phenylcyclopropane reacts with ethyleneplatinous chloride to give a phenylcyclopropane complex which on being heated gives $Pt_2Cl_2(1\text{-phenylallyl})_2$ [405]. Other substituted cyclopropanes react similarly. Vinylcyclopropane or spiropentane react with palladium chloride to give π-allylicpalladium chloride complexes [406].

The cyclopentadienyl ligand can be reduced to the cyclopentenyl ligand, C_5H_7, e.g. chromacene reacts with hydrogen and carbon monoxide to give $Cr(C_5H_7)(C_5H_5)(CO)_2$ and nickelocene on reduction with sodium amalgam in ethanol gives $Ni(C_5H_7)(C_5H_5)$ [376, 377].

3.1.2. Some Examples of π-Allylic Transition Metal Complexes

A summary of the various types of π-allylic complexes is now given with key references. Many of the compounds are discussed in more detail in the text. In this subsection "all" refers to allyl or a substituted allyl group.

Titanium

Ti(all)Cp$_2$ all = allyl, 2-methylallyl, 1,1-dimethylallyl, 1,2-di-methylallyl; paramagnetic[379].

Zirconium

$Zr(C_3H_5)_4$ Red, decomposes 0° [288, 407].

$Zr(C_3H_5)_2Cp_2$ One σ- and one π-allyl group[379].

[403] M. R. Churchill and J. Wormald, *Chem. Commun.* (1968) 1597.
[404] L. F. Dahl and W. E. Oberhansli, *Inorg. Chem.* **4** (1965) 629.
[405] W. J. Irwin and F. J. McQuillin, *Tetrahedron Letters* (1968) 1937.
[406] A. D. Ketley and J. A. Braatz, *Chem. Commun.* **16** (1968) 959.
[407] J. K. Becconsall, B. E. Job and S. O'Brien, *J. Chem. Soc.* A (1967) 423.

Hafnium

Hf(C$_3$H$_5$)$_4$ Orange red[407].

Thorium

Th(C$_3$H$_5$)$_4$ Light yellow[288].

Vanadium

V(C$_3$H$_5$)$_3$ Brown, decomposes $> -30°$ [288].

Niobium

Nb(C$_3$H$_5$)$_4$ Green, decomposes $>0°$ [288, 407].

Tantalum

Ta(C$_3$H$_5$)$_4$ Green, decomposes $>0°$ [288].

Chromium

Cr(C$_3$H$_5$)$_3$ Red black, m.p. 77–79° (decomp.)[288].
CrX(C$_3$H$_5$)$_2$ X = halogen[288].
Cr$_2$(C$_3$H$_5$)$_4$ [288]

Molybdenum

Mo(C$_3$H$_5$)$_4$ Green, allyl groups asymmetrically bonded[409].
Mo$_2$(C$_3$H$_5$)$_4$ Deep green[288].
Mo(all)Cp(CO)$_2$ all = allyl, or possibly ^3h-cycloheptatrienyl or indenyl[383].

[Mo$_2$X$_3$(all$_2$)(CO)$_4$]$^-$[NEt$_4$]$^+$ X = halogen, all = allyl or 2-methylallyl[387].
Mo$_2$(C$_3$H$_5$)$_2$(C$_8$H$_{10}$N$_4$)(CO)$_4$ C$_8$H$_{10}$N$_4$ is derived from allyl cyanimide[410].
Mo(C$_3$H$_5$)(CO)$_2$(dipyrazoylborate)   Orange red, sublimes[411].
Mo(*p*-methylbenzyl)(CO)$_2$Cp   One double bond of the phenyl ring forms part of a π-allylic system[412].

MoI(NO)(C$_5$H$_5$)$_2$?   One of the cyclopentadienyl rings is possibly bonded by 3–C atoms only[413].

[Mo$_2$X$_3$(all)$_2$(CO)$_4$][NEt$_4$]   X = halogen, all = allyl or 2-methylallyl[387].
[MoXall(CO)$_2$MeCN]$_2$ From Mo(CO)$_6$ + allX in MeCN. X = Cl, Br, I or NCS. all = allyl, 2-methylallyl[414].

Tungsten

W(C$_3$H$_5$)$_4$ Light brown, decomposes $>95°$ [288].
W(all)(CO)$_2$Cp [46]

Manganese

Mn(all)(CO)$_4$ all = allyl, 2-methylallyl, 1-methylallyl and various other substituted allyl ligands[382, 397].

[408] E. O. Fischer and K. Ulm, *Chem. Ber.* **94** (1961) 2413.
[409] K. C. Ramey, D. C. Lini and W. B. Wise, *J. Am. Chem. Soc.* **90** (1968) 4275.
[410] H. T. Dieck and H. Friedel, *J. Organomet. Chem.* **12** (1968) 173.
[411] S. Trofimenko, *J. Am. Chem. Soc.* **90** (1968) 4754.
[412] F. A. Cotton and M. D. LaPrade, *J. Am. Chem. Soc.* **90** (1968) 5418.
[413] R. B. King, *Inorg. Chem.* **7** (1968) 90.
[414] H. T. Dieck and H. Friedel, *J. Organomet. Chem.* **14** (1968) 375.

Iron

$Fe(C_3H_5)_3$	Golden orange, decomposes $> -40°$ [288].
$FeX(all)(CO)_3$ $[Fe(all)(CO)_3]^+$	} X = halogen, all = various π-allylic ligands[386, 396].
$Fe(all)Cp(CO)$	all = C_3H_5, 1-methylallyl[122].
$Fe_2(CO)_6C_6H_6Ph_2$ $Fe_2(CO)_6C_6H_8$ $Fe_2(CO)_7C_3H_4$	} From $Fe_2(CO)_9$ and allenes[402].
$Fe_2(CO)_5C_{12}H_8$	From acenaphthylene[403].

Ruthenium

$Ru(all)_2L_2$	} all = allyl or 2-methylallyl[380].
$Ru_2X_2(all)_2(diolefin)_2$	} $L_2 = (PPh_3)_2$ or COD or NBD [380].
$RuCl_2(C_{12}H_{18})$	From $RuCl_3$ + butadiene[415].
$RuCl_2(C_{10}H_{16})$	From $RuCl_3$ + isoprene[400].

Osmium

$OsC_8H_8(CO)_3$?	The C_8H_8 possibly contains a π-allylic system[416].

Cobalt

$Co(C_3H_5)_3$	Golden red, decomposes $> -40°$ [288].
$[Co(C_3H_5)Cp(CO)][PF_6]$ [416a]	
$Coall(CO)_3$	all = allyl or 1-methylallyl (*syn*- and *anti*-isomers) or many other substituted allylic groups[394].
$[Coall(CN)_4]^{2-}$	all = allyl, 1-methylallyl, 2-methylallyl[384].
$CoX(C_3H_5)Cp$	X = Br or I [388].
$Coall(COD)$	all = π-cyclo-oct-2-enyl[417].
$CoC_4H_6(C_8H_{13})$ [418]	

Rhodium

$Rhall(PPh_3)_2$	all = allyl, 2-methylallyl, cyclohexenyl, cyclo-octenyl[395].
$[RhC_3H_5(CO)_2]_x$	Evidence for a monomer ⇌ dimer equilibrium in solution[419].
$Rh(C_3H_5)_3$	Orange crystals. Volatile[385].
$Rh_2X_2(all)_4$ $RhCl(all)_2L$ $[Rh(all)_2py_2]^+$ $Rh(all)_2(acac)$ $[RhCl_2(all)(CO)]_x$	} X = Cl, Br, I or OAc. all = allyl, 2-methylallyl, 1-methylallyl. L = PR_3, AsR_3 or py, etc.[385].
$Rh(all)_2Cp$	one π- and one σ-allyl group[385].
$[Rh_2Cl_4(all)_2(butadiene)]$	all = allyl, 1- or 2-methylallyl[389].

[415] J. K. Nicholson and B. L. Shaw, *J. Chem. Soc.* A (1966) 807.
[416] M. I. Bruce, M. Cooke and M. Green, *Angew. Chem. Int. Edn.* **7** (1968) 639.
[416a] E. O. Fischer and R. D. Fischer, *Z. Naturforsch.* **16b** (1961) 475.
[417] S. Otsuka and M. Rossi, *J. Chem. Soc.* A (1968) 2630.
[418] G. Natta, U. Giannini, P. Pino and A. Cassata, *Chem. e Ind.* (*Milan*) **47** (1965) 524.
[419] S. O'Brien, *Chem. Commun.* (1968) 757.

$[RhCl_2(all)]_x$ [389, 420]

$RhX_2(all)L_2$ — X = Cl, Br. all = allyl or 2-methylallyl. L = PPh_3, $AsPh_3$ or $SbPh_3$ [389, 390].

$Rh_2Cl_2(1\text{-methylallyl})_2(H_2O)_2$ [421]

Iridium

$Ir(C_3H_5)_3$ — White. Sublimes easily. Decomposes > 65° [422].

$IrX_2all(CO)L$ — X = Cl, Br. all = allyl, 2-methylallyl, 1-phenylallyl. L = cyclo-octene, py, isoquinoline, PPh_3, $AsPh_3$, $AsMe_2Ph$ [76].

$IrCl_2allL_2$ — all = allyl, 2-methylallyl. L = PEt_3, PMe_2Ph, PEt_2Ph [381].

$[IrCl(all)(CO)L_2]^+$ — all = allyl or 2-methylallyl. L = PR_3 [18].

Nickel

$Ni(all)_2$ — Yellow. all = a variety of acyclic or cyclic allylic ligands. Exists as *cis*- and *trans*-isomers [288, 423].

$Ni_2X_2all_2$ — X = Cl, Br or I. all = allyl, 1-methylallyl, 2-methylallyl, cyclohexenyl, cyclohepetenyl, etc. [288].

$NiXallL$ — L = neutral ligand, e.g. PPh_3 [288].

$NiallCp$ — all = allyl, cyclopentenyl, etc. Red, volatile [382].

$NiC_{12}H_{18}$ — $C_{12}H_{18}$ = dodeca-2,5,10-trien-1,12-diyl [288].

$Ni(C_8H_{12})L$ — C_8H_{12} = octa-2,6-dien-1,8-diyl. L = phosphite [288].

$NiMe(C_3H_5)$ — Violet crystals. Decomposes > −35° [424].

$[Niall(thiourea)_2]^+$ — all = allyl or methylallyl [425].

$Ni(C_3H_5)COPR_3$ — From $NiX(C_3H_5)PR_3$ and CO [426].

Palladium

$Pd_2X_2all_2$ — X = Cl, Br, I, SCN, OAc, OBz, etc. "all" can be one of a large number of mono-, di-, tri- and tetra-substituted allylic ligands; see text [376, 392, 393, 398, 406, 427].

$Pdall_2$ — all = C_3H_5, C_4H_7. Exists in *syn*- and *anti*-forms [407].

$PdallCp$ — all = C_3H_5, red and volatile. Many analogues with different allyls [382].

$Pdall(acac)$ — Many different allylic ligands [427].

$PdXallL$ — X = Cl, Br or I. Many different allylic ligands. L = amine, phosphine, arsine, etc. [428].

$[PdallL_2]^+$ — L = tertiary phosphine or L_2 = chelating diamine [429].

[420] G. Paiaro, A. Musco and G. Diana, *J. Organomet. Chem.* 4 (1965) 466.
[421] H. E. Swift and R. J. Capwell, *Inorg. Chem.* 7 (1968) 620.
[422] P. Chini and S. Martinengo, *Inorg. Chem.* 6 (1967) 837.
[423] E. J. Corey, L. Hegedus and M. F. Semmelhack, *J. Am. Chem. Soc.* 90 (1968) 2417.
[424] B. Bogdanovic, H. Bönnemann and G. Wilke, *Angew. Chem.* 78 (1966) 839.
[425] F. Guerrieri, *Chem. Commun.* 16 (1968) 983.
[426] F. Guerrieri and G. P. Chiusoli, *J. Organomet. Chem.* 15 (1969) 209.
[427] S. D. Robinson and B. L. Shaw, *J. Chem. Soc.* (1963) 4806.
[428] J. Powell and B. L. Shaw, *J. Chem. Soc.* A (1967) 1839.
[429] J. Powell and B. L. Shaw, *J. Chem. Soc.* A (1968) 774.

Platinum

Ptall$_2$	all = C$_3$H$_5$ or alkyl-substituted allyl group. Pt(C$_3$H$_5$)$_2$ exists as *cis*- and *trans*-isomers[242].
Pt$_2$X$_2$all$_2$ Pt(acac)all	X = Cl, Br, or I. all = 2-methylallyl or various substituted allyl. The compound with all = C$_3$H$_5$ is not π-allylic but has a bridging allyl[242, 405].
PtC$_3$H$_5$Cp	Yellow, volatile[430].
[PtC$_3$H$_5$(PPh$_3$)$_2$]X	X = large anion[390].

3.1.3. Structures and Bonding in π-Allylic Complexes[98]

3.1.3a. *Structures*

The structures of many π-allylic metal complexes have been determined by X-ray diffraction.

The titanium compound Ti(1,2-dimethylallyl)Cp$_2$ is paramagnetic and one cannot use n.m.r. to prove the structure. There was doubt whether such compounds were σ- or π-allylic complexes. The X-ray structure shows it to be π-allylic with three allylic C–Ti distances of Ti–C$_1$ 2.34 Å, Ti–C$_2$ 2.43 Å and Ti–C$_3$ 2.35 Å [431]. The prediction that the benzyl complex Mo(CH$_2$Ph)Cp(CO)$_2$ would have three carbon atoms of the benzyl group bonded to the molybdenum in an allylic grouping has been confirmed by the X-ray structure of the corresponding *p*-methoxybenzyl complex[412], shown diagrammatically in [3.1.14].

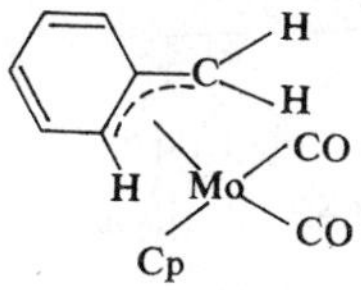

[3.1.14]

X-ray structural determinations on some organic derivatives of iron show that allylic–metal bonding can be preferred to olefinic–metal bonding, e.g. in azulene-di-iron pentacarbonyl[432], three of the carbon atoms form an allylic system [3.1.15].

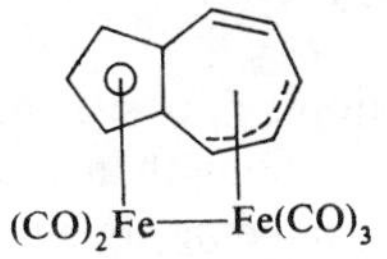

(CO)$_2$Fe——Fe(CO)$_3$

[3.1.15]

Allylic–iron bonding is also found in pentacarbonyl(bicyclo[3,2,1]octadienyl)iron.

The suggestion that dodeca-2,6,10-triene-1,12-diylnickel might be an intermediate in the nickel-catalysed cyclotrimerization of butadiene[214] was indirectly substantiated by the X-ray structural determination of RuCl$_2$(C$_{12}$H$_{18}$); see [3.1.16].

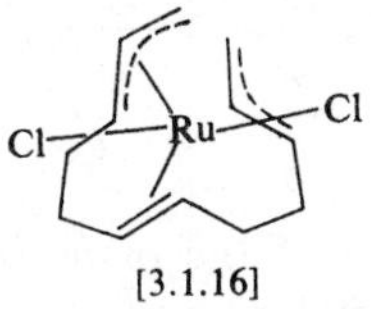

Cl—Ru—Cl

[3.1.16]

[430] B. L. Shaw and N. Sheppard, *Chem. and Ind. (London)* (1961) 517.
[431] R. B. Helmholdt, F. Jellinek, H. A. Martin and A. Vos, *Rec. trav. chim.* 86 (1967) 1263.
[432] M. R. Churchill, *Inorg. Chem.* 6 (1967) 190.

This compound was made from ruthenium trichloride and butadiene and has the postulated dodeca-2,6,10-triene-1,12-diyl ligand[415].

The X-ray structure of the cobalt complex $Co(C_8H_{13})(C_4H_6)$ is discussed in section 3.1.5.

The structure of $RhCl_2(2\text{-methylallyl})(AsPh_3)_2$ shows a symmetrically bonded π-2-methylallylic group with the methyl displaced towards the rhodium atom by 0.2 Å; see [3.1.17].

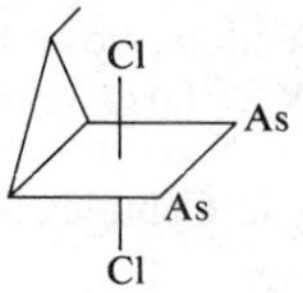

[3.1.17]

The three allylic C–Rh distances are 2.22, 2.26 (central) and 2.24 Å. The dihedral angle between the plane of the allyl ligand and the As–Rh–As plane is 127° [433].

The allyl ligands in $Rh_2Cl_2(C_3H_5)_4$ are asymmetrically bonded to the rhodium with the two mutually *trans*, apical (terminal) carbon atoms of the allylic system farther from the rhodium than the two terminal carbons opposite the bridging chlorine system; see the partial structure [3.1.18]

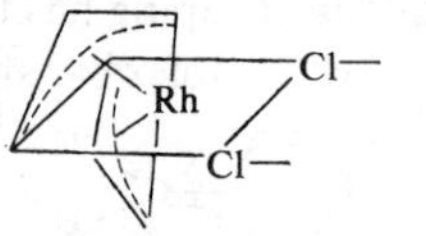

[3.1.18]

and the important bond lengths for the two allylic systems in [3.1.19] and [3.1.20] [434].

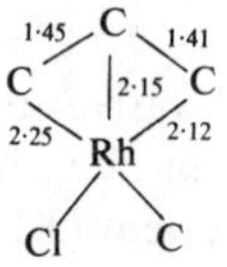

[3.1.19]

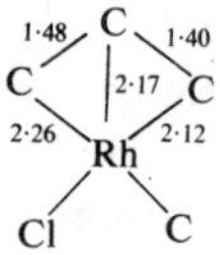

[3.1.20]

The two terminal allylic carbon atoms opposite the electro-negative chlorine ligand are thus more strongly bonded to the rhodium than the other two terminal carbon atoms. Carbon is frequently a *trans*-bond weakening ligand atom in platinum metal complexes, whilst bonds opposite chlorine are usually relatively strong.

Di-π-2-methylallylnickel is a sandwich compound with the nickel atom at the centre of symmetry of the two 2-methylallyl ligands; see [3.1.21].

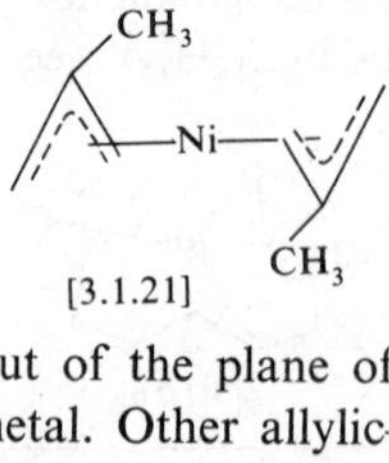

[3.1.21]

The methyl groups are displaced out of the plane of the three allylic–carbon atoms by approximately 0.5 Å towards the metal. Other allylic–nickel complexes whose structures

433 T. G. Hewitt and J. J. de Boer, *Chem. Commun.* (1968) 1413.
434 Mary McPartlin and R. Mason, *Chem. Commun.* (1967) 16.

have been determined by X-rays are 1-*exo*-cyclopentadienyltetramethylcyclobutenyl-π-cyclopentadienylnickel(II), 2-carbethoxyallylnickel bromide[435] and the five-coordinate NiBr(2-methylallyl)diphos.

The X-ray structure of dimeric π-allylpalladium chloride has been examined by four groups with one very accurate determination at −140°C and others at room temperature. The geometry is shown in [3.1.22].

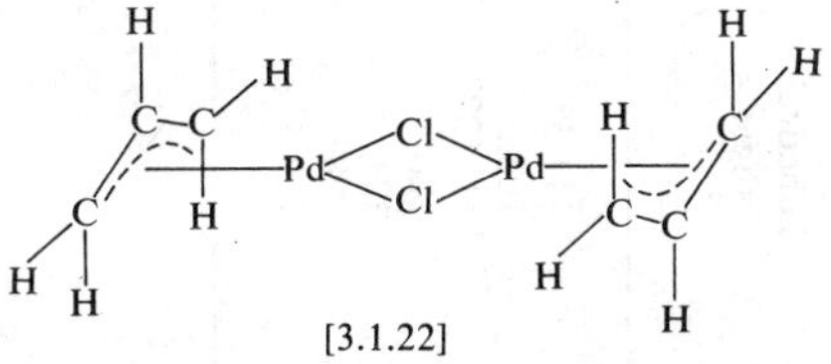

[3.1.22]

At −140° the average Pd–C bond length is 2.117 Å, but at room temperature the two terminal carbon atoms are 2.15 Å from the metal and the central one 2.02 Å away. Both C–C bonds are similar, averaging 1.376 Å at −140° and 1.36 Å at room temperature. The dihedral angle made by the plane of the three carbon atoms with the plane of the palladium and two chlorines is 111.5 ± 0.9° at −140° and 108° at room temperature.

Other allylic–palladium complexes whose structures have been determined by X-rays are Pd_2Cl_2(2-methylallyl)$_2$, Pd_2Cl_2(1,1,3,3-tetramethylallyl)$_2$, Pd_2Cl_2(1,3-dimethylallyl)$_2$ (which has a bent Cl-bridging system of dihedral angle 150°) the *endo*- and *exo*-forms of $Pd_2Cl_2(C_4Ph_4OEt)_2$, $Pd_2Cl_2(C_3H_4 \cdot C_3H_4Cl)_2$ [3.1.7], Pd(acac)(cyclo-octadienyl), Pd(SnCl$_3$)-(allyl)PPh$_3$, Pd_2B1_2(cycloheptenyl)$_2$ and Pd(acac)(dehydro Dewar hexamethylbenzene). Bond lengths and the dihedral angle formed by the allylic ligand and the PdCl$_2$ plane in some chlorine bridged allylic complexes of type Pd_2Cl_2(all)$_2$ are given in Table 16.

The structure of $Pd_2(OOCCH_3)_2(C_3H_5)_2$ [3.1.23] has a similarity to the structure of

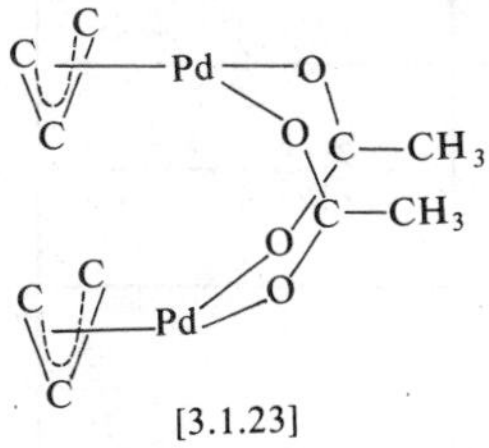

[3.1.23]

dimeric copper acetate. The two complexes PdCl(allyl)PPh$_3$ and PdCl(2-methylallyl)PPh$_3$ have asymmetrically bonded allyl groups. The asymmetry in PdCl(2-methylallyl)PPh$_3$ was first postulated from n.m.r. results and other considerations and confirmed by the X-ray structure[436]. The structure of PdCl(allyl)(PPh$_3$) has been determined very accurately at low temperatures[437] and the structure and bond lengths are given in [3.1.24] and [3.1.25].

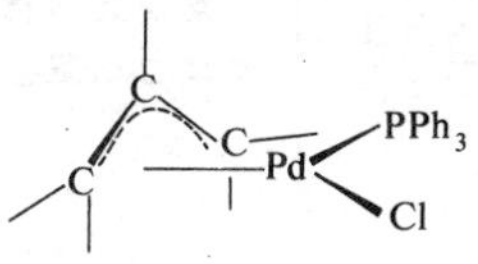

[3.1.24] 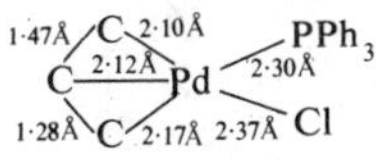   [3.1.25]

[435] M. R. Churchill and T. A. O'Brien, *Inorg. Chem.* **6** (1967) 1386.
[436] R. Mason and D. R. Russell, *Chem. Commun.* (1966) 26.
[437] A. E. Smith, Preprints American Chemical Society Meeting, New York, September 1969, Paper XIV-17.

TABLE 16. SOME INTERATOMIC DISTANCES (Å) AND DIHEDRAL ANGLES FORMED BETWEEN THE ALLYLIC CARBON ATOMS AND THE PLANE OF THE PdCl$_2$ SYSTEM, IN COMPLEXES OF TYPE Pd$_2$Cl$_2$(all)$_2$

Allyl substituent	Pd–C distances			C–C distances		Dihedral angle
	C$_1$	C$_2$	C$_3$	C$_1$–C$_2$	C$_2$–C$_3$	
—[a]	2.12	2.12	2.11	1.36	1.40	111.5°
2-Methyl[b]	2.08	2.10	2.06	1.37	1.36	108.5°
1,3-Dimethyl[c]	2.14	2.07	2.03			123, 127°
2-t-Butyl[e]	2.14	2.15	2.14	1.39	1.39	120°
1,1,3,3-Tetramethyl[d]	2.12	2.14	2.12	1.41	1.42	121.5°
1-Carbethoxy-2-hydroxyl[f]	2.13	2.19	2.10	1.43	1.33	108°
2-(3-Chloroprop-1'-en-2'-yl)[g]	2.19	2.12	2.17	1.47	1.39	117.5°

[a] At −140°C, A. E. Smith, Acta Cryst. 18 (1965) 331.
[b] R. Mason and A. G. Wheeler, J. Chem. Soc. A (1968) 2549.
[c] G. R. Davies, R. H. B. Mais, S. O'Brien and P. G. Owston, Chem. Commun. (1967) 1151.
[d] R. Mason and A. G. Wheeler, J. Chem. Soc. A (1968) 2543.
[e] M. K. Minasyan, S. P. Gubin and Yu T. Struchkov, Zh. Strukt. Khim. 8 (1967) 1108.
[f] K. Oda, N. Yasuoka, T. Ueki, N. Kasai, M. Kakudo, Y. Tezuka, T. Ogura and S. Kawaguchi, Chem. Commun. (1968) 989.
[g] A. D. Broadbent and G. E. Pringle J. Inorg. Nucl. Chem. 33 (1971) 2009. This is the product from allene and PdCl$_2$(PhCN)$_2$ (see section 3.1.6).

There has been a great deal of work on the rate processes which occur in solutions containing the species derived from allylicpalladium chlorides and tertiary phosphines or tertiary arsines. These two X-ray determinations[436, 437] therefore have a special significance.

3.1.3b. *Bonding*

A molecular orbital theory of bonding in allylic–metal complexes has been developed[438]. The allylic radical has three π molecular orbitals, ψ_1 (bonding), ψ_2 (non-bonding) and ψ_3 (anti-bonding). The first two can overlap with the metal orbitals as shown in Fig. 24.

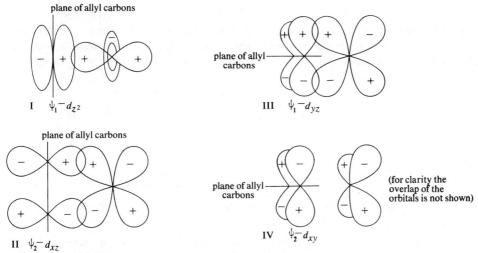

FIG. 24. Molecular orbital scheme for the bonding in π-allylic complexes, e.g. of type $Pd_2Cl_2(all)_2$. In I and II the allyl group would be perpendicular to the plane formed by the $PdCl_2$ system, i.e. a dihedral angle of 90°, whilst in III and IV it would be horizontal, i.e. a dihedral angle of 180°.

Two of the interactions would favour the allyl group being vertical to the xz-plane, but the other two would favour the allyl group lying in the xz-plane. One would therefore expect the allyl group to take up an orientation somewhere between these two extremes. Calculation of the overlap integrals for the various orientations (Fig. 24) and maximization of the bonding energy suggests that in allylicpalladium chloride complexes the dihedral angle formed between the allylic ligand and the plane of the $PdCl_2$-grouping would be about 110°, in good agreement with observation (see Table 16). Overlap of filled d-orbitals on the metal with the antibonding π^* orbital (ψ_3) on the allyl ligand is calculated to be very small, i.e. there is little "back bonding" in allylic complexes (contrast with olefin, allene or acetylene complexes).

3.1.4. Nuclear Magnetic Resonance Spectroscopy, Infrared Spectroscopy and Dipole Moments

3.1.4a. *Nuclear Magnetic Resonance Spectroscopy*

This is the most important physical technique for studying allylic complexes. The stereochemistry of the allylic substituents is usually determined by ^{1}H n.m.r. spectroscopy, and it

438 S. F. A. Kettle and R. Mason, *J. Organomet. Chem.* 5 (1966) 573.

has also been used a great deal for studying rapid rate processes such as σ-allylic $\leftrightarrow$ π-allylic conversions. The symmetrical nature of the bonding in π-allylic complexes was first demonstrated by n.m.r.

One would expect a symmetrically bonded allyl ligand [3.1.26] to give an AMM′XX′

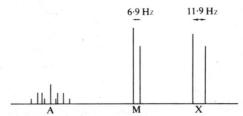

[3.1.26]

^{1}H n.m.r. pattern. J_{MX} and $J_{MX'}$ are small and under the usual running conditions not observed. Thus the reported resonance pattern for allylpalladium chloride consists of doublets for the M- and X-resonances and a nine-line pattern for the A resonance (Fig. 25).

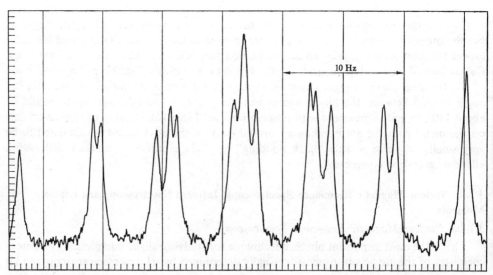

FIG. 25. ^{1}H n.m.r. pattern for $Pd_2Cl_2(C_3H_5)_2$ in $CDCl_3$. Diagrammatic.

However, with a more homogeneous magnetic field the M- and X-resonances can each be resolved into a doublet of triplets and the A-resonance can be resolved into fifteen lines (Fig. 26).

FIG. 26. ^{1}H n.m.r. pattern (60 MHz) for the central hydrogen A resonance of $Pd_2I_2(C_3H_5)_2$ in $CDCl_3$.

Thus J_{MX} and $J_{MX'}$ are not zero, and under good resolution second-order splittings become apparent. However, practically all the published ^{1}H n.m.r. data have been obtained from first-order spectra of the type shown in Fig. 25 and not Fig. 26.

It is known from a study of the n.m.r. spectra of olefinic hydrogens that *trans-vic*-hydrogens couple more strongly than *cis-vic*-hydrogens and that both coupling constants are positive. As can be seen from Fig. 25, $J_{AX} > J_{AM}$ and, mainly on this evidence, the high field resonance is assigned to transition of the *anti*-protons (XX'). The relative signs of coupling constants in π-allylic complexes can be determined by spin tickling. Thus for the π-1-phenylallyl complex, Pd(acac)(1-phenylallyl) [3.1.27]

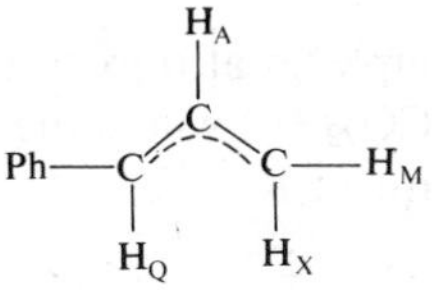

[3.1.27]

the values are $J_{AX} \pm 11.9$ Hz, $J_{AQ} \pm 11.2$ Hz, $J_{AM} \pm 6.9$ Hz, $J_{MX} \mp 1.0$ Hz, $J_{MQ} \mp 1.0$ Hz, $J_{XQ} \mp 0.4$ Hz; the upper set of coupling constants is almost certain to be correct, i.e. J_{AX} is positive, etc.

π-1-Methylallylcobalt tricarbonyl exists as two stereoisomers with *syn*- or *anti*-1-methylallyl ligands [3.1.28] or [3.1.29] respectively[439].

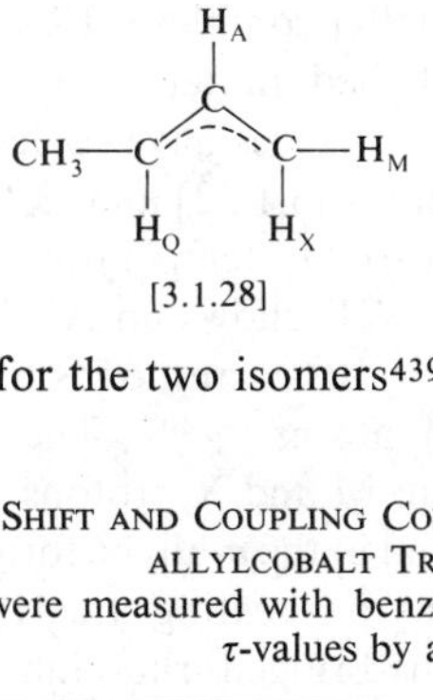

[3.1.28] 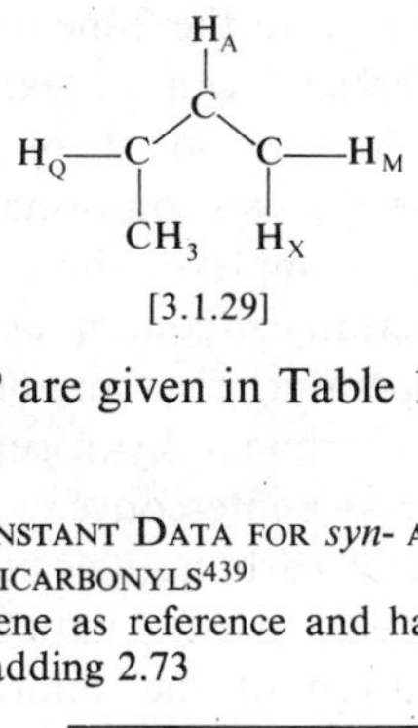   [3.1.29]

The τ- and J-values for the two isomers[439] are given in Table 17.

TABLE 17. CHEMICAL SHIFT AND COUPLING CONSTANT DATA FOR *syn*- AND *anti*-π-1-METHYL-ALLYLCOBALT TRICARBONYLS[439]
The chemical shifts were measured with benzene as reference and have been converted to τ-values by adding 2.73

	syn [3.1.28]	*anti* [3.1.29]		*syn* [3.1.28]	*anti* [3.1.29]
	τ-values			J-values (Hz)	
H_A	4.62	4.38	J_{AM}	6.3	7.1
H_M	6.46	6.08	J_{AX}	10.5	12.3
H_X	7.38	6.53	J_{AQ}	10.5	6.8
H_Q	6.22	5.11	J_{MX}	0.2	0.6
CH_3	7.76	8.11	J_{MQ}	0.2	1.6
			J_{Q-CH_3}	6.5	7.0

[439] D. W. Moore, H. B. Jonassen, T. B. Joyner and A. J. Bertrand, *Chem. and Ind.* (London) (1960) 1304.

The values of J_{AQ} are 10.5 and 6.8 Hz, corresponding to *trans-* and *cis*-vicinal couplings respectively. The stereochemistries of hundreds of π-allylic complexes have been determined from the n.m.r. spectra, assuming *trans-vic*-coupling constants > *cis-vic*-coupling constants.

The ¹H n.m.r. spectra of the diallylic complexes M(all)₂, where M = Ni, Pd or Pt, clearly show that *cis-* [3.1.30] and *trans*-isomers [3.1.31] are present in solution.

[3.1.30]   [3.1.31]

Other more complex allylic complexes also exist as *cis-* and *trans*-isomers in solution, e.g. complexes of type Mo(all)Cp(CO)₂ [440], where the proportion of each form [3.1.32] or [3.1.33] depends on the allyl substituent R.

[3.1.32]   [3.1.33]

For the allyl complexes (R = H) the two forms can be clearly seen at 5° and conformer [3.1.32] predominates, but with the 2-methylallyl complexes (R = Me) conformer [3.1.33] predominates. The differences can be explained in terms of non-bonding interactions between the ligands. At 86° the two sets of resonances have coalesced to one by some rapid rate process which causes the two conformations [3.1.32] and [3.1.33] to interconvert[440].

Many isoleptic allylic complexes show variable temperature n.m.r. spectra. Thus the ¹H n.m.r. spectrum of tetrallylzirconium at −74° shows an AMM′XX′ (AM₂X₂) pattern. This changes as the temperature rises and at −10° corresponds to an AX₄ pattern, i.e. with magnetically equivalent terminal hydrogen atoms[288, 407]. The equivalence is probably caused by a rapid π → σ → π interconversion. M and X protons would become equivalent by rotation about the metal–carbon single bond in the σ-allylic form. Other allylic complexes which change from AM₂X₂ to AX₄ patterns on heating are Hf(C₃H₅)₄ and Th(C₃H₅)₄. Also the ¹H n.m.r. pattern of the chlorine-bridged rhodium complex Rh₂Cl₂(C₃H₅)₄ changes from that of an asymmetrically bonded π-allyl to an AX₄ pattern on heating to 140°. In contrast, the n.m.r. spectrum of Mo(C₃H₅)₄ shows asymmetrically bonded allylic groups and, surprisingly, the n.m.r. spectrum does not change with temperature until decomposition sets in at 170°.

Trisallylrhodium and trisallyliridium also exhibit variable temperature n.m.r. patterns. At −65° trisallylrhodium shows three AM₂X₂(AMM′XX′) patterns (i.e. splitting by ¹⁰³Rh is small), corresponding to three symmetrically bonded π-allylic groups. At −10° two of the AM₂X₂ patterns have collapsed into one pattern which is now twice as intense as the remaining AM₂X₂ pattern. Preliminary X-ray diffraction results show that the geometry is essentially based on a trigonal prism.

440 J. W. Faller and M. J. Incorvia, *Inorg. Chem.* 7 (1968) 840.
441 J. K. Becconsall and S. O'Brien, *Chem. Commun.* (1966) 720.

At low temperatures the three allylic groups will be different; if one of them (say A) starts to rotate rapidly, i.e. [3.1.34] → [3.1.35] → [3.1.34] (Fig. 27), then allyl groups B and C will become equivalent and their resonances will coalesce.

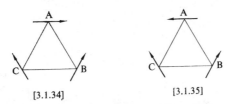

[3.1.34]   [3.1.35]

FIG. 27. Representation of the two forms of tris-π-allylrhodium(III) looking down the three-fold axis of the trigonal prism. Each arrow represents an allyl group, i.e. looking down the plane of the allyl group with the arrow head the central carbon atom.

Trisallyliridium is similar and at room temperature its ^{1}H n.m.r. spectrum consists of two AM_2X_2 patterns—one twice as intense as the other[442].

When treated with triphenylphosphine, the complexes of type $Pd_2Cl_2(all)_2$ give mononuclear species, $PdCl(all)PPh_3$, in which the allylic ligand is asymmetrically bonded to the palladium (see discussion of X-ray structures, section 3.1.3). The ^{1}H n.m.r. spectrum of $PdCl(2$-methylallyl$)PPh_3$ is shown diagrammatically in Fig. 28.

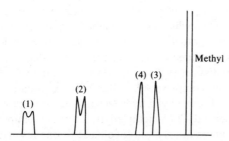

FIG. 28. ^{1}H n.m.r. spectrum of $PdCl(2$-methylallyl$)PPh_3$ at 100 MHz (diagrammatic). The assignment of the resonances is as shown in [3.1.36] (Fig. 29).

The splitting of the two resonances (1) and (2) is due to coupling with the phosphorus nucleus in *trans*-position. The addition of triphenylphosphine to solutions of this complex brings about changes in the n.m.r. spectrum due to a number of rapid rate processes. For example, the splittings of resonances (1) and (2) (Fig. 28) disappear on adding a small amount of triphenylphosphine to a solution of $PdCl(2$-methylallyl$)PPh_3$ in deuteriochloroform at room temperature. The loss of ^{31}P-coupling is due to rapid exchange of free and complexed triphenylphosphine, see the [3.1.36] ⇌ [3.1.37] interconversion (Fig. 29).

This interconversion also causes collapse of resonances (1) with (4) and (2) with (3) (Fig. 28), but these collapses require a slightly higher concentration of added triphenylphosphine than that required to lose the ^{31}P-coupling from resonances (1) and (2). Further addition of triphenylphosphine eventually causes equivalence of all four allylic hydrogens, i.e. (1), (2), (3) and (4), giving a single, sharp, resonance peak. Equivalence is probably

442 P. Chini and S. Martinengo, *Inorg. Chem.* **6** (1969) 837.

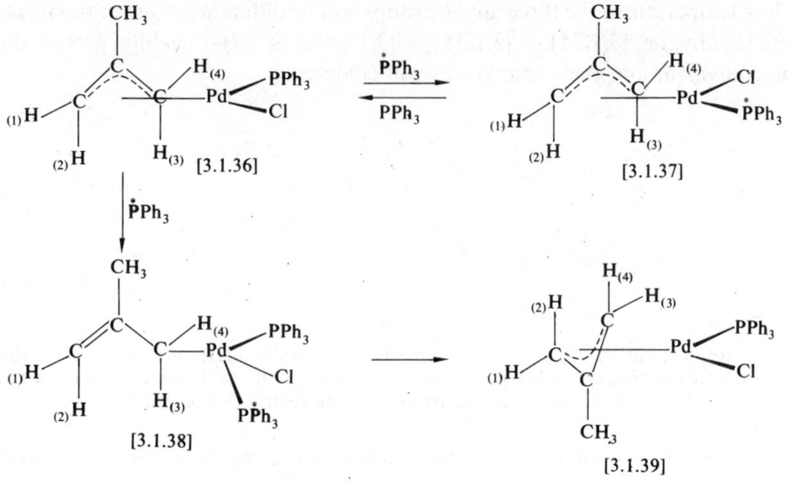

FIG. 29. Some of the rapid substitution processes and $\pi \rightarrow \sigma \rightarrow \pi$ processes which occur in solutions containing PdCl(2-methylallyl)PPh$_3$ and PPh$_3$.

caused by conversion of a π-allylic to a σ-allylic complex, i.e. [3.1.36] $\rightarrow$ [3.1.38] in Fig. 29. In [3.1.38] rotation about the C–Pd single bond can occur, followed by loss of triphenylphosphine, giving [3.1.39] in which protons (3) and (4) have interchanged positions, i.e. the rapid conversion [3.1.36] $\rightarrow$ [3.1.38] $\rightarrow$ [3.1.39] makes protons (3) and (4) equivalent on the n.m.r. time scale. A system which causes magnetic equivalence of all four hydrogens of a terminal allylic ligand by a rapid rate process is sometimes called "a dynamic σ-allylic system". Evidence for a σ-bonded intermediate in such systems comes from infrared data. The solutions show a band at *ca.* 1630 cm^{-1} due to a terminal $v(C{=}C)$ which is absent from the spectrum of π-allylic complexes[443]. Although the mononuclear complex PdCl(2-methylallyl)PPh$_3$ shows all four allylic hydrogens to be non-equivalent, in the presence of the bridged-chloro dimer Pd$_2$Cl$_2$(2-methylallyl)$_2$ at room temperature some rapid rate process makes the resonances of (3) and (4) collapse to one resonance, although the resonances of (1) and (2) and the coupling with ^{31}P remain. The nature of this interaction between PdCl(2-methylallyl)PPh$_3$ and Pd$_2$Cl$_2$(2-methylallyl)$_2$ is not known. However, in [3.1.36] the carbon atom in *trans*-position to the PPh$_3$ is less strongly bonded to the palladium than the other terminal carbon atom (see section 3.1.3). Hence some interaction with Pd$_2$Cl$_2$(2-methylallyl)$_2$ probably occurs to give an intermediate σ-2-methylallylic complex analogous to [3.1.38] but without the $\overset{*}{P}Ph_3$. This can invert to re-form [3.1.39], causing equivalence of protons (3) and (4) but without causing interchange of protons (1) and (2), which remain *trans* to the same PPh$_3$ (i.e. ^{31}P coupling is not lost).

There have been many studies on systems such as Pd$_2$Cl$_2$(2-methylallyl)$_2$/PPh$_3$ discussed above[428, 443, 444]. The systems studied are of the type Pd$_2$X$_2$all$_2$ with a ligand L, where X = Cl, Br, I, OAc, SCN, etc., all = allyl, 2-methylallyl, 2-phenylallyl, 1,3-dimethylallyl, 1,1-dimethylallyl, 1,2,3-trimethylallyl, and 1,1,3,3-tetramethylallyl and many others. The ligands L are various tertiary phosphines or tertiary arsines. There have been some very thorough kinetic studies on these systems using low-temperature n.m.r. spectroscopy[444].

443 F. A. Cotton, J. W. Faller and A. Musco, *Inorg. Chem.* **6** (1967) 179.
444 K. Vrieze, A. P. Praat and P. Cossee, *J. Organomet. Chem.* **12** (1968) 533.

Rate processes which cause magnetic equivalence of the terminal hydrogens due to rapid $\pi \to \sigma \to \pi$ interconversions have also been observed in rhodium complexes of the type $RhCl_2(all)L_2$ (L = PPh_3, $AsPh_3$ or $SbPh_3$) of configuration [3.1.40].

[3.1.40]

Although at $-80°$ the ^1H n.m.r. spectrum is consistent with the structure [3.1.40], at $+30°$ all four terminal hydrogens are magnetically equivalent. For L = PPh_3, however, coupling with the ^{31}P-nucleus is still present at $+30°$, so that even though a $\pi \to \sigma \to \pi$ conversion is occurring, the exchange of PPh_3 is not rapid. At $+50°$, however, exchange of triphenylphosphine is also rapid, and coupling with the ^{31}P-nucleus is lost. The rates of the $\pi \to \sigma \to \pi$ interconversions have been measured and decrease as the σ-donor strengths of L decrease, i.e. $PPh_3 > (p\text{-}Me_2NC_6H_4)_3As > AsPh_3 > SbPh_3$.

The compound $Mo(C_5H_5)_3NO$ formed from $[MoI_2(NO)(C_5H_5)]_2$ and thallium cyclopentadienide has a variable temperature ^1H n.m.r. spectrum[445]. Thus (1) at $25°$ one sharp line is observed, (2) at $-28°$ the spectrum of a slow moving $h^1\text{-}C_5H_5$ is observed, (3) at $-52°$ the spectrum consists of a well-defined AA'BB'X spectrum due to $h^1\text{-}C_5H_5$ and a single line due to the remaining ten protons, (4) at $-95°$ the single line has split into two lines each corresponding to five protons. These lines become sharp at $-110°$ and between $-52°$ and $-110°$ the AA'BB'X pattern has apparently changed into an ABCDX pattern. The suggested explanation is that the compound is $Mo(h^1\text{-}C_5H_5)(h^3\text{-}C_5H_5)(h^5\text{-}C_5H_5)NO$. The motion of the $h^1\text{-}C_5H_5$ ring is "frozen out" by $-52°$ and at $-110°$ the motion of the $h^3\text{-}C_5H_5$ ring is not stopped but the presence of the two rings $h^3\text{-}C_5H_5$ and $h^5\text{-}C_5H_5$ causes nonequivalence of the five protons in the $h^1\text{-}C_5H_5$ ring[445].

The "hapto" nomenclature refers to the number of carbon atoms of a particular ligand which are actually bonded to the metal atom, i.e. h^3—three carbon atoms as in a π-allylic ligand, h^5—five carbon atoms as in a π-cyclopentadienyl[446].

3.1.4b. Infrared Spectroscopy

Assignments have been made to the various vibrations of the allylic ligands in π-allylic complexes of palladium and nickel[447]. In the C–H stretching region a band at 2849 cm⁻¹ has been assigned to $v(C\text{-}H)$ for the *anti*-hydrogens of $Pd_2Br_2(C_3H_5)_2$; in $Ni_2Cl_2(C_3H_5)_2$ the band is at 2865 cm⁻¹. A band in the region 1450–1500 cm⁻¹ has been assigned to the C–C asymmetric stretching vibration of a coordinated π-allyl ligand. Thus in chloroform solution $Pd_2Cl_2(C_3H_5)_2$ has a band at 1458 cm⁻¹ and the corresponding π-1-methylallyl complex a band at 1500 cm⁻¹. A band at *ca*. 1020 cm⁻¹ has been assigned to the symmetric stretch $v_s(C\text{-}C)$. A band due to the bending mode of the allylic ligand, $\delta(C\text{-}C\text{-}C)$ in nickel(II) or palladium(II) complexes occurs at *ca*. 510 cm⁻¹ for unsubstituted allyl, from 520 to 567 cm⁻¹ for π-2-methylallyl or π-1-methylallyl, and at *ca*. 585 cm⁻¹ for cyclic allylic ligands[447].

[445] F. A. Cotton and P. Legzdins, *J. Am. Chem. Soc.* **90** (1968) 6232.
[446] F. A. Cotton, *J. Am. Chem. Soc.* **90** (1968) 6230.
[447] H. P. Fritz, *Chem. Ber.* **94** (1961) 1217.

There have been several studies on metal–chlorine stretching frequencies in π-allylic–metal chloride complexes. Bridged chloro-complexes such as $Pd_2Cl_2(all)_2$ show two very intense bands at low frequencies, characteristic of the bridging system, e.g. with $Pd_2Cl_2(C_3H_5)_2$ they are at 253 and 244 cm⁻¹. Terminal metal–chlorine stretching frequencies are higher, e.g. for $PdCl(C_3H_5)PPh_3$ at 279 cm⁻¹, and for some rhodium(III) complexes *ca.* 300 cm⁻¹. All bands due to metal–chlorine stretch are intense and easily identified by comparing the spectrum with that of the corresponding bromide or iodide. Since metal–chlorine stretching frequencies in platinum metal complexes depend mainly on the nature of the *trans*-ligand, the low value of 279 cm⁻¹ for $v(Pd–Cl)$ in $PdCl(C_3H_5)PPh_3$ suggests that the π-allyl ligand has quite a strong *trans*-bond weakening effect towards chlorine (comparable with a tertiary phosphine).

3.1.4c. *Dipole Moments*

Although halogeno-bridged π-allylic complexes of nickel and palladium are often symmetrical in the crystalline state, in solution they have non-zero dipole moments; 1.3–1.6 D for nickel and 2.1–2.2 D for palladium. This has been explained in various ways such as partial dissociation into a polar monomer, an angular displacement of the bridging halogens or partial conversion to a *cis*-arrangement of the allylic ligands, i.e. [3.1.41] → [3.1.42].

[3.1.41]   [3.1.42]

The low dipole moments of π-allyl-π-cyclopentadienylnickel(II) (0.78 D) and π-cyclopentadienyl-π-cyclopentenylnickel (1.16 D) are in agreement with sandwich structures. The appreciable dipole moment of π-cyclopentadienyl-π-cyclopentenyldicarbonylchromium (3.5 D) suggests that the two organic ligands are not mutually *trans*.

3.1.5. Reactions

3.1.5a. π–σ-*Interconversions*

One of the more important reactions of allylic complexes is the interconversion of π- and σ-forms. Examples have been given in sections 3.1.1 and 3.1.4a. $\pi \to \sigma$ conversion is an important step in many catalytic reactions involving allylic intermediates (see section 3.1.5). Here we describe a few more examples.

The π-allylrhodium complex $RhCl_2(\pi$-$C_3H_5)(PPh_3)_2$ reacts with liquid sulphur dioxide at 20° to give $RhCl_2(\sigma$-$C_3H_5)SO_2(PPh_3)_2$ [448], and the π-2-methylallyliridium complex $IrCl_2(C_4H_7)(PMe_2Ph)_2$ when treated with carbon monoxide gives a σ-2-methylallyl complex $IrCl_2(C_4H_7)CO(PMe_2Ph)_2$. $IrCl_2(\pi$-$C_3H_5)CO(cyclo-octene)$ when boiled in chloroform/methanol gives an insoluble σ-allyl complex $[IrCl_2(\sigma$-$C_3H_5)CO(cyclo-octene)]_x$ (x is presumably 2).

3.1.5b. *Allylic Transfer and Disproportionation Reactions*

There are several examples of the transfer of allylic ligands between metals[385, 449], e.g.

$$Rh(C_3H_5)_3 + PdCl_4^{2-} \to \tfrac{1}{2}[Rh_2Cl_2(C_3H_5)_4] + \tfrac{1}{2}[Pd_2Cl_2(C_3H_5)_2] + 2Cl^-$$

$$\downarrow {}^{20°}_{RhCl_3}$$

$$Rh_2Cl_2(C_3H_5)_4 \xrightarrow[70°]{RhCl_3} [RhCl_2(C_3H_5)]_x$$

[448] H. C. Volger and K. Vrieze, *J. Organomet. Chem.* **13** (1968) 479.
[449] R. F. Heck, *J. Am. Chem. Soc.* **90** (1968) 317.

$$Pd_2Cl_2(C_3H_5)_2 + Fe_2(CO)_9 \rightarrow FeCl(C_3H_5)(CO)_3 + 2Pd + 3CO$$
$$\tfrac{1}{2}[Pd_2Cl_2(C_3H_5)_2] + NaCo(CO)_4 \rightarrow \pi\text{-}C_3H_5Co(CO)_3 + Pd + NaCl + CO$$

Allylicpalladium halides react with mercury to give allylicmercuric halides and palladium metal in quantitative yield. Di(allyl)platinum reacts with mercury to give diallylmercury and platinum metal[450].

These allylic transfer reactions probably go via a bridged or μ-bonded intermediate [3.1.43]. The binuclear compound $Pt_2(acac)_2(C_3H_5)_2$ is bridged by the allyl groups in this way[242].

[3.1.43]

The disproportionation of allylicnickel halides is the basis of a method of synthesis of di(allylic)nickel complexes. In solutions containing allylicnickel bromides the following equilibrium is set up rapidly:

$$\underset{\text{red}}{Ni_2Br_2all_2} \rightleftharpoons \underset{}{NiBr_2} + \underset{\text{yellow}}{Niall_2}$$

The allylic group is probably transferred via an allylic-bridged intermediate. In donor solvents such as dimethylformamide or N-methylpyrrolidine, the equilibrium is driven over to the RHS and the yellow diallylicnickel complex, Ni(all)$_2$, can be distilled out of the solution in yields of 60%, or greater than 80% if triethylenetetramine is also added to the solution[423].

3.1.5c. *Pyrolysis*

Allylpalladium chloride on heating to 160–220° gives palladium metal and allyl chloride. 2-Methylallylpalladium chloride gives 2-methylallyl chloride and some 1-chloro-2-methyl-prop-1-ene. Higher homologues tend to give more complicated products, e.g. conjugated dienes.

3.1.5d. *Alkoxide Attack*

In the presence of an alcohol and a base, allylicpalladium halides decompose very rapidly to give palladium metal and olefin(s) (allyl-H). The products from a large number of such reactions have been studied[451] (see Table 18 for a few examples). Some allylic–rhodium and –iridium halides react similarly with sodium methoxide in methanol to give olefins.

TABLE 18. REDUCTIVE DECOMPOSITION OF BRIDGED CHLORO π-ALLYLIC-PALLADIUM COMPLEXES USING POTASSIUM HYDROXIDE IN METHANOL[451]

Substituents on allyl ligand	Olefin formed (yield)
1,3-Dimethyl-2-ethyl	3-Ethylpent-2-ene (75%)
1,1,2-Trimethyl	{ 2,3-Dimethylbut-1-ene (23%)
	2,3-Dimethylbut-2-ene (55%)
1-tert-Butyl-2-methyl	{ 2,4,4-Trimethylpent-2-ene
	2,4,4-Trimethylpent-1-ene
1,2-Pentamethylene	1-Methylcyclohept-1-ene

450 A. N. Nesmeyanov, A. Z. Rubezhov, L. A. Leites and S. P. Gubin, *J. Organomet. Chem.* 12 (1968) 187.
451 H. Christ and R. Hüttel, *Angew. Chem.* 75 (1963) 921.

The mechanism of this reaction with methoxide ion has been studied in some detail for the complex Pd_2Cl_2(2-neopentylallyl)$_2$ [452]. This complex was treated with the following combinations of reactants: (a) $D_2O/CD_3OD/NaOD$, (b) $H_2O/CD_3OH/NaOH$ and (c) $D_2O/CH_3OD/NaOD$, and the deuterium content of the γ-position in the product [3.1.44] estimated by n.m.r.

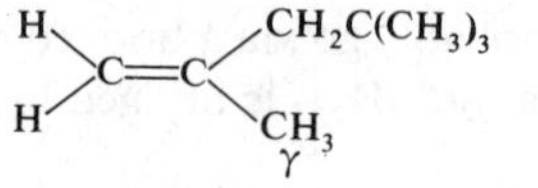

$$\underset{H}{\overset{H}{\diagdown}}C=C\underset{\underset{\gamma}{CH_3}}{\overset{CH_2C(CH_3)_3}{\diagup}}$$

[3.1.44]

Combinations of reactants (a) and (c) gave approximately 1 deuterium in position γ but (b) gave very little. Thus the mechanism shown in Fig. 30 was suggested.

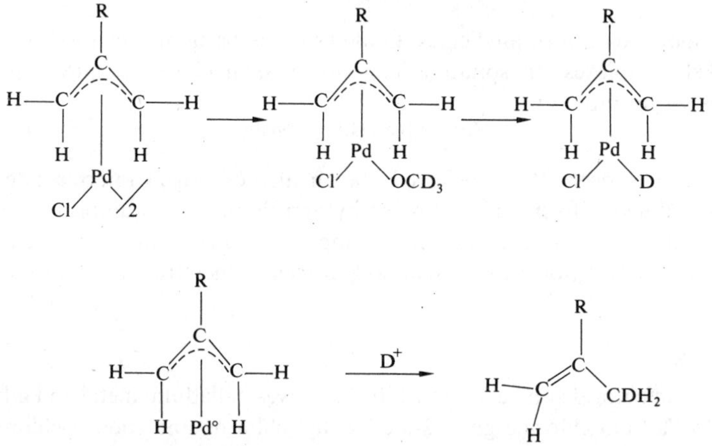

FIG. 30. Proposed mechanism for the decomposition of Pd_2Cl_2(2-neopentylallyl)$_2$ to 2,4,4-trimethylpent-l-ene by sodium methoxide in aqueous methanol (specifically deuteriated; see text). Combination (a), i.e. $D_2O/CD_3OD/NaOD$, used as the example (R = neopentyl).

3.1.5e. *Reduction*

Dodeca-2,6,10-trien-1,12-diylnickel reacts with hydrogen to give nickel metal and dodecane. Sodium borohydride has also been used as a reducing agent, e.g. the π-allylic complex dichloro(cyclododeca-1,5-dienyl)rhodium gives cyclododecane with sodium borohydride.

3.1.5f. *Hydrogen Transfer*

In the catalytic dimerization of butadiene to 5-methylhepta-1,3,6-triene by $Co(C_3H_5)_3$, hydrogen transfer probably occurs in the conversion of the known intermediate 5-methyl-hepta-2,6-dienyl(butadiene)cobalt(I) to 1-methylallyl-(5-methylhepta-1,3,6-triene)cobalt (see section 3.1.4).

An allyliciron hydride intermediate may be involved in the iron carbonyl catalysed isomerization of allyl alcohol to propionaldehyde, since $CH_2{:}CHCD_2OH$ gives CH_2DCH_2CDO [453].

[452] R. Hüttel and P. Kochs, *Chem. Ber.* **101** (1968) 1043.
[453] W. T. Hendrix, F. G. Cowherd and J. L. von Rosenberg, *Chem. Commun.* **2** (1968) 97.

3.1.5g. *Oxidative Hydrolysis*

Allylicpalladium halides react with water to give (mainly) olefin, aldehyde and palladium metal.

$$Pd_2Cl_2(2\text{-methylallyl})_2 + H_2O \rightarrow CH_2\!\!=\!\!CH \cdot MeCHO + C_4H_8 + 2HCl + Pd$$

3.1.5h. *Alkoxy Exchange*[427]

The alkoxyallylic complexes [3.1.45] and [3.1.46], formed from $[PdCl_4]^{2-}$ and conjugated dienes in alcohols, undergo acid catalysed alkoxy exchange very readily.

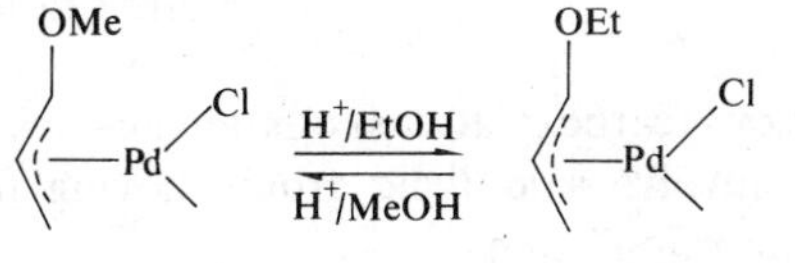

[3.1.45]   [3.1.46]

3.1.5i. *Fission with Acid or by Compounds with Active Hydrogen, giving Olefin*

The rhodium complex $RhCl(\pi\text{-}C_3H_5)Cp$ reacts with concentrated hydrochloric acid to give $Rh_2Cl_4Cp_2$ and propene. Iridium complexes $IrCl_2(\pi\text{-all})L_2$ (L = PEt_3, PMe_2Ph or PEt_2Ph; all = allyl or 2-methylallyl) react with dilute hydrochloric acid to give bridged chloro-complexes, $Ir_2Cl_6L_4$ or $Ir_2Cl_6L_3$, and presumably the olefins, allH.

With a system containing a reduced nickel–triethyl phosphite catalyst and an amine such as diallylamine or morpholine, butadiene is dimerized almost exclusively to *cis-*, *trans-* and *trans, trans*-octa-1,3,6-triene. The suggested mechanism is shown below; it involves hydrogen (proton) addition to an octa-2,6-dien-1,8-diylnickel complex and subsequent elimination of hydrogen (proton)[452]. See paragraph (2) of section 3.1.5j.

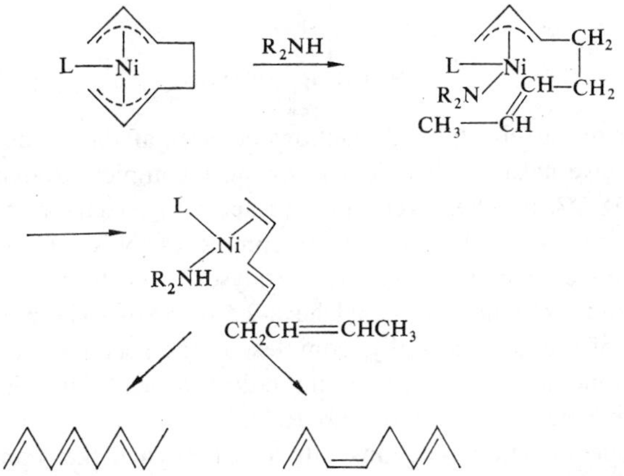

3.1.5j. *Insertion Reactions*

As discussed in section 1.1.5, unsaturated molecules such as carbon monoxide, olefins, acetylenes, sulphur dioxide, etc., can be inserted into metal–carbon σ-bonds. Allylic–metal complexes probably undergo insertion reactions via a σ-allylic intermediate.

(1) *Carbon monoxide.* Carbon monoxide can be inserted between the metal and the allylic group. Sometimes the carbonyl inserted complex is isolable, but it frequently reacts further to give metal-free organic products.

Carbon monoxide reacts with π-allylcobalt tricarbonyl to give the unstable σ-allylcobalt tetracarbonyl; this with triphenylphosphine gives a butenoyl complex, presumably via a coordinatively unsaturated butenylcobalt tricarbonyl[276, 277].

$$\begin{array}{c} CH_2 \\ \| \\ HC\!\!-\!\!-Co(CO)_3 \longrightarrow CH_2\!:\!CHCH_2Co(CO)_4 \rightleftharpoons CH_2\!:\!CHCH_2COCo(CO)_3 \\ \diagdown CH_2 \end{array}$$

$$\Big\downarrow PPh_3$$

$$CH_2\!:\!CHCH_2COCo(CO)_3PPh_3$$

The various carbonylations (carboxylations) of π-allylic–nickel or –palladium complexes described below probably involve a σ-allylic group moving from metal to coordinated carbon monoxide in the "insertion" step.

Allylic halides react with tetracarbonylnickel to give π-allylicnickel halides, hence the following is a probable mechanism for the carboxylation of allyl bromide, catalysed by tetracarbonylnickel[282].

$$\begin{array}{ccc} \left[\!\!\begin{array}{c} Br \\ \diagup\!\!-Ni\diagdown \end{array}\!\!\right]_2 & \xrightarrow{\;+CO\;} & CH_2\!:\!CHCH_2Ni(CO)_2Br \\ & & \\ \Big\uparrow CH_2\!:\!CHCH_2Br & & \Big\downarrow \begin{array}{c} CO \\ insertion \end{array} \\ & & \\ Ni(CO)_4 & \xleftarrow[\text{elimination of}]{+2CO} CH_2\!:\!CHCH_2CONi(CO)_3Br \\ & CH_2\!:\!CHCH_2COBr & \\ & & \\ & \Big\downarrow H_2O & \\ & CH_2\!:\!CHCH_2COOH & \end{array}$$

There are a number of nickel-catalysed reactions between allylic halides, carbon monoxide and acetylene(s) to give hexa-2,5-dienoic acid or more complex products such as phenols, ketonic acids, etc.[275, 282]. π-Allylnickel halides react with tertiary phosphines and carbon monoxide to give air-sensitive, five-coordinate complexes, $NiX(\pi\text{-}C_3H_5)(CO)(PR_3)$. These complexes in toluene/methanol at $0°$ and 1 atm catalyse the reaction of allyl halide, acetylene, carbon monoxide and methanol to methyl hexa-2,5-dienoate. The reaction is believed to involve conversion of the π- to a σ-allyl complex by the acetylene followed by acetylene insertion, carbon monoxide insertion and elimination of acid chloride, which then reacts with methanol to give methyl hexa-2,5-dienoate[455].

Carbon monoxide insertion reactions into other allylicnickel complexes include the conversion of the α,ω-diallylic complex $NiC_{12}H_{18}$ by carbon monoxide at $-60°$ into vinyl-cycloundecadienone and tetracarbonylnickel; this probably goes via carbon monoxide insertion into a σ-allylic intermediate followed by elimination of the ketonic product. Bis-π-cyclo-octenylnickel(II) reacts with carbon monoxide to give a 92% yield of di(cyclo-octenyl)ketone. π-Cyclo-octenylnickel chloride (dimer) gives cyclo-oct-2-enoyl chloride with carbon monoxide.

[454] P. Heimbach, *Angew. Chem. Int. Edn.* 7 (1968) 882.
[455] F. Guerrieri and G. P. Chiusoli, *Chem. Commun.* (1967) 781.

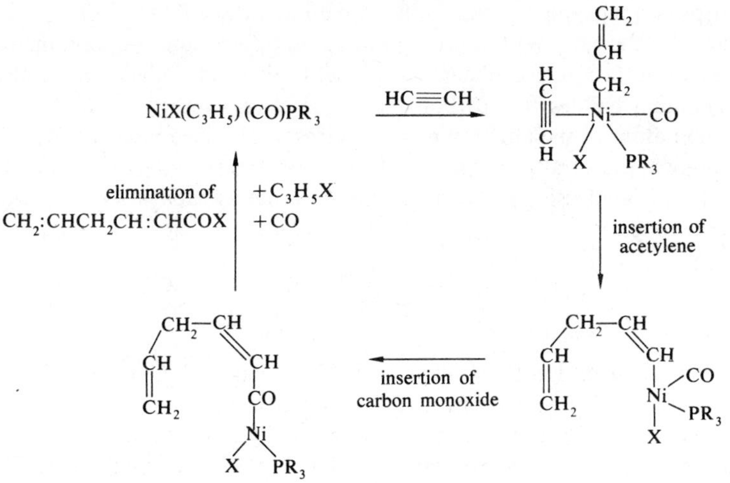

$$NiX(C_3H_5)(CO)PR_3 \xrightarrow{HC\equiv CH}$$

elimination of $\quad$ +C$_3$H$_5$X
CH_2:CHCH$_2$CH:CHCOX $\quad$ +CO

insertion of acetylene

insertion of carbon monoxide

[3.1.47]

A small amount of phenol is also formed, probably by an internal insertion reaction of the intermediate [3.1.47].

Similar mechanisms have been proposed for the carbonylation of π-allylicpalladium chlorides, although more vigorous conditions are required than with nickel systems. π-Allylpalladium chloride is carbonylated at 100° in ethanol to ethyl but-3-enoate, but in benzene a butenoyl chloride is formed[354].

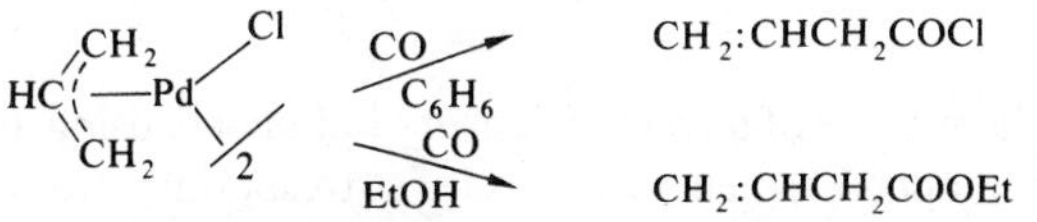

$$\xrightarrow[\text{EtOH}]{\text{CO}} \quad CH_2\text{:}CHCH_2COCl$$
$$\xleftarrow{\text{CO}}{C_6H_6} \quad CH_2\text{:}CHCH_2COOEt$$

The carbonylation of some allylic compounds can be carried out catalytically in the presence of palladium chloride or allylpalladium chloride. With allyl chloride and carbon monoxide at 90° and 85 atm, using allylpalladium chloride as a catalyst, there is 90% conversion to but-3-enoyl chloride. The reaction is first order in allylpalladium chloride concentration, and the rate is proportional to the square of the carbon monoxide pressure. The reaction is thought to go via a σ-allylpalladium dicarbonyl intermediate[456].

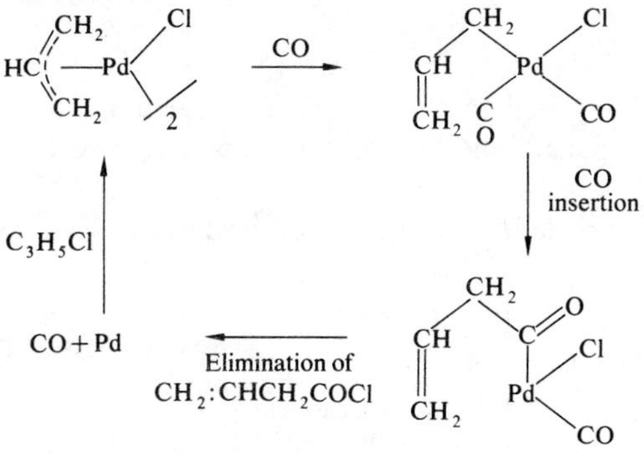

C_3H_5Cl

$CO + Pd$

Elimination of
CH_2:CHCH$_2$COCl

CO insertion

---

456 D. Medema, R. van Helden and C. F. Kohll, *Inorganica chimica acta* (1969) 255.

Added triphenylphosphine enhances the rate up to a PPh_3/Pd ratio of 0.55, then further addition decreases the rate. With a ratio up to about 0.5 the rate is approximately proportional to the concentration of triphenylphosphine. In these carbonylations of allylic halides (or allylicpalladium halides) the carbon monoxide always inserts at the least-substituted terminal carbon atom of the allyl fragment (contrast with diene insertions).

Several other types of allylic compounds, e.g. allyl ethers, allyl esters and allyl alcohol, have been carbonylated using π-allylpalladium chloride as catalyst. The reactions can be summarized as follows:

$$R_1CH:C\cdot CH_2X \xrightarrow[\text{catalyst}]{CO + Pd_2Cl_2(C_3H_5)_2} R_1CH:C\cdot CH_2COX$$
$$\quad\quad | \quad\quad\quad\quad\quad\quad\quad\quad\quad\quad\quad\quad | $$
$$\quad\quad R_2 \quad\quad\quad\quad\quad\quad\quad\quad\quad\quad\quad R_2$$

where R_1 = Cl, alkyl or H; R_2 = alkyl or H; X = halogen, OH, OR, OCOR. Cycloalkenyl chlorides can also be carbonylated. The relative rates of carbonylation decrease in the order allyl > crotyl > cyclo-octenyl > 1-chloroallyl. At the end of the reaction π-allylicpalladium carbonyl complexes can be isolated, e.g. 2-methylallyl chloride gives the complex [3.1.48].

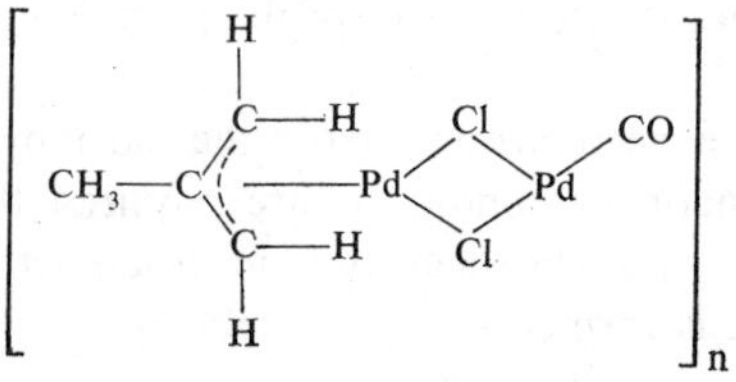

[3.1.48]

The carbonylation of a mixture of allyl chloride and butadiene, using allylpalladium chloride as catalyst, gives some octa-3,7-dienoyl chloride, corresponding to successive insertion of butadiene and carbon monoxide.

Butadiene reacts with palladium chloride to give π-4-chlorocrotylpalladium chloride. This allylic complex reacts with carbon monoxide in benzene as follows:

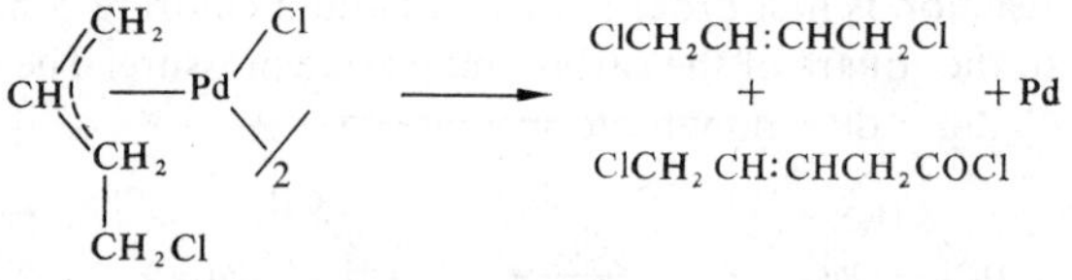

Butadiene can be carbonylated catalytically to ethyl pent-3-enoate in ethanol containing hydrogen chloride and palladium chloride. Presumably the reaction goes via a π-allylic–palladium complex[457].

Some of the π-allylic–palladium complexes derived from allene can also be carbonylated, e.g.

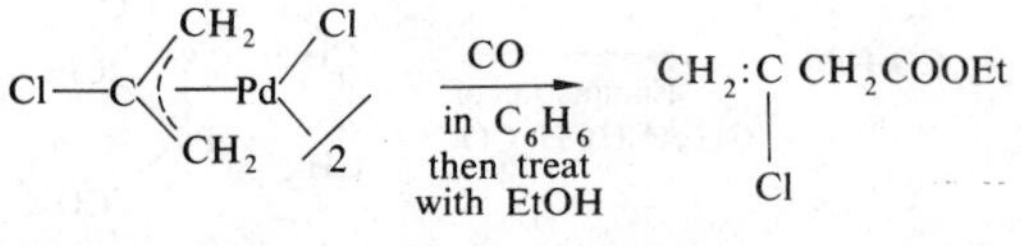

457 S. Brewis and P. R. Hughes, Chem. Commun. (1965) 157.

The carbonylation of $[PdCl(C_6H_8Cl)]_2$ (prepared from allene) gives products corresponding to attack by one, two or three carbon monoxides per palladium, e.g. one of the products is [3.1.49]

[3.1.49]

Relatively little has been done on the carbonylation of allylic–rhodium complexes. Treatment of π-allyl– or π-2-methylallyl–rhodium complexes of type $RhCl_2(all)(PPh_3)_2$ with carbon monoxide gives eventually quantitative yields of *trans*-$RhCl(CO)(PPh_3)_2$ and the allylic chloride. However, an infrared study of the intermediates shows various labile complexes containing σ-allylic ligands and a carbonyl-inserted product to be present. The carbonyl inserted intermediate is probably of the type

$$RhCl_2\{COCH_2C(CH_3):CH_2\}(PPh_3)_2 \text{ }[448].$$

As discussed in section 3.1.5k, many bis(allylic) complexes eliminate 1,5-dienes when treated with carbon monoxide.

(2) *Insertion of mono-olefins, diolefins or acetylenes.* Conjugated dienes or olefins can be inserted into allylic–nickel bonds very rapidly. Many of the oligomerization reactions of butadiene and related co-oligomerizations, discovered by Wilke and coworkers, involve insertion as a key step[214, 288, 458]. Some of the products formed and the proposed π- or σ-allylic intermediates are summarized in Fig. 31.

The nickel catalyst can be introduced into the system either as bis(allyl)nickel or as the product of treating nickel acetylacetonate with two-thirds of a molar quantity of triethyl-aluminium. The presence of one mole of a ligand, Lg, such as a tertiary phosphine or a phosphite, blocks a coordination site and prevents trimerization of butadiene, i.e. it prevents conversion of the octadiendiylnickel [3.1.50] to dodeca-2,6,10-trien-1,12-diylnickel [3.1.51] in which the nickel has an inert gas electron configuration and no available orbital through which Lg could remain coordinated.

When butadiene is oligomerized in the presence of a nickel catalyst and one mole of tris-2-phenylphenyl phosphite, the initial product is mainly *cis,cis*-divinylcyclobutane. This could be formed either via the di-σ-allylic complex [3.1.52] or by the fusion of two double bonds, each from a butadiene molecule as in olefin dismutation (see section 2.1.5e). As the reaction proceeds, however, more vinylcyclohexene and particularly more *cis,cis*-cyclo-octa-1,5-diene are formed via the π-allylic complexes [3.1.53] and [3.1.54] respectively. It has been suggested that in the early stages of the reaction the large concentration of butadiene present forces the equilibria [3.1.54] ⇄ [3.1.53] ⇄ [3.1.52] in favour of [3.1.52], possibly by co-ordinating to the nickel, but as the butadiene is used up the proportion of the bis-π-allylic structure [3.1.54] increases and cyclo-octa-1,5-diene is the preferred product. The nickel catalyst will also readily convert *cis,cis*-divinylcyclobutane into *cis,cis*-cyclo-octa-1,5-diene.

In the absence of added ligand Lg the bisallylic complex [3.1.50] reacts with butadiene to give the complex [3.1.51]. In the presence of butadiene this can eliminate the various cyclo-dodecatrienes shown and the resultant "bare" nickel atom is presumably again converted into the octadiendiyl complex [3.1.50]. Sometimes the olefinic bond in the product

[458] P. Heimbach and W. Brenner, *Angew. Chem. Int. Edn.* **6** (1967) 800.

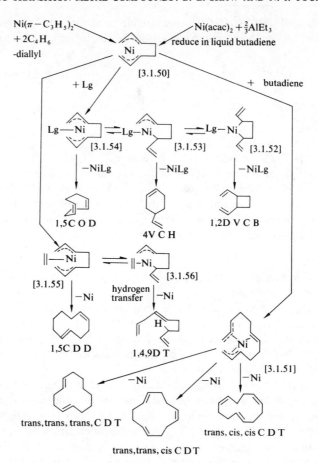

FIG. 31. Some nickel(0) catalysed oligomerization reactions of butadiene.

will have a *cis*- and sometimes a *trans*-stereochemistry. The *cis*- or *trans*-stereochemistry depends on whether the π-allylic ligand has an *anti*- or a *syn*-configuration.

Ethylene and butadiene can be co-oligomerized. Thus the π-allylic octa-2,6-dien-1,8-diylnickel complex [3.1.50] can react with ethylene to give either *cis,trans*-cyclodeca-1,5-diene or *trans*-deca-1,4,9-triene, presumably via [3.1.55] or [3.1.56] respectively.

Butadiene and acetylenes can also be co-oligomerized. Thus disubstituted acetylenes react with butadiene to give cyclodecatrienes[459].

$$R_1C\text{:}CR_2 + 2C_4H_6 \xrightarrow[\text{catalyst}]{Ni/PPh_3}$$

These products are presumably formed via the octa-2,6-dien-1,8-diylnickel complex [3.1.50].

[459] P. Heimbach and R. Schimpf, *Angew. Chem.* **80** (1968) 704.

Dipyridyliron complexes {e.g. $FeEt_2(dipyridyl)_2$ or $Fe(dipy)_2$} are also effective catalysts for the cyclodimerization of butadiene (and isoprene). Butadiene gives mixtures of cyclo-octa-1,5-diene and 4-vinylcyclohexene (ratio *ca.* 3:1). 1,1,4,4-Tetradeuteriobutadiene has been dimerized by this catalyst and the cyclo-octa-1,5-diene formed was specifically deuteriated 3,3,4,4,7,7,8,8-d_8 [460], i.e.

No significant amount of hydrogen transfer occurs in the oligomerization.

The complexes $Fe(NO)_2(CO)_2$ or $Fe(C_3H_5)(NO)_2CO$ will catalyse the dimerization of butadiene or isoprene, especially in the presence of sunlight. A feature is the specific dimerization of butadiene to 4-vinylcyclohex-1-ene. Cyclic isoprene dimers are also formed readily.

A complex $C_{12}H_{19}Co$ is formed by treating cobaltous chloride with butadiene and sodium borohydride in ethanol. The complex will catalyse the dimerization of butadiene to 3-methylhepta-1,4,6-triene. Triallylcobalt will also catalyse the dimerization of butadiene, but is probably rapidly converted into the "true" catalyst, $C_{12}H_{19}Co$ in the process.

The X-ray structure of $C_{12}H_{19}Co$ shows it to be (5-methylhepta-2,6-dienyl)(butadiene)-cobalt(I)[461], drawn diagrammatically in [3.1.57].

[3.1.57]   [3.1.58]

It is suggested that this complex [3.1.57] is in tautomeric equilibrium with a (π-1-methyl-allyl)(methylheptatriene) complex [3.1.58]. The conversion of [3.1.57] to [3.1.58] requires a hydrogen migration from C_4 to C_8. This migration could occur via a cobalt hydride species since hydrogen linked to C_4 is calculated to be only 3.1 Å from the cobalt. It is suggested that an attacking butadiene molecule coordinates to the cobalt and displaces the vinyl grouping C_6–C_7. The butadiene is inserted into the 1-methylallyl–cobalt bond to give a new 5-methylhepta-2,6-dienyl ligand. Butadiene then displaces the 5-methylhepta-1,3,6-triene to give [3.1.57] and the cycle can be repeated.

Although triallylcobalt catalyses the branched-chain dimerization of butadiene, di(allyl)-cobalt iodide catalyses the polymerization of butadiene to 1,4-*cis*-polybutadiene containing a little 1,2-polybutadiene. In the presence of aluminium bromide the reactivity is increased and the product is essentially 1,4-*cis*-polybutadiene. A chain growth mechanism has been suggested[288] which involves insertion of butadiene to give an α,ω-diallyliccobalt iodide

[460] A. Yamamoto, K. Morifuji, S. Ikeda, T. Saito, Y. Uchida and A. Misono, *J. Am. Chem. Soc.* **90** (1968) 1878.
[461] G. Allegra, F. Lo Giudice, G. Natta, U. Giannini, F. Fagherazzi and P. Pino, *Chem. Commun.* (1967) 1263.

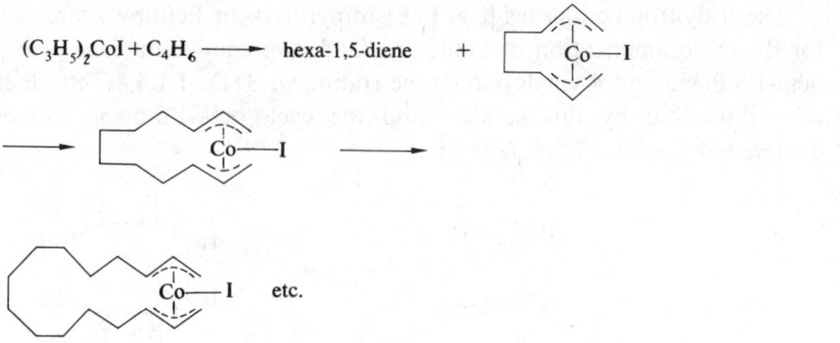

$(C_3H_5)_2CoI + C_4H_6 \longrightarrow$ hexa-1,5-diene   +   [structure] Co—I

$\longrightarrow$ [structure] Co—I $\longrightarrow$

[structure] Co—I   etc.

It is possible that the end of a chain could be transferred to another cobalt atom, i.e. the two ends of the growing polymer chain could be on different metal atoms.

Triallylchromium(III) causes extremely rapid conversion of butadiene to 1,2-polybutadiene. It will also polymerize methyl methacrylate; a key step is thought to be the insertion of the double bond of the methacrylate into the chromium–allyl σ-bond. Butadiene reacts with rhodium trichloride to give $Rh_2Cl_4(C_4H_7)_2C_4H_6$, the structure of which is possibly [3.1.59], although the detailed stereochemistry is not known[389]. Ethylene will displace the butadiene from this complex reversibly to give a π-1-methylallyl(ethylene)rhodium complex [3.1.60].

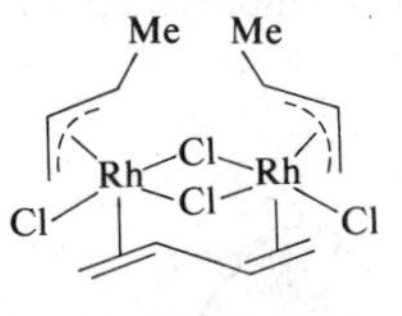

[structure]   Me   Me
Rh—Cl—Rh
Cl   Cl
Cl   Cl
[3.1.59]

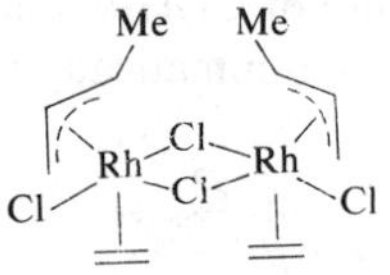

[structure]   Me   Me
Rh—Cl—Rh
Cl   Cl   Cl
[3.1.60]

The codimerization of butadiene and ethylene to hexa-1,4-diene by rhodium trichloride has been shown to occur via an intermediate π-1-methylallylrhodiumethylene complex such as [3.1.60] [462]. In the following reaction scheme (Fig. 32) only the organic ligands are shown.

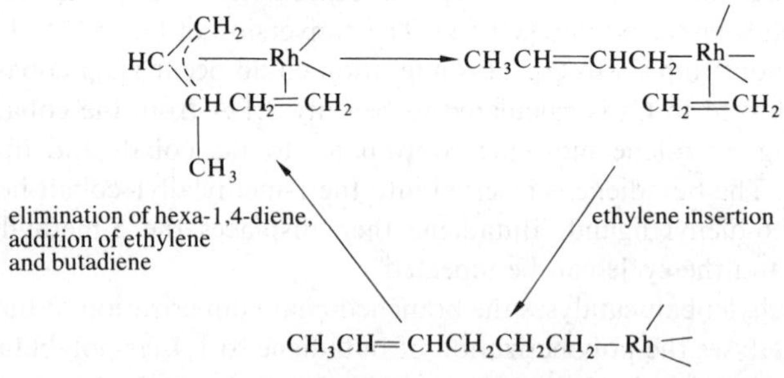

$$HC \underset{CH}{\overset{CH_2}{\Big\langle}} —Rh \underset{CH_2=CH_2}{\big\langle} \longrightarrow CH_3CH=CHCH_2—Rh— \underset{CH_2=CH_2}{\big\langle}$$

CH₃

elimination of hexa-1,4-diene,
addition of ethylene
and butadiene

ethylene insertion

$CH_3CH= CHCH_2CH_2CH_2— Rh \big\langle$

FIG. 32. Proposed reaction scheme (simplified) for the codimerization of ethylene and butadiene, catalysed by rhodium trichloride[462].

Hexa-1,4-diene is of industrial interest since when copolymerized with ethylene and propylene it gives an elastomer which can be vulcanized.

[462] R. Cramer, J. Am. Chem. Soc. **89** (1967) 1633.

A catalyst system formed from $CoCl_2(diphos)_2$ and triethylaluminium is also very efficient for the codimerization of ethylene and butadiene to hexa-1,4-diene[463]. It is thought that a cobalt hydride species is formed which adds to the butadiene to give a π-1-methylallyl-cobalt complex. This is then attacked by the ethylene, etc.; the overall sequence of reactions is similar to that postulated for the rhodium system (above). $Fe(C_2H_4)(diphos)_2$ will also catalyse codimerization of ethylene and butadiene, but various hexadienes are formed.

Treatment of butadiene with the palladium(0) complexes $Pd(maleic\ anhydride)(PPh_3)_2$ or $Pd(PPh_3)_4$ in methanol at 70° gives 1-methoxyocta-2,7-diene (85%), together with a little octa-1,3,7-triene. In CH_3OD as solvent the product is 1-methoxy-6-deuterio-octa-2,7-diene and it is believed that the reaction goes via the α,ω-diallylic complex [3.1.61] formed by oligomerizing two butadienes followed by attack by OMe⁻ (or MeOH) and H⁺ (or D⁺)[464].

$$Ph_3P\text{——}Pd \quad \longrightarrow \quad MeOCH_2CH:CHCH_2CH_2CHDCH:CH$$

[3.1.61]

In benzene, tetrahydrofuran or acetone as solvent the main product is octa-1,3,7-triene.

A related reaction is the formation of 1-phenoxyocta-2,7-diene from butadiene and phenol using a variety of palladium complexes as catalysts. From a mixture of palladium chloride, sodium phenoxide, butadiene and phenol essentially quantitative yields (*ca.* 97%) of 1-phenoxyocta-2,7-diene are obtained[465]. The mechanism of this reaction is probably similar to that of the methoxyoctadiene synthesis described above. π-Allylpalladium chloride gives an even more active catalyst than palladium chloride for the synthesis of phenoxy-octadiene.

Rhodium trichloride reacts with butadiene in ethanol to give ethoxybutenes and ethoxy-octadienes. The reactions very probably involve allylic–rhodium complexes as inter-mediates[466]. Similar reactions occur with isoprene.

Solutions of palladium salts, e.g. the acetate or nitrate in glacial acetic acid, catalyse various reactions of allene, e.g. at 50° a product mixture consisting of allyl acetate, 2,3-dimethylbuta-1,3-diene, 3-methyl-2-acetoxymethylbuta-1,3-diene and 2,3-diacetoxymethyl-buta-1,3-diene in approximate ratio of 7:19:100:14. However, if one mole of triphenyl-phosphine per palladium atom is added, allene is absorbed at a faster rate and only very small amounts of these products are formed. Instead a polymer of allene is produced by head-to-head union, with a molecular weight of 1000–2000. This polymer probably has the structure [3.1.62] with acetate or hydrogen end groups.

$$\left(\begin{array}{c} CH_2 \qquad CH_2 \\ C\text{—}C \\ -CH_2 \qquad CH_2 \end{array}\right)_n$$

[3.1.62]

463 M. Iwamoto and S. Yuguchi, *Bull. Chem. Soc. Japan* **41** (1968) 150.

464 S. Takahashi, H. Yamazaki and N. Hagihara, *Bull. Chem. Soc. Japan* **41** (1968) 254.

465 E. J. Smutny, H. Chung, K. C. Dewhirst, W. Keim, T. M. Shryne and H. E. Thyret, Division of Petroleum Chemistry Preprints, American Chemical Society Meeting, April 1969, p. B 100.

466 K. C. Dewhirst, *J. Org. Chem.* **32** (1967) 1297.

It is very probably formed via a π-allylic intermediate of the type discussed in section 3.1.1f.

When halo- or carboxylato-bridged allylic complexes of palladium are treated with 1,3-dienes, insertion of the diene occurs with the formation of new π-allylic–palladium complexes [3.1.63]. The reaction can be represented in a general form as

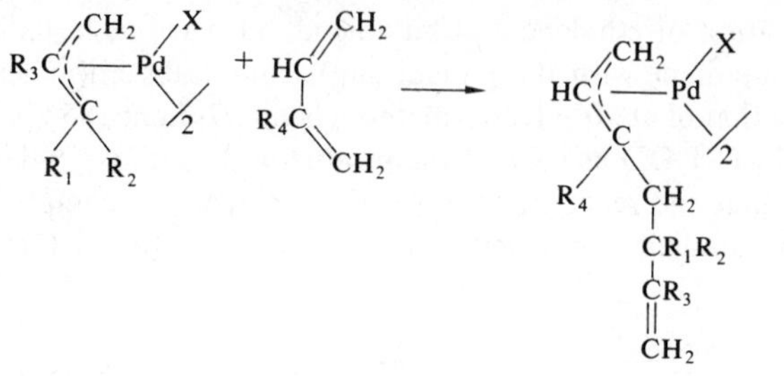

[3.1.63]

where R_1, R_2, R_3 = H, alkyl, Cl or COOR; R_4 = H, alkyl or Cl; X = Cl, Br, I, OOCCH$_3$, OOCCF$_3$, CNS.

Using 1,1,4,4-tetradeuteriobutadiene and allylpalladium chloride the product has the structure and stereochemistry [3.1.64], as is readily shown by n.m.r. spectroscopy[469].

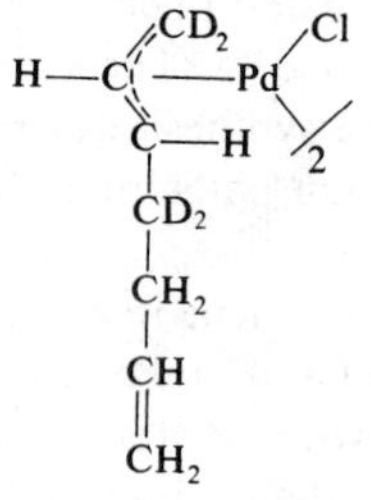

[3.1.64]

At 20° reactions can take as little as a few minutes for completion or even weeks, depending on the nature of $R_1 \rightarrow R_4$, X, etc. The following generalizations can be made about the relative rates of these diene insertion reactions[468]:

(1) Dependence on diene:

butadiene > isoprene > 2:3-dimethylbutadiene

(2) Dependence on allyl substituent:

2-chloro > 2-carboxymethyl $\gg$ 2-hydro > 2-methyl > 1,1,2-trimethyl

(3) Dependence on bridging group:

CF$_3$COO > CH$_3$COO > NO$_2$ > Cl > Br > NCS $\sim$ I

[467] G. D. Shier, J. Organomet. Chem. 10 (1967) 15.

[468] R. van Helden, C. F. Kohll, D. Medema, G. Verberg and T. Jonkhoff, Rec. trav. chim. 87 (1968) 961.

[469] A. Bright, B. L. Shaw and G. Shaw, Division of Petroleum Chemistry Preprints, American Chemical Society Meeting, April 1969, p. 81.

The reaction between Pd_2Br_2(2-methylallyl)$_2$ and isoprene has been studied in detail; it is first order in both reactants. Insertion of the diene occurs at the more heavily alkyl-substituted allylic terminal carbon atom, e.g. from 1,1,4,4-tetradeuteriobutadiene and 1,1,2-trimethylallylpalladium chloride the product is [3.1.65] [469].

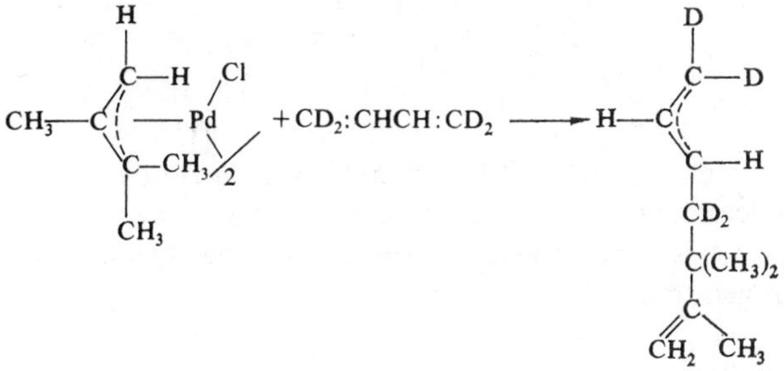

[3.1.65]

This behaviour is similar to that of allylic Grignard reagents, e.g. in the carbonation of crotylmagnesium bromide. Crotylmagnesium bromide is at least 98 % in the linear form, i.e. $CH_3CH:CHCH_2MgBr$ and yet on carbonation the branched-chain carboxylic acid $CH_2:CH \cdot CH(COOH)CH_3$ is produced. A cyclic mechanism has been proposed to explain this. A similar cyclic mechanism is possibly involved in the insertion of dienes into allylic–palladium complexes, i.e. the mechanism can be written as shown in Fig. 33.

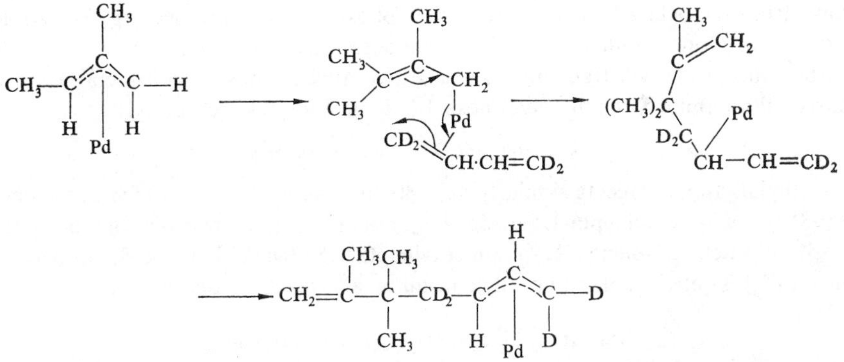

FIG. 33. Proposed cyclic mechanism for diene insertions into π-allylic–palladium complexes (the mechanism has been written in a simplified form, omitting many of the ligands, etc.).

Allene inserts as follows into π-allylic–palladium complexes[468]:

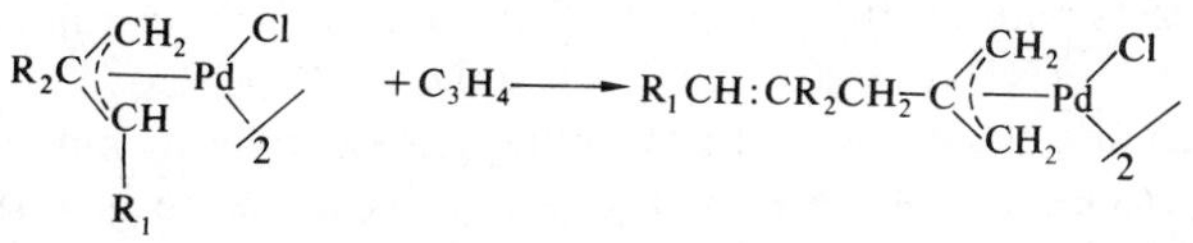

where R_1 = Cl, $COOCH_3$, CH_3; R_2 = Cl, H, CH_3.

Although allylpalladium acetate and butadiene react at ca. 20° to give the insertion product mentioned above, with longer reaction times (24 hr) a complex $C_{12}H_{18}Pd_2(OAc)_2$ is formed. The suggested structure for this compound [3.1.66] is an α,ω-C_{12} allylic ligand bonded to two palladium atoms.

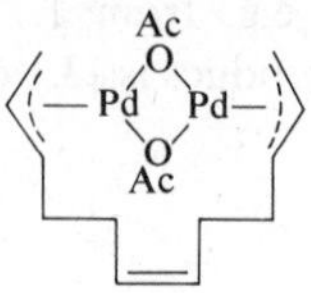

[3.1.66]

At a slightly higher temperature (50°) this complex will catalyse the linear trimerization of butadiene to n-dodeca-1,3,6,10-tetraene and can be recovered at the end of the reaction. This tetraene product is very probably formed from the α,ω-diallylic ligand shown in [3.1.66] by a hydrogen-transfer reaction[468].

3.1.5k. *Elimination Reactions*

A generalized elimination reaction may be represented as follows:

$$M^{n+}\!\!-\!\!A \rightarrow M^{(n-2)+} + A\!\!-\!\!B$$
$$\overset{|}{B}$$

i.e. two ligands are eliminated (joined by a covalent bond) and the formal valence state of the metal goes down by two. In the process more ligands may become coordinated to the metal and the $M^{(n-2)+}$ valence state may only have a transient existence. An elimination reaction is frequently the final step in a catalytic process, e.g. the formation of cyclo-octa-1,5-diene from the octa-2,6-dien-1,8-diylnickel–phosphite complex (see Fig. 31). Attack on the nickel with more butadiene presumably induces the elimination.

Carbon monoxide will frequently displace two allyl ligands from a metal as hexa-1,5-diene, e.g. di(allyl)nickel readily gives hexa-1,5-diene and nickel carbonyl:

$$Ni(C_3H_5)_2 + 4CO \rightarrow Ni(CO)_4 + C_6H_{10}$$

Di-(1-methylallyl)nickel reacts similarly at $-40°$ to give a mixture of *trans,trans*-octa-2,6-diene (98%) and 3-methylhepta-1,5-diene (2%), but at higher temperatures a typical product composition is octa-2,6-diene (58%), 3-methylhepta-1,5-diene (38%) and 3,4-dimethylhexa-1,5-diene (4%). Triallylcobalt and triallylchromium also react rapidly with carbon monoxide.

$$Co(C_3H_5)_3 \xrightarrow{\text{CO}} Co(C_3H_5)(CO)_3 + \text{hexa-1,5-diene}$$
$$Cr(C_3H_5)_3 \xrightarrow{\text{CO}} Cr(CO)_6 + \text{hexa-1,5-diene}$$

Other ligands, e.g. tertiary phosphines, pyridine, cyclo-octa-1,5-diene, will also induce elimination of allylic ligands. Some examples with allylic–rhodium complexes are shown in Fig. 34. Other examples are the following:

Triallyliron reacts with triethylphosphine under nitrogen to give hexa-1,5-diene and a nitrogen compound, $v(N\!:\!N) = 2038$ cm^{-1}, possibly $Fe(N_2)(PEt_3)_4$ [470]. The α,ω-diallylic complex of ruthenium, $C_{12}H_{18}Ru$, reacts with pyridine to give butadiene trimers and $RuCl_2py_4$. π-Allylnickel halides react with triphenylphosphine to give the nickel(I) complexes $NiX(PPh_3)_3$ (X = halogen) and presumably hexadiene[471].

[470] C. H. Campbell, A. R. Dias, M. L. H. Green, T. Saito and M. G. Swanwick, *J. Organomet. Chem.* **14** (1968) 349.

[471] L. Porri, M. C. Gallazzi and G. Vitulli, *Chem. Commun.* **5** (1967) 228.

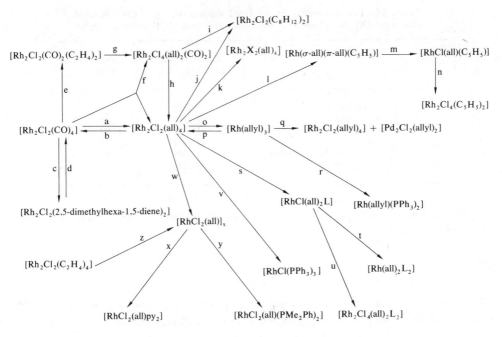

Fig. 34. Some reactions of allylic–rhodium(III) complexes.

^a allCl + H₂O + MeOH (allCl = allylic chloride).
^b CO.
^c 2-methylallyl chloride + H₂O + MeOH at 60°, at 0°, or in the presence of base the product is [Rh₂Cl₂(2-methylallyl)₄].
^d CO.
^e + [Rh₂Cl₂(C₂H₄)₄].
^f mix[Rh₂Cl₂(CO)₄] and [Rh₂Cl₂(all)₄] in allCl.
^g allCl.
^h allCl + H₂O + MeOH.
ⁱ and ^j + C₈H₁₂(= Cyclo-octa-1,5-diene).
^k alkali metal halides or silver acetate, X = Br, I, OAc.
^l TlC₅H₅.

^m dil. HCl.
ⁿ concHCl, heat.
^o allylHgCl.
^p dil.HCl or allCl or RhCl₃ in the cold.
^q Na₂PdCl₄.
^r PPh₃.
^s +L = py, PR₃, AsR₃ etc.
^t L₂ = 2py or 2,2′-dipyridyl.
^u dil.HCl.
^v PPh₃.
^w RhCl₃, boil.
^x py.
^y PMe₂Ph.
^z allCl.

3.1.5l. *Reaction with Organic Halides*

Di(allylic)nickel bromides, $Ni_2Br_2(all)_2$, in polar coordinating media such as dimethyl formamide or N-methylpyrolidine react readily with organic iodides (preferably) or bromides to give the coupled product, e.g. $Ni_2Br_2(2$-methylallyl$)_2$ and methyl iodide give 2-methylbut-1-ene. Remarkably, the reaction goes with aryl or vinyl halides, and even hydroxyl functions in the alkyl halide do not interfere. Some of the halides which react in this way with $Ni_2Br_2(2$-methylallyl$)_2$ are given in Table 19.

Organic dihalides also react, e.g. 1,6-di-iodohexane gives 2,11-dimethyl-1,11-dodecadiene (95% yield) and 1,4-dibromobenzene gives 1,4-dimethylallylbenzene in 97% yield. Using similar reactions and di(π-1,1-dimethylallyl)nickel bromide, α-santalene and *epi*-α-santalene have been synthesized.

TABLE 19. THE COUPLING OF Ni_2Br_2(2-METHYLALLYL)$_2$ WITH ORGANIC HALIDES (RX) IN DIMETHYL FORMAMIDE TO GIVE $RCH_2C(CH_3):CH_2$[472]

RX	Reaction time {hr (temp. °C)}	Yield (%)
Cyclohexyl iodide	3 (22)	91
t-Butyl iodide	24 (22)	25
Iodobenzene	1 (22)	98
Vinyl bromide	13 (22)	70
Phenyl α-chloromethyl ether	18 (60)	50
Chloroacetone	2 (22)	46

The reactions possibly go via an oxidative addition of the alkyl (aryl) halide followed by an elimination step.

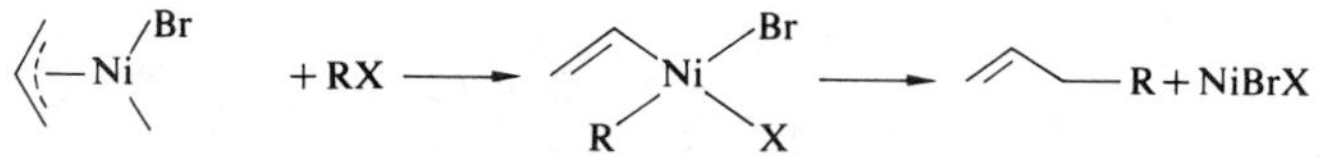

A related reaction is the synthesis of large ring compounds from α,ω-diallylic bromides and tetracarbonylnickel in dimethyl formamide or N-methylpyrolidine.

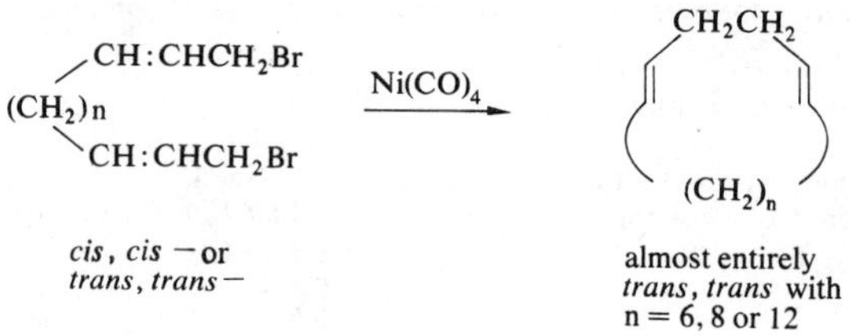

With α,ω-diallylic bromides $CH_2Br \cdot CH:CH(CH_2)_nCH:CHCH_2Br$, $n = 2$ or 4, six-membered rings are also formed, e.g. with $n = 2$, the products are 4-vinylcyclohexene and cyclo-octa-1,5-diene, and with $n = 4$, cis- and trans-1,2-divinylcyclohexane. Humulene (eleven-membered ring) has been synthesized using this cyclization coupling reaction as a key step.

3.1.5m. Reactions with Aldehydes or Ketones

Di(allylic)nickel bromides react with aldehydes or ketones to give carbinols, e.g.

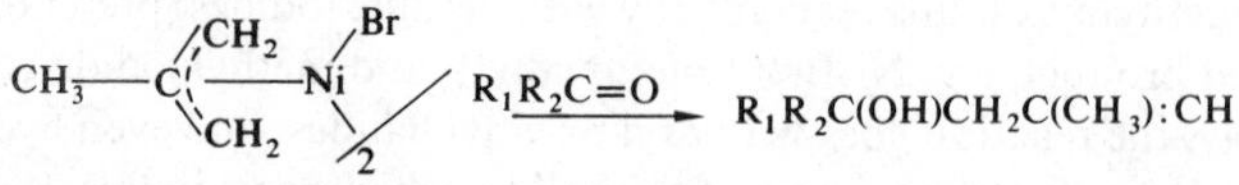

where $R_1R_2C=O$ = acetone, cyclopentanone, benzaldehyde or acrolein.

These reactions go much more slowly than the reactions with organic iodides and require more drastic conditions (50° for ca. 24 hr)[473].

[472] E. J. Corey and M. F. Semmelhack, J. Am. Chem. Soc. 89 (1967) 2755.
[473] L. Cassar and G. P. Chiusoli, Tetrahedron Letters (1965) 3295.

3.1.5n. *Reactions of the Halogen or Carboxylato Bridges*

Many allylic–metal complexes contain two metals bridged by halogens or carboxylate groups. They therefore have many of the reactions of bridging systems such as metathesis and bridge splitting. Examples of reactions of bridged allylic–rhodium complexes have been summarized in Fig. 34; the great majority of other examples come from palladium chemistry, e.g. metathesis:

$$\text{allPd} \underset{Cl}{\overset{Cl}{<>}} \text{Pdall} \quad \xrightarrow{MX} \quad \text{allPd} \underset{X}{\overset{X}{<>}} \text{Pdall}$$

where MX = LiBr, NaI, AgOAc, KSCN, etc.

A useful reaction of halo-bridged π-allylic–metal complexes is their conversion to mononuclear acetylacetonato or cyclopentadienyl complexes,

$$[\text{all MCl}]_2 \quad \overset{Tl\ acac}{\underset{or\ TlCp}{\overset{\longrightarrow}{\underset{NaCp}{\longleftarrow}}}} \quad \begin{array}{l}(\text{all})\text{Macac}\\[1em](\text{all})\text{MCp}\end{array}$$

since these products are often easily purified and the acac or C_5H_5 ligands do not complicate the interpretation of n.m.r. spectra.

Ligands L such as tertiary phosphines, tertiary arsines, amines, etc., will split halogen bridges to give mononuclear products. A second molecule of L can displace the halide to give ionic species. The reactions are often reversible.

$$\text{all Pd} \underset{2}{\overset{Cl}{<>}} \quad \underset{}{\overset{L(1)}{\rightleftharpoons}} \quad \text{all Pd} \underset{Cl}{\overset{L}{<}} \quad \overset{L(2)}{\rightleftharpoons} \quad [\text{all Pd } L_2]^+$$

The formation of an ionic species, step (2), is favoured by an ionizing solvent such as aqueous acetone. Chelating ligands such as α,α'-dipyridyl readily give ionic species of type [allPd(chelate)]$^+$. Allylicnickel halides can give five-coordinate species, however, e.g. NiBr(π-2-methylallyl)(diphos).

The bridge-splitting reactions of allylicpalladium halides by neutral ligands L and subsequent formation of ionic species can be followed conductimetrically in acetone or aqueous acetone. The tendency towards salt formation decreases in the order of ligands L = PMe$_2$Ph > PPh$_3$ > AsPh$_3$ > SbPh$_3$ > py and in the order of bridging ligands X, Cl > Br > I. Increasing alkyl substitution on the allylic ligand reduces the tendency towards salt formation[429].

3.1.5o. *Miscellaneous Reactions*

Reduction of allylicnickel bromides with a zinc/copper couple in dimethyl formamide gives very high yields (> 90%) of diallylic–nickel complexes. Examples of diallylic–nickel complexes prepared in this way include di(-2-methylallyl)nickel and di(-2-carboxyethylallyl)-nickel[423]. Allylpalladium chloride reacts with chloro(methylcarboxy)mercury to give methyl vinylacetate.

Some π-allylic complexes react with diphenylphosphine complexes to give the olefin, allH, and compounds containing a metal–metal–phosphorus three-membered ring[474].

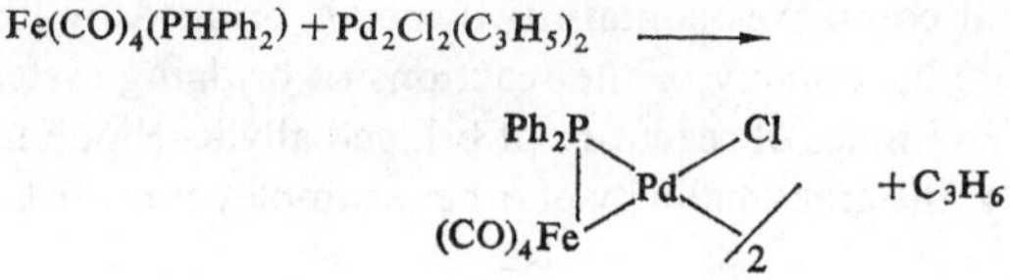

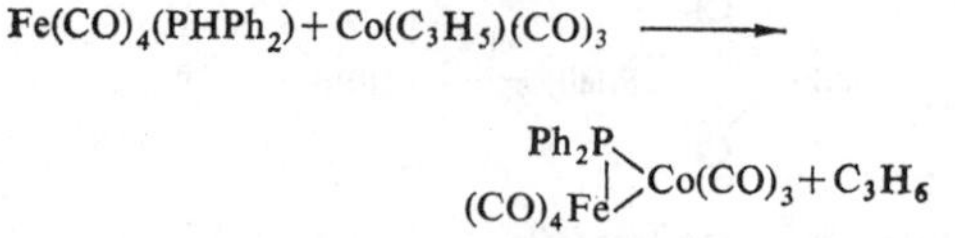

$$Fe(CO)_4(PHPh_2) + Pd_2Cl_2(C_3H_5)_2 \longrightarrow$$

$$Fe(CO)_4(PHPh_2) + Co(C_3H_5)(CO)_3 \longrightarrow$$

π-Allylnickel bromide reacts with methylmagnesium chloride at −78° to give π-C_3H_5NiMe as deep violet crystals (96% yield). This methyl complex decomposes at −35°, giving butenes, propene, ethane, methane and nickel metal. It gives adducts with tertiary phosphines $NiMe(\pi\text{-}C_3H_5)L$, where L = PEt_3, PPh_3 or $P(\text{cyclohexyl})_3$, which are stable at 20° and have the allylic group asymmetrically bonded[424].

3.2.1. Fluoroallyl Complexes

Very few fluoroallyl complexes are known. Treatment of $Fe(CO)_3C_6F_8$ [3.2.1] with caesium fluoride gives the anionic π-allylic complex [3.2.2].

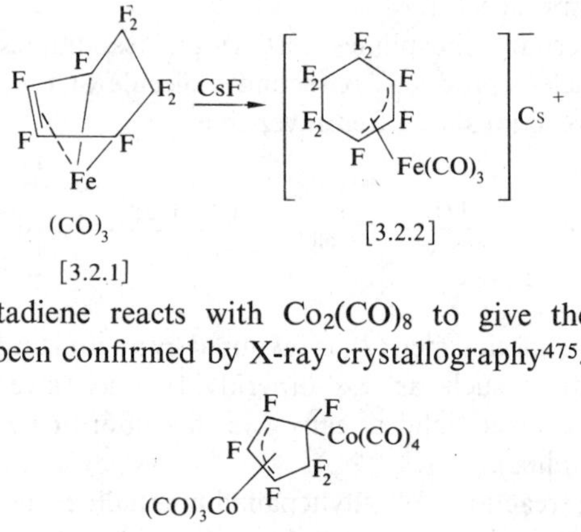

[3.2.1]   [3.2.2]

Hexafluorocyclopentadiene reacts with $Co_2(CO)_8$ to give the complex [3.2.3], the structure of which has been confirmed by X-ray crystallography[475].

[3.2.3]

4. CONJUGATED DIOLEFIN AND RELATED COMPLEXES

4.1. COMPLEXES WITH CONJUGATED DIOLEFINS INCLUDING FLUORODIOLEFINS BUT EXCLUDING CYCLOBUTADIENES[13, 149, 151, 476, 477]

4.1.1. Preparation

The most commonly used method is to treat a metal complex with a conjugated diolefin, but electrophilic attack on a dienyl–metal complex can also be used.

[474] B. C. Benson, R. Jackson, K. K. Joshi and D. T. Thompson, *Chem. Commun.* (1968) 1506.
[475] P. B. Hitchcock and R. Mason, *Chem. Commun.* (1966) 503.
[476] R. Pettit and G. F. Emerson, *Advances Organomet. Chem.* **1** (1964) 1.
[477] R. Pettit, G. Emerson and J. Mahler, *J. Chem. Educ.* **40** (1963) 175.

4.1.1a. *From the Diolefin and a Metal Halide, Metal Carbonyl, etc.*

As described in section 2, there are several examples of butadiene forming a bridge between two metal atoms. Compounds where only one of the double bonds is coordinated are also known. However, when rhodium trichloride is treated with butadiene at $-5°$ for several days, the 5-coordinate rhodium(I) complex $RhCl(C_4H_6)_2$, [4.1.1], is formed. The structure has been shown by X-ray diffraction[478].

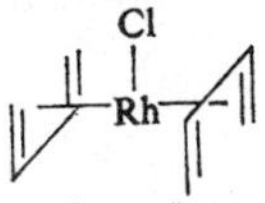

[4.1.1]

Most conjugated diolefin–metal complexes have been prepared from metal carbonyls; those of iron, being the most extensively studied, will be discussed first. Tricarbonyl-(butadiene)iron was made by Reihlen in 1930 from pentacarbonyliron and butadiene, but it was incorrectly formulated as the tetracarbonyl until 1958. A large number of complexes of the type $Fe(CO)_3$(conjugated diolefin) are known[13, 476, 477]. They can usually be made by treating $Fe(CO)_5$ with the diolefin, either in di-n-butyl ether under reflux or by irradiation with ultraviolet light. Less vigorous conditions such as heating under reflux in benzene are sufficient when starting with the more reactive dodecacarbonyltri-iron or enneacarbonyldi-iron. Under the vigorous conditions necessary using pentacarbonyliron, the complex may be of a rearranged olefin. Thus *cis*- or *trans*-piperylene give the *trans*-piperylene complex [4.1.2].

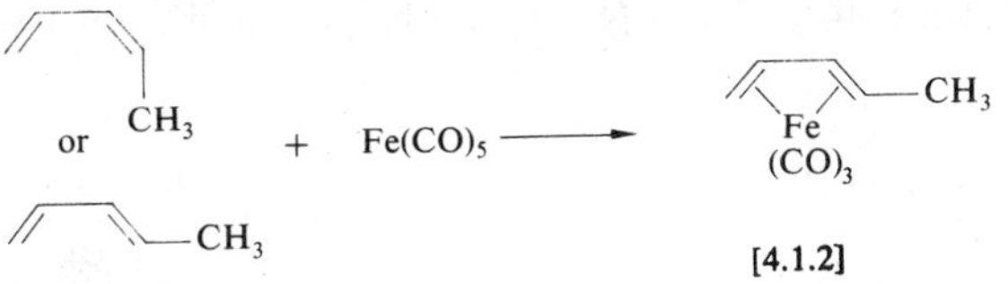

[4.1.2]

The *cis*-piperylene complex [4.1.3] is formed under milder conditions from $Fe_2(CO)_9$ [479].

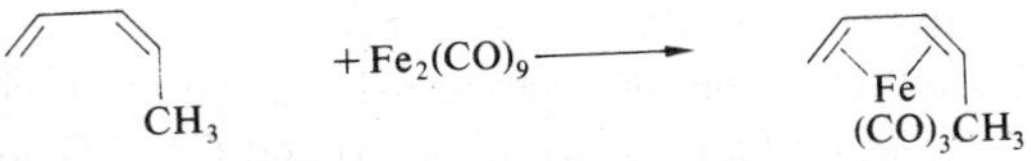

[4.1.3]

Rearrangement of the diolefin by hydrogen transfer is common. Thus cyclohexa-1,4-diene gives tricarbonyl(cyclohexa-1,3-diene)iron. Other examples are[13, 480]:

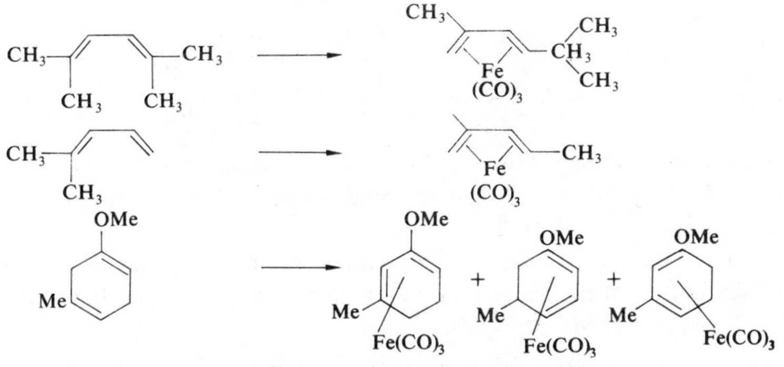

478 L. Porri, A. Lionetti, G. Allegra and A. Immirzi, *Chem. Commun.* (1965) 336.
479 R. K. Kochhar and R. Pettit, *J. Organomet. Chem.* 6 (1966) 272.
480 A. J. Birch, P. E. Cross, J. Lewis, D. A. White and S. B. Wild, *J. Chem. Soc.* A (1968) 332.

A cyclopropane ring may be opened,

$$\underset{CH_2=C-CH}{\overset{C_6H_4OMe}{|}} \longrightarrow \underset{Fe(CO)_3}{MeOC_6H_4-\!\!\!-CH_3}$$

but also ring closure may occur, e.g. cyclo-octa-1,3,5-triene gives some of the bicyclo derivative [4.1.4].

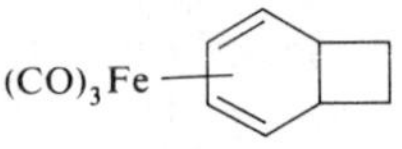

$$(CO)_3Fe$$

[4.1.4]

Complexes involving part of an extended conjugated system are known, e.g. the complex [4.1.5] formed from Vitamin A aldehyde and $Fe_3(CO)_{12}$ [481].

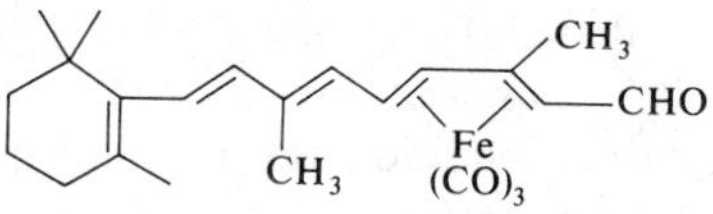

$$\underset{CH_3}{\overset{CH_3}{\underset{Fe(CO)_3}{}}}-CHO$$

[4.1.5]

The tricarbonyliron grouping can move along a conjugated system but not readily. Thus two distinct species [4.1.6] and [4.1.7] can be prepared from 1-phenyl-6-p-tolylhexa-1,3,5-triene.

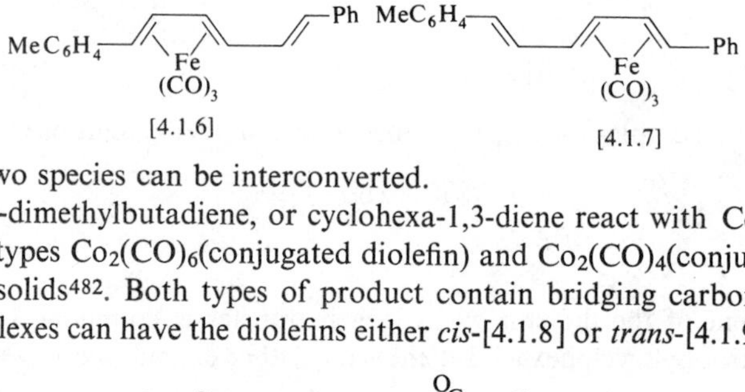

$$MeC_6H_4\underset{Fe(CO)_3}{}\!\!\!-Ph \qquad MeC_6H_4\underset{Fe(CO)_3}{}\!\!\!-Ph$$

[4.1.6]   [4.1.7]

On heating, the two species can be interconverted.

Butadiene, 2,3-dimethylbutadiene, or cyclohexa-1,3-diene react with $Co_2(CO)_8$ to give complexes of the types $Co_2(CO)_6$(conjugated diolefin) and $Co_2(CO)_4$(conjugated diolefin)$_2$, as red or yellow solids[482]. Both types of product contain bridging carbonyl groups. The bis(diolefin) complexes can have the diolefins either cis-[4.1.8] or trans-[4.1.9] and in hexane

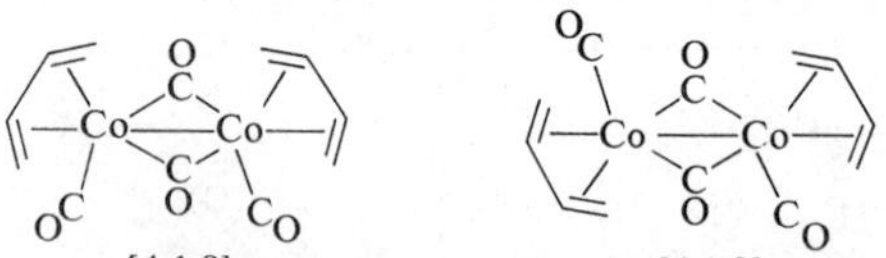

$$\underset{[4.1.8]}{} \qquad \underset{[4.1.9]}{}$$

[4.1.8]   [4.1.9]

solution infrared spectroscopy shows the bis(2,3-dimethylbuta-1,3-diene) complex to be the trans-isomer and the cyclohexa-1,3-diene complex to be a cis-/trans-mixture, with possibly some monomeric species as well[483].

Butadiene or cyclohexa-1,3-diene reacts either thermally or, better, photochemically with the hexacarbonyls of the Group VI metals to give complexes of type $M(CO)_2$(conjugated diolefin)$_2$ (M = Cr, Mo or W)[484].

[481] A. J. Birch, H. Fitton, R. Mason, G. B. Robertson and J. E. Stangroom, Chem. Commun. (1966) 613
[482] G. Winkhaus and G. Wilkinson, J. Chem. Soc. (1961) 602.
[483] P. A. McArdle and A. R. Manning, Chem. Commun. (1968) 1020.
[484] E. O. Fischer, H. P. Kögler and P. Kuzel, Chem. Ber. 93 (1960) 3006.

4.1.1b. *Nucleophilic Attack on Dienyl Complexes*

Many dienyl–metal complexes are attacked by nucleophiles to give diene–metal complexes. Most of the examples occur with derivatives of iron carbonyls. Tricarbonyl(cyclohexa-1,3-diene)iron when treated with trityl fluorborate is converted into the cyclohexa-2,4-dienyl(tricarbonyl)iron cation. This reacts with nucleophiles Y^- to give derivatives of tricarbonyl(cyclohexa-1,3-diene)iron.

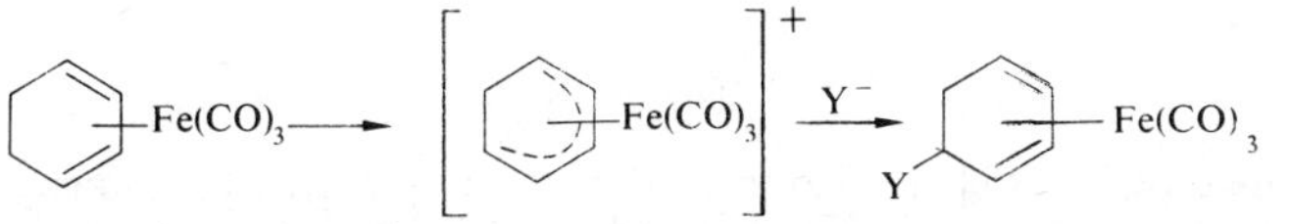

The nucleophilic reagents include water+sodium hydrogen carbonate, cyanide ion, methoxide ion, pyrrolidine, morpholine, diethyl malonate, acetylacetone, dimedone and methyllithium, the corresponding Ys being OH, CN, OMe, NC_4H_8, NC_4H_8O, $CH(COOEt)_2$, $CHAc_2$, $C_8H_{11}O_2$ and Me [13, 480, 485]. Cycloheptatrienyl(tricarbonyl)iron behaves similarly with nucleophiles to give cyclohepta-1,3-diene-iron derivatives.

The methoxycyclohexa-1,3-diene complex [4.1.10] reacts with the trityl cation to give [4.1.11]. This cyclohexadienyl cation reacts with water to give the cyclohexadienone complex [4.1.12], which is in effect a complex of phenol in its keto-form [13, 480].

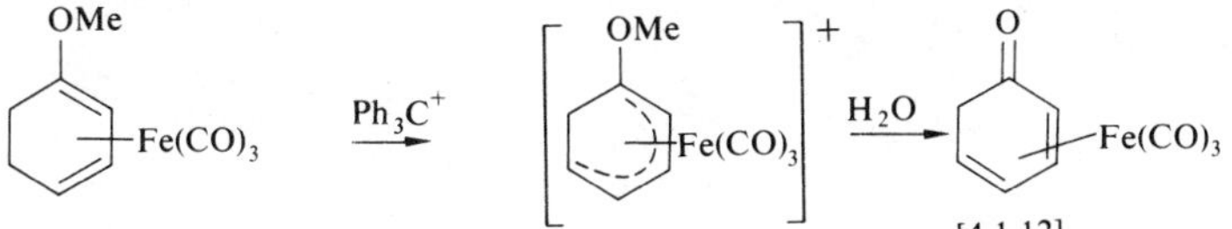

[4.1.10]   [4.1.11]   [4.1.12]

The diene carbinol complex [4.1.13] reacts with acid to give the hexadienyl complex [4.1.14]; this with sodium borohydride gives a mixture of the *cis*- and *trans*-(tricarbonyl)penta-1,3-dieneiron complexes.

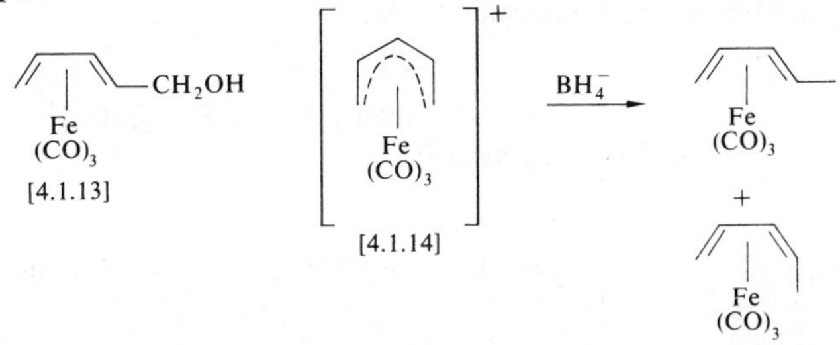

[4.1.13]   [4.1.14]

In the reduction of tricarbonyl(benzene)manganese by sodium borohydride the main product is the cyclohexadienyl complex [4.1.15], but a small amount of the hydro(cyclohexadiene) complex [4.1.16] is also formed [486].

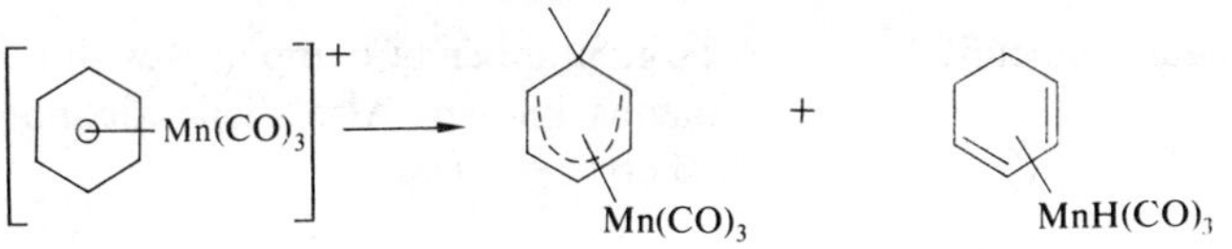

[4.1.15]   [4.1.16]

485 G. F. Grant and P. L. Pauson, *J. Organomet. Chem.* **9** (1967) 553.
486 G. Winkhaus, *Z. anorg. allgem. Chem.* **319** (1963) 404.

The cobalticenium ion is attacked by nucleophiles ($Y^- = H^-$, D^- or Ph^-) to give substituted cyclopentadienecobalt complexes [4.1.17].

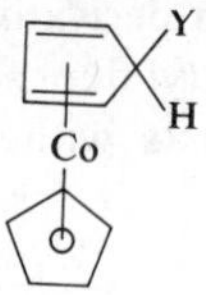

[4.1.17]

Complexes of the type [4.1.17] are also formed by treating dicyclopentadienylcobalt(II) with halides, YX, where X = halogen and Y = alkyl, fluoroalkyl or acyl; or with C_2F_4, giving Y = CF_2CF_2H. These substituted cyclopentadiene complexes were first formulated with the group Y in the *endo*-position, but it has now been definitely established that the group is *exo* (see section 4.1.3). $RhCp_2^+$ also reacts with nucleophiles to give the rhodium analogues of [4.1.17].

6,6'-Diarylfulvene derivatives of iron carbonyl are known. The compounds of composition $Fe(CO)_3$(fulvene) presumably have the structure [4.1.18] [489].

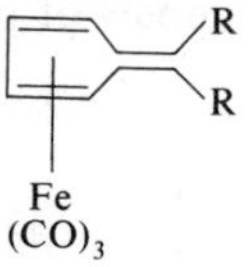

[4.1.18]

Some 6,6'-diphenylfulvene derivatives of type [$M(6,6'$-diphenylfulvene)$_2$]PF_6 (M = Co, Rh or Ir)[490] probably have one of the fulvenes bonded to the metal in an analogous fashion to the fulvene of [4.1.18].

4.1.2. Complexes of Metals with Conjugated Diolefins

Vanadium

$VCp(CO)_2$(diolefin)	Buta-1,3-diene, 2,3-dimethylbuta-1,3-diene or cyclohexa 1,3-diene[484].

Molybdenum

$Mo(CO)_2$(diolefin)$_2$	Buta-1,3-diene[169], cyclohexa-1,3-diene[173].

Manganese

$MnH(CO)_3$(cyclohexa-1,3-diene)[486]

Iron

$Fe(CO)_3$(conjugated diolefin)	A large number of complexes with acyclic and cyclic olefins known. Also diene alcohols, esters, acids, ethers[13, 480, 485].

[487] M. L. H. Green, L. Pratt and G. Wilkinson, *J. Chem. Soc.* (1959) 3753.
[488] E. O. Fischer and G. E. Herberich, *Chem. Ber.* **94** (1961) 1517.
[489] E. Weiss and W. Hübel, *Chem. Ber.* **95** (1962) 1186.
[490] E. O. Fischer and B. J. Weimann, *J. Organomet. Chem.* **8** (1967) 535.

Ruthenium

$Ru(CO)_3(cyclohexa-1,3-diene)$[491]

Cobalt

$Co_2(CO)_6$(conjugated diolefin)	2,3-Dimethylbuta-1,3-diene[482].
$Co_2(CO)_4$(conjugated diolefin)$_2$	Buta-1,3-diene, cyclohexa-1,3-diene, 2,3-dimethyl-buta-1,3-diene[482, 483].
$CoCp(C_5H_5Y)$	C_5H_5Y = substituted cyclopentadiene. Y = H, D, Me, CCl_3, $CHCl_2$, CF_3, etc.[487].

Rhodium

$Rh(Ph_2PC_6H_4)(PPh_3)(C_4H_6)$ [492]
$RhCl(C_4H_6)_2$ [478]
$RhCp(C_5H_6)$ [487]

Iridium

$IrCp(C_5H_6)$ [493]

4.1.3. Structures and Bonding in Conjugated Diolefin–Metal Complexes

4.1.3a. *Structures*

The structures of cyclopentadienone–metal complexes and the hexakis(trifluoromethyl)-benzene–rhodium complex have been discussed in section 2.4.5. The structure of tricarbonyl-(butadiene)iron, as determined by X-ray diffraction, is shown in [4.1.19].

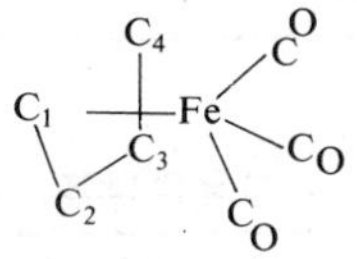

[4.1.19]

Bond angles are 118° for C_1–C_2–C_3 and 70° for Fe–C_1–C_2. The iron atom is approximately equidistant (2.1 Å) from each diene carbon. The C–C distances are 1.45 Å (Table 20). An electron diffraction study gives slightly different C–C (1.41 Å) and Fe–C (2.08 Å) distances.

The structures of several cyclopentadienone and cyclopentadiene complexes have been determined by X-ray diffraction. In all of them the five-membered ring is folded, e.g. in π-cyclopentadienyl-1-benzoylcyclopentadienecobalt [4.1.20] and π-cyclopentadienyltetra-methylcyclopentadienonecobalt [4.1.21].

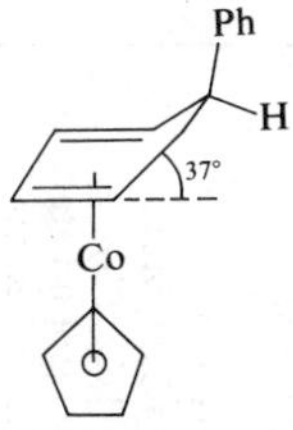

[4.1.20] 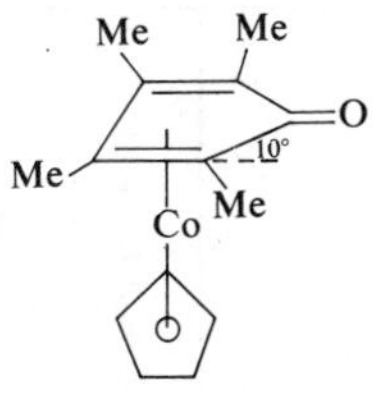   [4.1.21]

491 B. F. G. Johnson, R. D. Johnston, P. L. Josty, J. Lewis and I. G. Williams, *Nature* **213** (1967) 901.
492 W. Keim, *J. Organomet. Chem.* **16** (1969) 191.
493 E. O. Fischer and U. Zahn, *Chem. Ber.* **92** (1959) 1624.

TABLE 20. CARBON–CARBON DISTANCES (Å) IN CONJUGATED DIOLEFIN COMPLEXES

	C_1–C_2	C_2–C_3	C_3–C_4	Reference
Fe(CO)$_3$(C$_4$H$_6$) (crystal)	1.45	1.45	1.45	O. S. Mills and G. Robinson, Acta Cryst. 18 (1962) 562
(gas)	1.41	1.41	1.41	C. S. Speed, Dissertation Abs. 28 (1968) 4033
Fe(CO)$_3$(C$_6$F$_8$)	1.40	1.38	1.40	M. R. Churchill and R. Mason, Proc. Chem. Soc. (1964) 226
Fe(CO)$_3${(CF$_3$)$_4$C$_5$O}	1.41	1.40	1.37	M. Gerloch and R. Mason, Proc. Roy. Soc. A, 279 (1964) 170
Fe(CO)$_3$(C$_8$H$_8$)	1.42	1.42	1.42	B. Dickens and W. N. Lipscomb, J. Am. Chem. Soc. 83 (1964) 4862
Fe$_2$(CO)$_6$(C$_8$H$_8$)	1.44	1.39	1.44	B. Dickens and W. N. Lipscomb, J. Chem. Phys. 37 (1962) 2084
RhCp{(CF$_3$)$_6$C$_6$}	1.48	1.42	1.53	M. R. Churchill and R. Mason, Proc. Roy. Soc. A, 292 (1966) 61
CoCp(C$_5$H$_5$Ph)	1.49	1.36	1.54	M. R. Churchill and R. Mason, Proc. Roy. Soc. A, 279 (1964) 191
ReCpMe$_2$(C$_5$H$_5$Me)	1.45	1.31	1.45	N. W. Alcock, Chem. Commun. (1965) 177
CoCp(C$_5$H$_5$COPh)				M. R. Churchill, J. Organomet. Chem. 4 (1965) 258
Butadiene, ground state	1.36	1.45	1.36 ⎱	M. R. Churchill and R. Mason, Advances in Organomet. Chem. 5
Butadiene, excited state	1.45	1.39	1.45 ⎰	(1967) 91

The distortion from planarity is much greater in the cyclopentadiene complexes than in the cyclopentadienone complexes. As discussed in sections 2.3.3 and 4.1.3b, the distortion from planarity is caused by back donation of electrons from the metals to the ψ_3 MO of the diene system.

The central C–C distance C_2–C_3 of butadiene in its ground state (1.45 Å) is longer than the two outer C–C distances, C_1–C_2 and C_3–C_4 (1.36 Å). Other conjugated dienes have similar distances. However, on complexation C_2–C_3 becomes shorter and C_1–C_2 and C_3–C_4 become longer (Table 20). It is also found that with carbonyls as the *trans*-ligands the three C–C bonds are about equally long, but with cyclopentadienyl as the *trans*-ligand the central C–C distance, C_2–C_3, is markedly shorter than the other two (Table 20). An MO scheme to explain these effects has been proposed (see section 4.1.3b).

4.1.3b. *Bonding*

The π-molecular orbitals of a 1,3-diene are shown pictorially in Fig. 35.

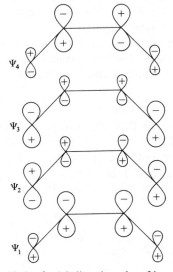

FIG. 35. π-Molecular orbitals of a 1,3-diene in order of increasing energy $\psi_1 \rightarrow \psi_4$.

In the ground state of the free diene, ψ_1, ψ_2 will be occupied and ψ_3, ψ_4 unoccupied. On complexation overlap of ψ_1 with the metal orbitals s, p_z and d_{z^2}; and ψ_2 with p_x and d_{xz} corresponds to electron donation from the butadiene to the metal orbitals. In addition, the lowest unoccupied level of the butadiene can overlap with the p_y- and d_{zy}-orbitals of the metal and would correspond to back donation from metal to butadiene. There are two important consequences of back donation into ψ_3. (1) The central bond C_2–C_3 becomes shorter and C_1–C_2 and C_3–C_4 longer, i.e. in valence bond terms the bonding corresponds to the σ,π-structure [4.1.22] for a cyclic 1,3-diene.

M

[4.1.22]

The extent of back coordination into ψ_3 depends on the *trans*-ligand(s). With three carbonyls as *trans*-ligands back coordination into ψ_3 is less than with cyclopentadienyl as *trans*-ligand because the electron acceptor power of three carbonyls is greater than that of a cyclopenta-dienyl group. Hence, with cyclopentadienyl as *trans*-ligand the length C_2–C_3 is less than C_1–C_2 or C_3–C_4, but it is of comparable length when tricarbonyl is the *trans*-grouping.

(2) Back coordination into ψ_3 distorts the cyclopentadiene, cyclopentadienone or other types of ring containing a diene system from planarity. The cyclopentadienone ring only distorts a little (10–20°) because the ring carbonyl can accommodate the electrons from the metal.

4.1.4. Physical Properties

4.1.4a. *Nuclear Magnetic Resonance Spectra*[476]

With tricarbonyl(butadiene)iron the ^{1}H resonances fall into two widely separated regions, the central hydrogens absorbing at much lower τ-value than the outer ones. Generally, for acyclic conjugated dieneiron tricarbonyls the central hydrogens absorb 4.6–4.9 τ or lower with unsaturated substituents. With no substituents on C_1 or C_2 the terminal hydrogens absorb within the range *anti* 9.5–9.8 τ and *syn* 8.24–8.4 τ. Substituents move the τ-values down somewhat. Nuclear magnetic resonance spectroscopy is very useful in the determination of the structure of dieneiron tricarbonyl complexes[480].

4.1.4b. *Infrared Spectroscopy*

Many compounds of the type $Fe(CO)_3$(conjugated diolefin) have an intense carbonyl absorption band at *ca.* 1980 cm^{-1}. With $Fe(CO)_3(C_4H_6)$ the low frequency absorption is resolvable into two bands—at 1985 and 1975 cm^{-1}. With electro-negative substituents on the diene system the bands move to higher frequencies.

The infrared absorption spectrum of $CoCp(C_5H_6)$ shows a band at *ca.* 2750 cm^{-1}.

[4.1.23]

This band was originally assigned to the vibration of the *endo*-hydrogen of the cyclopenta-diene ring [4.1.23], but more recent structural determinations by X-rays on compounds with substituted cyclopentadiene rings make it almost certain that this vibration frequency is associated with the *exo*-hydrogen[98].

4.1.4c. *Electronic Effects*

trans,trans-Hexa-2,4-dienoic acid is a slightly stronger acid when free (pK_a 7.00) than when complexed (pK_a 7.25) to tricarbonyliron. This makes an interesting contrast with $Cr(CO)_3$(benzoic acid) which is a stronger acid than free benzoic acid; the pK_as are 4.77 and 5.68 respectively.

4.1.5. Reactions

Several reactions have already been discussed which probably involve conjugated diolefin–metal complexes as intermediates, e.g. oligomerization, insertion reactions, protonation and nucleophilic attack (see section 3.1). Here a few other reactions are discussed.

4.1.5a. *Hydride Ion Abstraction*

Trityl fluoroborate abstracts hydride ion from tricarbonyl(*cis*-penta-1,3-diene)iron to give a cation with all five carbon atoms of the pentadienyl ligand bonded to the iron [4.1.24].

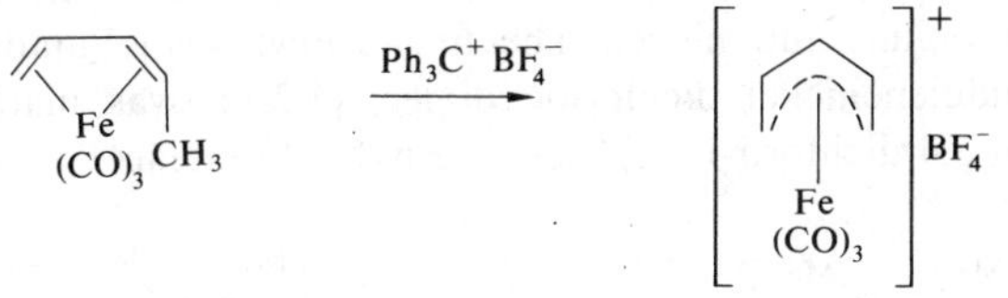

[4.1.24]

In contrast, trityl fluoroborate will not react with tricarbonyl(*trans*-penta-1,3-diene)iron, presumably because the methyl group is not suitably orientated with respect to the iron atom[476].

4.1.5b. *Reaction of Substituents*

A carbethoxy substituent on the coordinated diene system, e.g. as in

$$Fe(CO)_3\{CH_3(CH:CH)_2COOEt\},$$

can be hydrolysed to the corresponding carboxylic acid, and an aldehyde substituent converted into a carbinol, either by methylmagnesium iodide or by reduction with lithium aluminium hydride, e.g. [4.1.25] → [4.1.26].

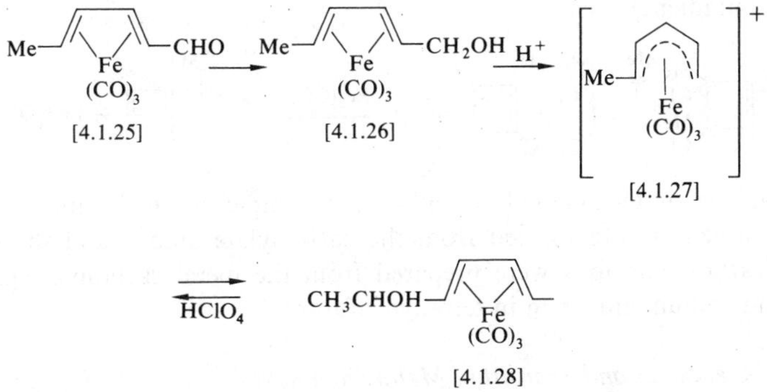

The resultant carbinol [4.1.26] reacts with a strong acid to give the hexadienyl complex [4.1.27]. This can be hydrated with water to give the secondary carbinol [4.1.28], i.e. the hydroxyl group adds to the more heavily substituted carbon atom.

4.1.5c. *Displacement of Conjugated Diolefin*

The diolefin can be displaced by triphenylphosphine (and presumably by some other ligands).

$$Fe(CO)_3(diolefin) + PPh_3 \rightarrow Fe(CO)_3(PPh_3)_2 + diolefin$$

4.1.5d. *Oxidation by* Ce^{4+} *or* Fe^{3+}

This causes breakdown of the complex and release of the diolefin.

4.2. CYCLOBUTADIENE–METAL COMPLEXES[13, 316, 494]

4.2.1. Preparation

4.2.1a. *From Dihalocyclobutenes*

This method has been used for cyclobutadiene complexes of chromium, molybdenum, tungsten, iron, ruthenium and nickel. The first known cyclobutadiene–metal complex, tetramethylcyclobutadienenickel dichloride dimer, [4.2.1], was made by treating tetracarbonylnickel with 3,4-dichloro-1,2,3,4-tetramethylcyclobutene[495].

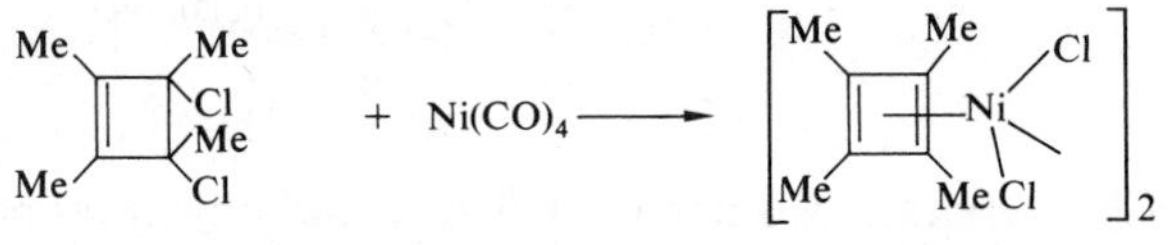

[4.2.1]

Tricarbonyl(cyclobutadiene)iron [4.2.2] is formed from 3,4-dichlorocyclobutene and enneacarbonyldi-iron

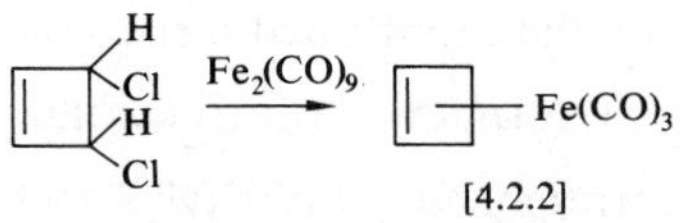

[4.2.2]

Di-, tri- and tetra-methylcyclobutadiene analogues of [4.2.2] are made similarly. A mixture of three isomeric dichlorocyclobutenes with enneacarbonyldi-iron gives tricarbonyl(1,2-dimethylcyclobutadiene)iron[496].

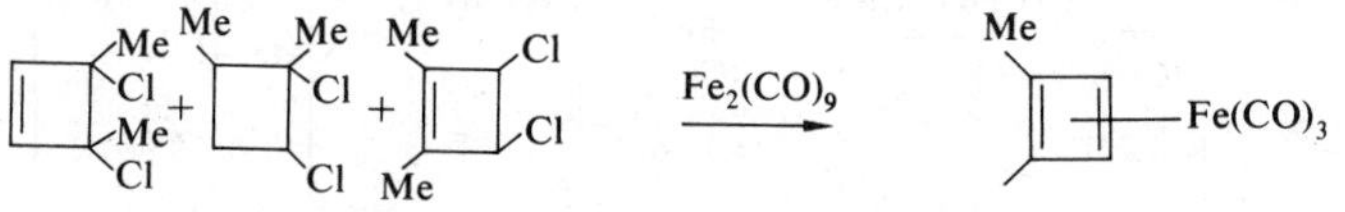

Cyclobutadiene or tetramethylcyclobutadiene complexes of chromium, molybdenum, tungsten or ruthenium are formed from the carbonylate anions and the dichlorocyclobutene. The carbonylate ions were prepared from the metal carbonyl, e.g. $Mo(CO)_6$ or $Ru_3(CO)_{12}$ and sodium amalgam in tetrahydrofuran[497].

4.2.1b. *From Acetylenes and Transition Metal Complexes*

Examples of the formation of cyclobutadiene–iron and –molybdenum complexes from acetylenes have been given in section 2.3.5. Another example already mentioned in an earlier section is tetraphenylcyclobutadienepalladium dichloride dimer, formed from tolan indirectly via ethoxy(tetraphenylcyclobutenyl)palladium chloride dimer (section 3.1.5).

[494] P. M. Maitlis, *Advan. Organomet. Chem.* **4** (1966) 94.
[495] R. Criegee and G. Schröder, *Justus Liebigs Ann. Chem.* **623** (1959) 1.
[496] H. A. Brune, W. Eberius and H. P. Wolff, *J. Organomet. Chem.* **12** (1968) 485.
[497] R. G. Amiet, P. C. Reeves and R. Pettit, *Chem. Commun.* (1967) 1208.

Cyclopentadienyl(cyclo-octa-1,5-diene)cobalt and tolan react to give cyclopentadienyl-tetraphenylcyclobutadienecobalt [4.2.3] [498].

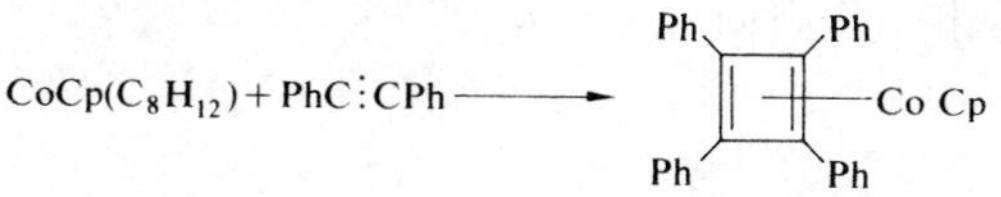
CoCp(C8H12) + PhC:CPh
Ph
Ph
Co Cp
Ph
Ph

[4.2.3]

4.2.1c. *From 1-Metallacyclopentadienes*

1,4-Dilithiotetraphenylbuta-1,3-diene reacts with dichlorodimethyltin to give the stannole [4.2.4]. This with nickel bromide in triglyme gives a good yield of tetraphenylcyclobutadienenickel dibromide [4.2.5] [499].

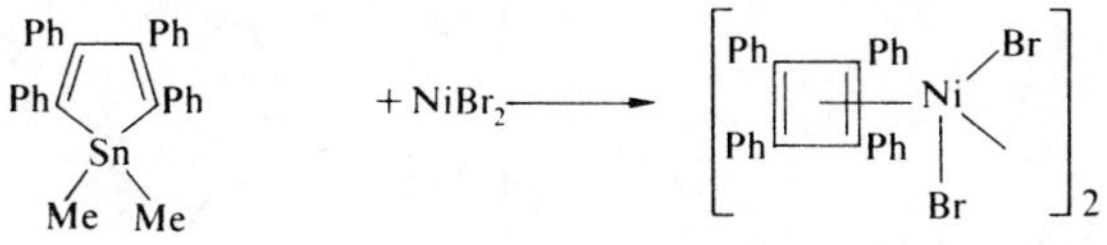
Ph Ph
Ph Ph
Sn
Me Me
+ NiBr2
Ph Ph
Br
Ni
Ph Ph
Br
2

[4.2.4] [4.2.5]

4.2.1d. *By Ligand Exchange from Other Cyclobutadiene–Metal Complexes*

Tetraphenylcyclobutadienepalladium halides are readily prepared from diphenylacetylene. The tetraphenylcyclobutadiene group can be transferred to another metal[500]. Some examples are outlined in Fig. 36.

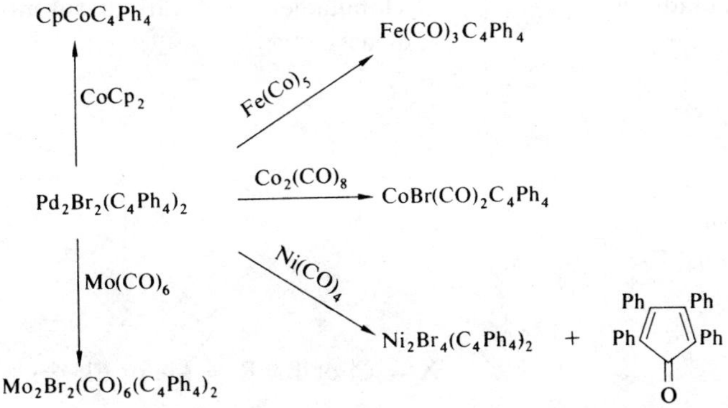
CpCoC4Ph4
CoCp2
Fe(CO)3C4Ph4
Fe(CO)5
Co2(CO)8
Pd2Br2(C4Ph4)2
CoBr(CO)2C4Ph4
Mo(CO)6
Ni(CO)4
Ph Ph
Ni2Br4(C4Ph4)2 + Ph Ph
O
Mo2Br2(CO)6(C4Ph4)2

FIG. 36. Synthesis of some tetraphenylcyclobutadiene–metal complexes by ligand transfer.

Cobalt complexes of type $CoX(CO)_2(C_4R_4)$ (X = Cl, Br or I; R = Ph or *p*-tolyl) are made similarly from $Pd_2X_4(C_4R_4)_2$ and octacarbonyldicobalt.

[498] A. Nakamura and N. Hagihara, *Bull. Chem. Soc. Japan* **24** (1961) 452.
[499] H. H. Freedman, *J. Am. Chem. Soc.* **83** (1961) 2194.
[500] A. Efraty and P. M. Maitlis, *J. Am. Chem. Soc.* **89** (1967) 3744.

4.2.1e. *Miscellaneous*

Photo-α-pyrone reacts with pentacarbonyliron to give tricarbonylcyclobutadieneiron and tricarbonyl(α-pyrone)iron in a combined yield of 10–15%. At $-15°$ the cyclobutadiene complex is the principal product[501].

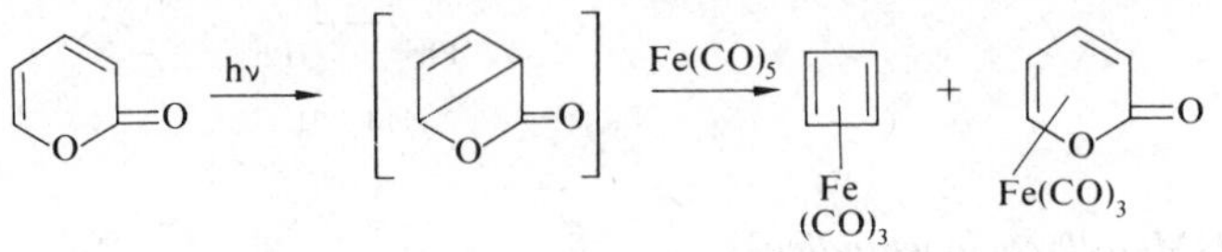

4.2.2. Cyclobutadiene–Metal Complexes

Chromium
$Cr(CO)_4(C_4Me_4)$ [497]

Molybdenum
$Mo(CO)_4(C_4R_4)$ [497]   R = H or Me.
$Mo(CO)_2(C_4Ph_4)_2$ [316]
$Mo(CO)_4(PhC\!:\!CPh)(C_4Ph_4)_2$ [316]
$Mo(CO)_2(tetracyclone)(C_4Ph_4)$ [316]
$Mo_2Br_2(CO)_4(C_4Ph_4)_2$ [503]

Tungsten
$W(CO)_4(C_4R_4)$ [497]   R = H or Me.

Iron
$Fe(CO)_3(cyclobutadienes)$ Cyclobutadiene and various substituted cyclobuta-
                                   dienes[13, 476].

Ruthenium
$Ru(CO)_3(C_4H_4)$ [497]

Cobalt
$CoCp(C_4Ph_4)$ [501a]
$CoBr(CO)_2(C_4Ph_4)$ [501b]

Nickel
$Ni_2X_4(C_4R_4)$ X = Cl or Br. R = Me or Ph [494, 495, 499, 501a].

Palladium
$Pd_2X_4(C_4Ph_4)_2$ X = Cl or Br [494, 501a].

[501] M. Rosenblum and C. Gatsonis, *J. Am. Chem. Soc.* **89** (1967) 5074.
[501a] M. L. Games and P. M. Maitlis, *J. Am. Chem. Soc.* **85** (1963) 1887.
[501b] P. M. Maitlis and A. Efraty, *J. Organomet. Chem.* **4** (1965) 175.
[502] J. D. Dunitz, H. C. Mez, O. S. Mills and H. M. M. Shearer, *Helv. chim. Acta* **45** (1962) 647.
[503] M. Mathew and G. L. Palenik, *Can. J. Chem.* **47** (1969) 705.

4.2.3. Structures and Bonding

The structure of $Ni_2Cl_4(C_4Me_4)_2$ is shown schematically in [4.2.6].

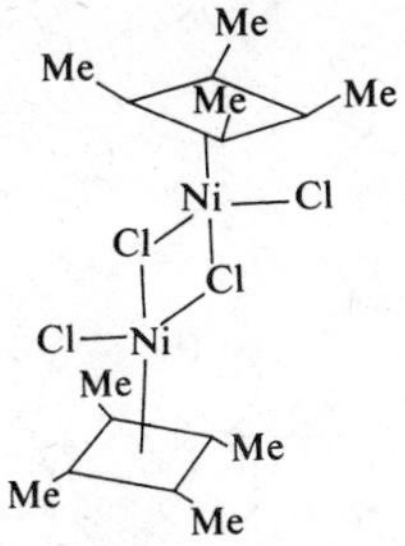

[4.2.6]

The crystals contain benzene of crystallization which makes the determination of accurate bond lengths and angles difficult. The cyclobutadiene ring C–C distances are 1.40–1.45 Å and the Ni–C distances are 2.00–2.05 Å. The methyl carbons are folded back and 0.12–0.18 Å away from the plane of the cyclobutadiene ring[502].

The structure of $Fe(CO)_3(C_4Ph_4)$ has been mentioned in section 2.4.5d. In this complex the cyclobutadiene ring is square with a C–C distance of 1.46 Å and the phenyl substituents bent 11° away from the iron. The Fe–C distances (2.06 Å) are the same as in ferrocene.

The structure of the complex $[MoBr(CO)_2(C_4Ph_4)]_2$ has been determined by X-ray diffraction[503] and is shown in [4.2.7].

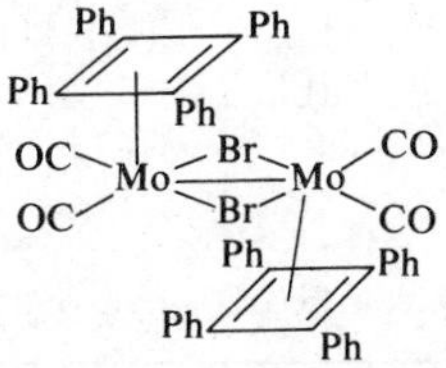

[4.2.7]

The Mo–Mo bond distance is 2.954 Å and the phenyl groups are bent away from the metal.

4.2.3a. Bonding

The π-molecular orbitals of cyclobutadiene, ψ_1, ψ_2, ψ_3 and ψ_4, are represented in Fig. 37. These orbitals can overlap with metal orbitals of the correct symmetry, as shown in Fig. 37. The contribution to the bonding of the overlap between ψ_4 and $d_{x^2-y^2}$ is probably very small. The overlaps of the orbitals ψ_2 and ψ_3 with the metal orbitals are probably the main contributors to the bonding. As with diene systems in five-, six-, seven- and eight-membered rings, complexation to the metal causes substituents on the diene system to bend away from the metal.

4.2.4. Physical Measurements

4.2.4a. Nuclear Magnetic Resonance Spectroscopy

A nuclear magnetic resonance study of tricarbonyl(cyclobutadiene)iron in a liquid crystal shows that the ratio of the length of the sides is 0.9977 ± 0.0045, i.e. probably a square[504].

[504] C. S. Yannoni, G. D. Caesar and B. P. Dailey, *J. Am. Chem. Soc.* **89** (1967) 2833.

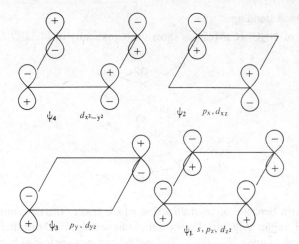

FIG. 37. π-Molecular orbitals of cyclobutadiene $\psi_1 \rightarrow \psi_4$. The metal orbitals with which they overlap are shown.

Some chemical shifts and coupling constants for tricarbonyl(cyclobutadiene)iron and its methyl derivatives have been measured very accurately. The coupling constants $^1J(^{13}C-H)$ for the ring hydrogens are large. For tricarbonyl(cyclobutadiene)iron the satellites due to splitting by ^{13}C are doublets, $^4J(H_1-H_3) = ^4J(H_2-H_4) = 8.9$ Hz; vicinal coupling constants, i.e. $^3J(H_1-H_2)$, must be small (Table 21).

TABLE 21. SOME N.M.R. DATA FOR Fe(CO)$_3$-CYCLOBUTADIENES IN CCl$_4$ SOLUTION[505]

Cyclobuta-1,3-diene substituents	Olefin protons		Methyl protons	
	τ	$J(^{13}C-H)$ Hz	τ	$J(^{13}C-H)$ Hz
—	6.089	191		
1,2-diMe	6.104	189.2	8.238	128.6
1,2,3-triMe	6.081	186.7	(1,3)8.256	128.3
			(2)8.234	
1,2,3,4-tetraMe	—	—	8.324	128.6

4.2.4b. Infrared Spectra

Table 22 gives some values of $v(C\vdots O)$ for Fe(CO)$_3$-cyclobutadienes[505]. It can be seen that as the number of methyl groups increases $v(CO)$ values decrease, probably because of increased back donation to the carbonyl groups from the iron.

505 H. A. Brune, H. P. Wolff and H. Hüther, Chem. Ber. 101 (1968) 1485.

TABLE 22. CARBONYL STRETCHING FREQUENCIES OF THE
CARBONYL GROUPS IN TRICARBONYLIRON–CYCLOBUTADIENE
COMPLEXES

Cyclobutadiene substituents	$\nu(CO)\,cm^{-1}$	
—	2055	1985
1,2-diMe	2045	1970
1,2,3-triMe	2040	1960
1,2,3,4-tetraMe	2035	1955

4.2.5. Reactions of Cyclobutadiene–Metal Complexes

4.2.5a. *Thermal Decomposition*

The products formed by thermal decomposition of cyclobutadiene–metal complexes are varied. Some of the products formed by thermal decomposition of $Ni_2Cl_4(C_4Me_4)_2$ are shown in Fig. 38.

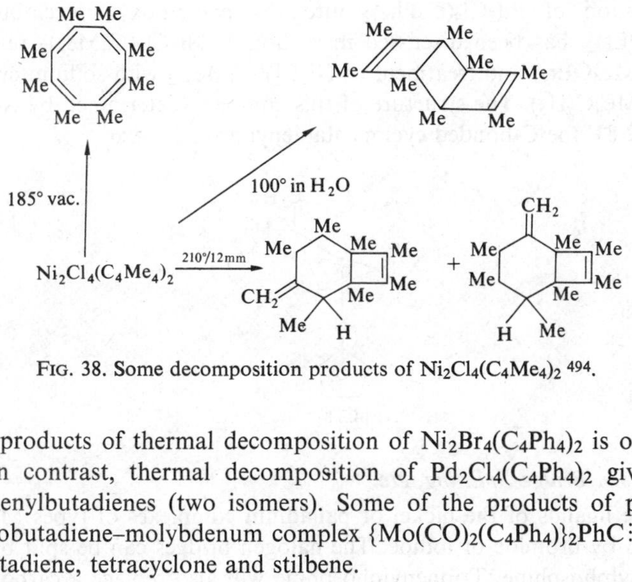

FIG. 38. Some decomposition products of $Ni_2Cl_4(C_4Me_4)_2$ [494].

One of the products of thermal decomposition of $Ni_2Br_4(C_4Ph_4)_2$ is octaphenylcyclo-octatetraene. In contrast, thermal decomposition of $Pd_2Cl_4(C_4Ph_4)_2$ gives mainly 1,4-dichlorotetraphenylbutadienes (two isomers). Some of the products of pyrolysis of the tetraphenylcyclobutadiene–molybdenum complex $\{Mo(CO)_2(C_4Ph_4)\}_2PhC\vdots CPh$ are penta-phenylcyclopentadiene, tetracyclone and stilbene.

4.2.5b. *Reduction*

Catalytic hydrogenation of $Ni_2Cl_4(C_4Me_4)_2$ gives *cis*-1,2,3,4-tetramethylcyclobutane in high yield. Lithium aluminium hydride reduction of $Fe(CO)_3(C_4Ph_4)$ gives *cis,cis*-1,2,3,4-tetraphenylbutadiene, whilst reduction with sodium in liquid ammonia gives 1,2,3,4-tetraphenylbutane. The products of reduction of several other cyclobutadiene–metal complexes have been identified[494].

[506] G. F. Emerson, K. Ehrlich, W. P. Giering and P. C. Lauterbur, *J. Am. Chem. Soc.* **88** (1966) 3172.

4.2.5c. *Oxidation*

Ceric ion oxidation of $Fe(CO)_3(C_4H_4)$ gives free cyclobutadiene which with alkynes gives Dewar-benzene derivatives; these on heating give arenes.

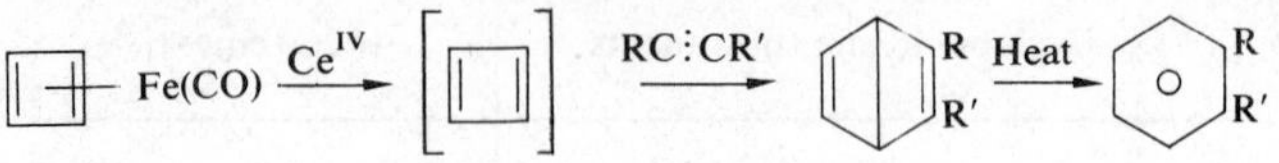

If the oxidation by ceric ion is carried out in the presence of dimethyl maleate the product is *endo-cis*-dicarbomethoxy-bicyclohexene.

4.2.5d. *Electrophilic Attack*

The cyclobutadiene ring of $Fe(CO)_3(C_4H_4)$ readily undergoes attack by electrophilic reagents. Thus the following substituents Y can be introduced to give compounds $Fe(CO)_3(C_4H_3Y)$; the reagents used are in parentheses. CHO (PhNMeCHO, $POCl_3$), RCO ($RCOCl$, $AlCl_3$), CH_2Cl (CH_2O, HCl), HgCl {$Hg(OAc)_2$, NaCl} and CH_2NMe_2 (CH_2O, Me_2NH).

4.2.5e. *Nucleophilic Attack on the Cyclobutadiene Ring*

The conversion of $Pd_2Cl_4(C_4Ph_4)_2$ into the *exo*-ethoxy π-cyclobutenyl complex, $Pd_2Cl_2(C_4Ph_4OEt)_2$, has been discussed in section 3. $Ni_2Cl_4(C_4Me_4)_2$ reacts similarly to give $Ni_2Cl_2(C_4Me_4OEt)_2$ and treatment of $Ni_2Cl_4(C_4Me_4)_2$ with sodium cyclopentadienide gives $NiCp(C_4Me_4C_5H_5)$. The structure of this complex (determined by X-ray diffraction) is shown in [4.2.8]; the C-bonded cyclopentadienyl group is *exo*.

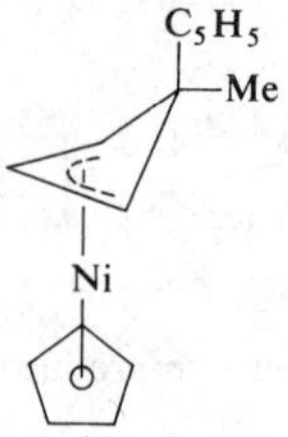

[4.2.8]

4.2.5f. *Metathesis, Bridge Splitting, etc.*

The chloride ligands of the nickel or palladium complexes of types $M_2Cl_4(C_4R_4)_2$ are readily replaced by bromide or iodide. The halogen bridges can be split by donor ligands such as triphenylphosphine. Triphenylphosphine will also replace a carbonyl ligand from some cyclobutadiene–metal–carbonyl complexes. The complex $CoBr(CO)_2(C_4Ph_4)$ reacts with arenes in the presence of aluminium chloride to give $[Co(arene)(C_4Ph_4)]^+$, or with cycloheptatriene to give a low yield of $[Co(cycloheptatriene)(C_4Ph_4)]^+$ [500].

4.3.1. Trimethylenemethaneiron Tricarbonyls

Tricarbonyl(trimethylenemethane)iron was first synthesized from 3-chloro-(2-chloromethyl)propene and $Fe_2(CO)_9$. The structure has been determined by gas-phase electron diffraction and is shown in [4.3.1].

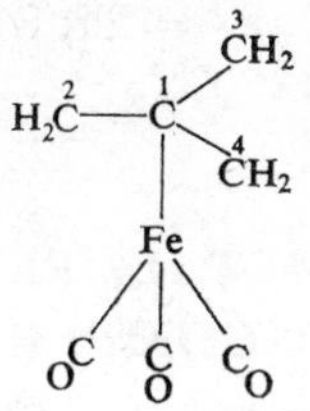

[4.3.1]

The distance Fe–C(1), 1.938 Å, is less than Fe–C(2), Fe–C(3) or Fe–C(4), which are all essentially equal at 2.123 Å. The molecule has C_{3v} symmetry with an anti-primatic arrangement of the six carbon atoms[507].

A more convenient synthesis of tricarbonyl(trimethylenemethane)iron is by thermal decomposition of chloro(2-methylallyl)tricarbonyliron:

$$FeCl(C_4H_7)(CO)_3 \rightarrow Fe(C_4H_6)(CO)_3$$

Other products formed in the reaction include ferrous chloride, isobutene and carbon monoxide. A one-step synthesis from commercially available materials is by treating enneacarbonyldi-iron with an excess of 2-methylallyl chloride and isolation by fractional distillation. The monomethyl and monophenyl derivatives $Fe(CO)_3(C_4H_5R)$ (R = Me or Ph) are made by a similar route. The structure of the phenyltrimethylenemethane complex has been determined by X-ray diffraction and is shown in [4.3.2].

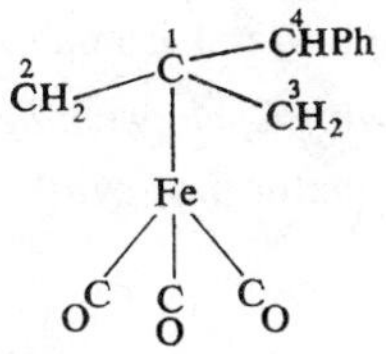

[4.3.2]

As in [4.3.1] the tricarbonyl group and the trimethylenemethane group adopt a mutually staggered conformation. The central carbon atom, C(1), is displaced by 0.31 Å from the plane of C(2)–C(3)–C(4). Some distances are Fe–C(1) 1.929 Å, Fe–C(2) 2.096 Å, Fe–C(3) 2.119 Å and Fe–C(4) 2.160 Å, C(1)–C(2) 1.406 Å, C(1)–C(3) 1.406 Å, C(1)–C(4) 1.434 Å.

Some methylenecyclopropanes coordinated to tricarbonyliron have been made from enneacarbonyldi-iron and methylenecyclopropanes [4.3.3].

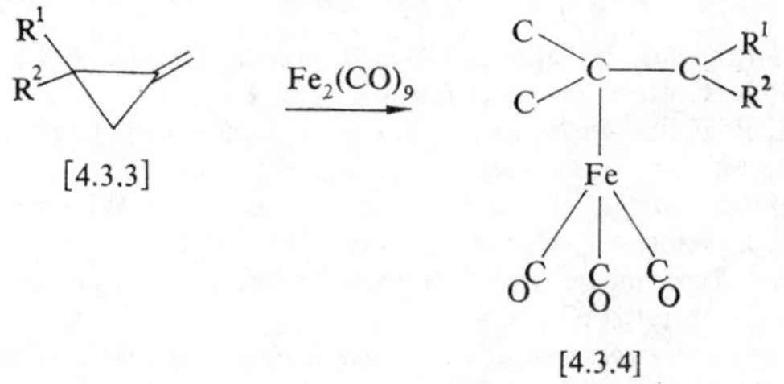

[4.3.3]

[4.3.4]

[507] A. Almenningen, A. Haaland and K. Wahl, *Chem. Commun.* (1968) 1027.
[508] K. Ehrlich and G. F. Emerson, *Chem. Commun.* (1969) 59.

Complexes of type [4.3.4] with R^1 = H, R^2 = Ph; R^1 = Me, R^2 = Ph and R^1 = R^2 = Ph were made[509].

5. CYCLOPENTADIENYL–METAL, PENTADIENYL–METAL AND RELATED COMPLEXES[510-516]

5.1. CYCLOPENTADIENYL–METAL COMPLEXES

A very important event in the history of organometallic chemistry was the discovery of dicyclopentadienyliron (ferrocene) in 1951 [517, 518]. Subsequently most metals have been shown to give cyclopentadienyl or substituted cyclopentadienyl derivatives. Cyclopentadienyl is a very useful ligand for stabilizing transition metal–hydride, –alkyl and –aryl bonds. There are many examples of cyclopentadienylmetal carbonyls or nitrosyls and of compounds containing metal–metal bonds, in which the bonded metals have cyclopentadienyl groups as ligands.

Many of these compounds are better discussed in other chapters, since the cyclopentadienyl group is serving as a stabilizing ligand and the more interesting chemistry is associated with the other ligands or with the metal–metal bond. Many cyclopentadienyl complexes are discussed elsewhere in our chapter; see, for example, section 2.2.5. In this section (5.1) we shall concentrate on aspects of cyclopentadienylmetal chemistry not covered elsewhere in this book.

5.1.1. Preparation of Cyclopentadienyl–Metal Complexes[510-516]

5.1.1a. *By Treating a Metal Halide with Cyclopentadiene in the Presence of a Base*

This is the most commonly used method of synthesizing ferrocene.

$$C_5H_6 + \text{Base} \rightleftharpoons \text{Base } H^+ \begin{Bmatrix} \\ \end{Bmatrix} C_5H_5^- \xrightarrow{FeCl_2}$$

Many different bases have been used, ranging from the weak, such as diethylamine or pyridine, to the strong, such as metallic sodium (which reacts with cyclopentadiene to give sodium cyclopentadienide). The method involving diethylamine[519] is probably the best laboratory synthesis, although another good procedure is to make sodium cyclopentadienide from cyclopentadiene and sodium in tetrahydrofuran or 1,2-dimethoxyethane, and then to

509 R. Noyori, T. Nishimura and H. Takaya, *Chem. Commun.* (1969) 89.
510 G. Wilkinson and F. A. Cotton, *Progress Inorg. Chem.* **1** (1959) 1.
511 E. O. Fischer and H. P. Fritz, *Advances Inorg. Chem. Radiochem.* **1** (1959) 55.
512 J. M. Birmingham, *Advances Organomet. Chem.* **2** (1964) 365.
513 M. Rosenblum, *Chemistry of the Iron Group Metallocenes*, John Wiley, New York (1965).
514 D. E. Bublitz and K. L. Rinehart, *Org. Reactions* **17** (1969) 1.
515 R. B. King, *Ferrocene, Ruthenocene and Osmocene in Organometallic Chemistry Reviews*, Section B, Annual Surveys, Elsevier, Lausanne (1967), p. 67.
516 M. I. Bruce, *Ferrocene, Ruthenocene and Osmocene in Organometallic Chemistry Reviews*, Section B, Annual Surveys, Elsevier, Lausanne (1968), p. 379.
517 T. J. Kealy and P. L. Pauson, *Nature* **168** (1951) 1039.
518 S. A. Miller, J. A. Tebboth and J. F. Tremaine, *J. Chem. Soc.* (1952) 632.
519 G. Wilkinson, *Org. Synth.* **36** (1956) 31.

add ferrous chloride. Many different substituted ferrocenes have been made from a substituted cyclopentadiene, a base and ferrous (or ferric) halide. Some of the substituted cyclopentadienes used are t-butyl, allyl-, acetyl-, trimethylsilyl-, phenylazo-, 1,2-dimethyl- or 1,3-diphenyl-cyclopentadiene. Polymethylene bridged ferrocenes of the type [5.1.1] can be made from polymethylene bridged cyclopentadienes [5.1.2].

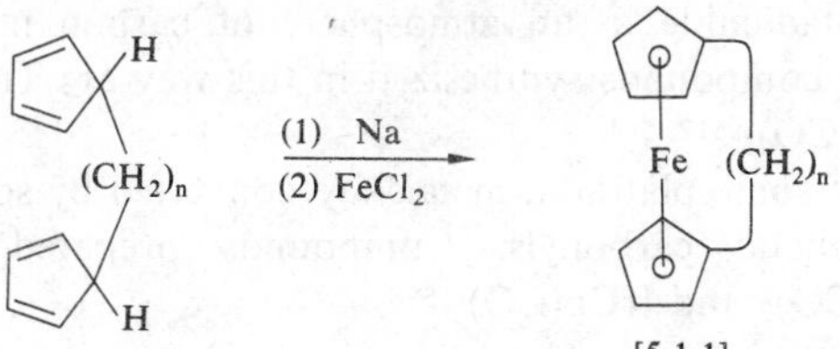

[5.1.2]   (1) Na   (2) FeCl$_2$ [5.1.1]

3-Methylcyclopent-2-enone when treated successively with sodamide, ferrous chloride and benzoyl chloride gives the dimethyl-dibenzoato-ferrocene [5.1.3], as a mixture of *meso*- and *racemic*-forms.

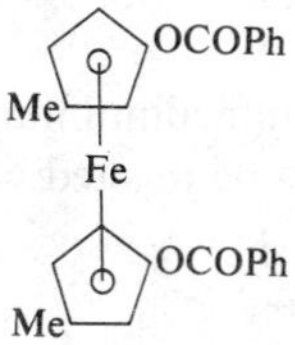

[5.1.3]

Ruthenocene (RuCp$_2$) is formed by treating ruthenium(III) acetylacetonate with cyclopentadienylmagnesium bromide, or anhydrous ruthenium trichloride with cyclopentadienylsodium[520]. Osmocene (OsCp$_2$) or manganocene, MnCp$_2$, are similarly made from cyclopentadienylsodium and osmium(IV) chloride or manganese(II) bromide respectively[521].

Nickelocene and cobalticene can be made from cyclopentadiene and the metal halide in the presence of diethylamine. Cyclopentadienyl derivatives of other metals are prepared by treating the metal halide with sodium cyclopentadienide, lithium cyclopentadienide or cyclopentadienylmagnesium halide. With an excess of the cyclopentadienide reduction to a lower valence state may occur, e.g.[510, 511, 522]

$$TiCl_4 + CpMgBr \rightarrow TiBr_2Cp_2$$
$$TiCl_2 + NaCp \rightarrow TiCp_2 \text{ } [37]$$
$$TiCl_2Cp_2 + NaCp \rightarrow TiCp_3$$
$$MoCl_5 + NaCp \rightarrow MoCl_2Cp_2$$
$$VCl_4 + CpMgBr \text{ (excess)} \rightarrow VCp_2$$

Sometimes mixed π-cyclopentadienyl-σ-cyclopentadienyl complexes may be formed, e.g.

$$MoCl_5 \xrightarrow{NaCp} Mo(\pi\text{-}C_5H_5)(\sigma\text{-}C_5H_5)_3$$
$$NbCl_5 \xrightarrow{NaCp} Nb(\pi\text{-}C_5H_5)_2(\sigma\text{-}C_5H_5)_2$$

[520] E. O. Fischer and H. Grubert, *Chem. Ber.* **92** (1959) 2302.
[521] G. Wilkinson and F. A. Cotton, *Chem. and Ind. (London)* (1954) 307.
[522] A. K. Fischer and G. Wilkinson, *J. Inorg. Nucl. Chem.* **2** (1956) 149.

or a hydrido complex

$$ReCl_5 + NaCp \rightarrow ReHCp_2$$

or a cyclopentenyl complex

$$NiCl_2 + NaCp \rightarrow NiCp_2 + NiCp(C_5H_7)$$

Cyclopentadienylmetal carbonyls can also be made by reacting a transition metal halide with sodium cyclopentadienide in an atmosphere of carbon monoxide (usually under pressure). Some of the compounds synthesized in this way are $TiCp_2(CO)_2$, $MnCp(CO)_3$, $TiCp(CO)_3$ and $ReCp(CO)_3$ [512, 514].

Carbonyl halides of some platinum metals are converted by sodium cyclopentadienide into cyclopentadienylmetal carbonyls. Compounds prepared in this way include $Ru_2Cp_2(CO)_4$, $RhCp(CO)_2$ and $IrCp(CO)_2$ [523].

5.1.1b. *From Cyclopentadienide Ions, Prepared from Fulvenes*

Fulvenes are readily prepared from cyclopentadiene and aldehydes or ketones.

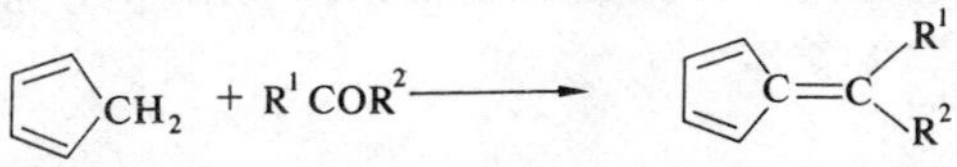

Sodium, organolithium reagents or lithium aluminium hydride add to fulvenes to give cyclopentadienide ions and these in turn can be reacted with ferrous chloride to give substituted ferrocenes. Some examples are:

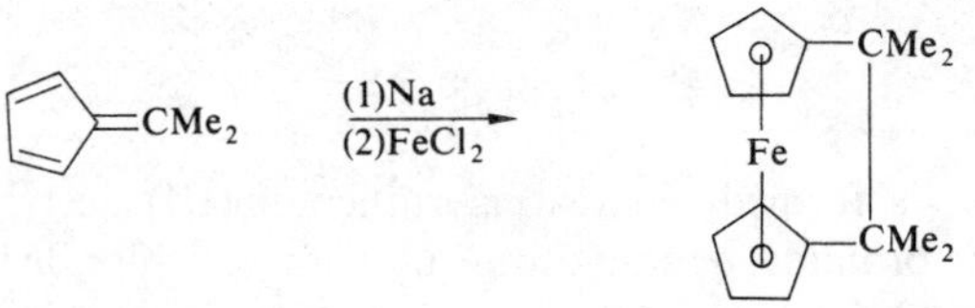

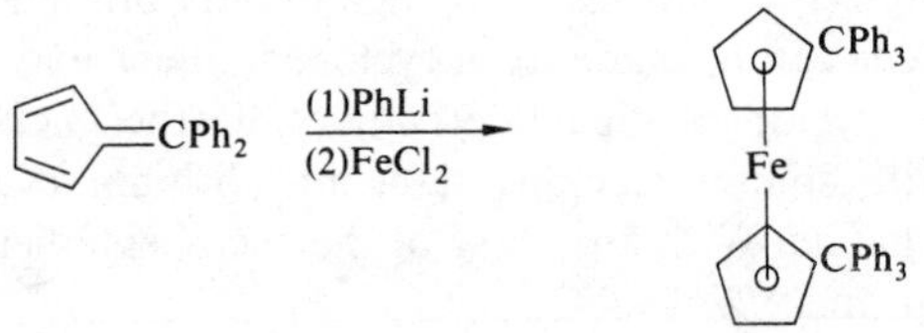

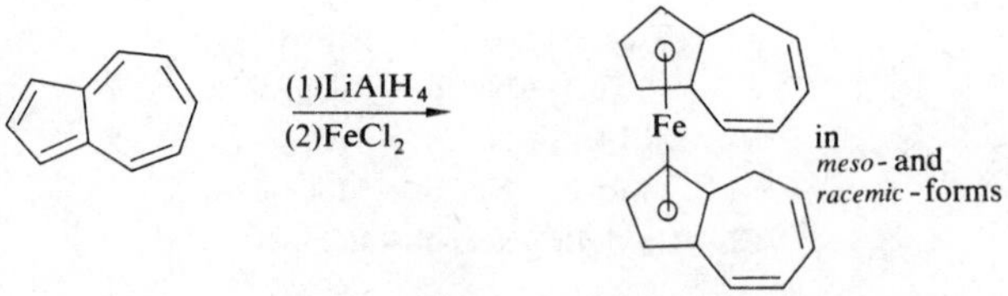

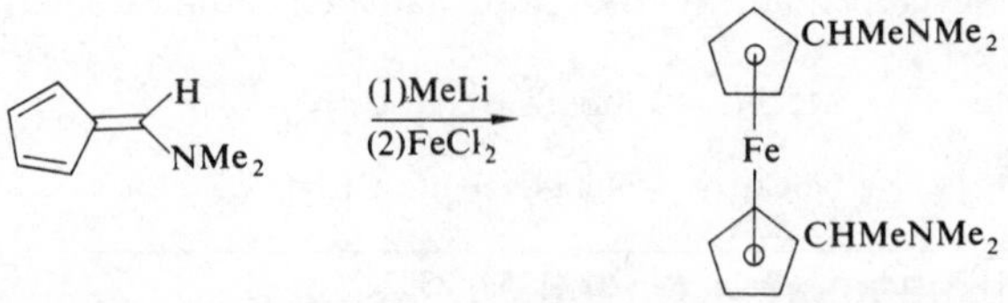

[523] E. O. Fischer and K. S. Brenner, *Z. Naturforsch.* **17B** (1962) 774.

There are many other examples of syntheses of substituted ferrocenes starting from fulvenes or related compounds[513–516].

Ring substituted derivatives of titanium, molybdenum, cobalt and nickel cyclopentadienyls have also been made using a substituted cyclopentadienide formed from a fulvene[524].

5.1.1c. *Formation from Cyclopentadienes or Fulvenes, and Metal Carbonyls*

These compounds are also discussed in the chapter on metal carbonyls. Treatment of metal carbonyls with alkali metal cyclopentadienides gives π-cyclopentadienyl–metal carbonyl complexes

$$[Nb(CO)_6]^- + NaCp \rightarrow NbCp(CO)_4$$
$$M(CO)_6 + NaCp \rightarrow [MCp(CO)_3]Na$$

where M = Cr, Mo or W.

A related synthesis is the conversion of titanium tetrachloride to $TiCp_2(CO)_2$ by treatment with NaCp in the presence of carbon monoxide[525].

Cyclopentadiene reacts with pentacarbonyliron to give $Fe_2Cp_2(CO)_4$ and with octacarbonyldicobalt to give $CoCp(CO)_2$. Under more vigorous conditions pentacarbonyliron and cyclopentadiene give ferrocene; similarly hexacarbonylchromium and cyclopentadiene at *ca.* 300° give chromocene, via $Cr_2Cp_2(CO)_6$ as an intermediate.

The Group VI metal carbonyls react with fulvenes to give ring substituted cyclopentadienyl–metal tricarbonyl complexes [5.1.4]. A hydrogen atom presumably comes from the solvent[205].

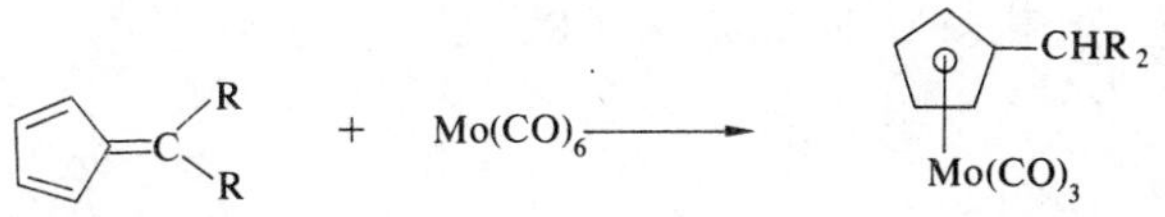

[5.1.4]

Fulvenes also react with pentacarbonyliron or with octacarbonyldicobalt to give substituted cyclopentadienyl–metal carbonyl complexes.

5.1.1d. *Miscellaneous Synthesis of Cyclopentadienyl–Metal Complexes*

Several examples of the formation of cyclopentadienyl–metal complexes from acetylenes are given in section 2.2.1.

Isobutene reacts with titanium tetrachloride at 300° and 30–60 atm to give trichloro-(pentamethylcyclopentadienyl)titanium[526].

$$CH_2:CMe_2 + TiCl_4 \rightarrow TiCl_3(C_5Me_5) + H_2 + CH_4 + C_2H_6 + C_3H_8, \text{ etc.}$$

Dewar hexamethylbenzene reacts with rhodium trichloride trihydrate or with iridium trichloride tetrahydrate in methanol to give a dimeric dichloropentamethylcyclopentadienyl complex, [5.1.5][527].

[524] G. R. Knox and P. L. Pauson, *J. Chem. Soc.* (1961) 4610.
[525] J. G. Murray, *J. Am. Chem. Soc.* **83** (1961) 1287.
[526] H. Röhl, E. Lange, T. Gössl and G. Roth, *Angew. Chem. Int. Edn.* **1** (1962) 117.
[527] J. W. Kang and P. M. Maitlis, *J. Am. Chem. Soc.* **90** (1968) 3259.

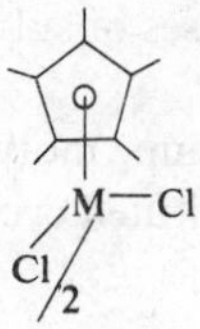

[5.1.5]

It is thought that in acidic methanol the Dewar hexamethylbenzene is converted into a chloro- (or methoxy-)ethylpentamethylcyclopentadiene, which then reacts with the rhodium or iridium trichloride.

5.1.2. Some Important Types of Cyclopentadienyl–Metal Complexes

As mentioned already, the cyclopentadienyl group is commonly used as a stabilizing ligand and its complexes are frequently best discussed in terms of the other ligands bonded to the metal. This list of the types of cyclopentadienyl–metal complexes is restricted to simple derivatives such as those where cyclopentadienyl or substituted cyclopentadienyl is the only ligand, and to cyclopentadienylmetal carbonyls, nitrosyls, hydrides and halides. Other examples of cyclopentadienyl–metal complexes are described in other chapters. There are many reviews on cyclopentadienyl–metal complexes[510–16]. In the following list Cp refers to cyclopentadienyl and frequently to substituted cyclopentadienyl.

5.1.2a. Complexes of Types M_xCp_y and $M_xCp_y^{n+}$

MCp_2	Zr?, V, Cr, Mn, Fe, Co, Ni, Mo, Ru, Rh, Os, Ir.
MCp_2^+	V, Cr, Fe, Co, Ni, Mo, Ru, Rh, Ir.
MCp_2^{2+}	Ni?, Os.
MCp_3	Ti.
MCp_4	Zr, Hf, Th.
M_2Cp_4	Ti.

5.1.2b. Cyclopentadienyl–Metal Carbonyls

$MCp(CO)_2$	Co, Rh, Ir.
$MCp(CO)_3$	Mn, Tc, Re.
$MCp(CO)_3^+$	Fe.
$MCp(CO)_3^-$	Cr, Mo, W.
$MCp(CO)_4$	V, Nb, Ta.
$MCp(CO)_4^+$	Cr, Mo, W.
$MCp_2(CO)_2$	Ti.
$M_2Cp_2(CO)_2$	Ni, Pt.
$M_2Cp_2(CO)_3$	Rh.
$M_2Cp_2(CO)_4$	Fe, Ru, Os.
$M_2Cp_2(CO)_6$	Cr, Mo, W.
$M_3Cp_3(CO)_3$	Co, Rh.
Miscellaneous	$Ni_3Cp_3(CO)_2$, $WMoCp_2(CO)_6$, $FeCoCp(CO)_6$, $FeNiCp_2(CO)_3$, $MoFeCp_2(CO)_5$, $MnFeCp(CO)_7$.

5.1.2c. Cyclopentadienyl–Metal Nitrosyls

$MCp(NO)$	Ni, Pd, Pt.

5.1.2d. *Cyclopentadienyl–Metal Hydrides*

$MHCp_2$	Re, Tc.
MH_2Cp_2	Mo, W.
MH_3Cp_2	Ta.
$MHCp_2^+$	Fe, Ru.
$MH_2Cp_2^+$	Re.
$MH_3Cp_2^+$	Mo, W.
$M_2H_2Cp_2$	Ti.
$M_nH_{2n}Cp_n$	Zr.

5.1.2e. *Cyclopentadienyl–Metal Halides*

X = halogen, most commonly chlorine, but frequently bromine or iodine and very occasionally fluorine.

MX_2Cp	Ti.
MX_3Cp	Ti, V, Cr, Zr.
MX_4Cp	Mo.
$MXCp_2$	Ti, V.
MX_2Cp_2	Ti, V, Zr, Mo, W, Hf.
MX_3Cp_2	Nb, Ta.
Miscellaneous	$[RhCpBr_2]_x$, $Pd_2Cl_2Cp_2$, $[MX_2Cp_2]^+$ (M = Mo, W, Re), $[OsICp_2]^+$.

5.1.3. Structures and Bonding

5.1.3a. *Structures*[513, 528]

A number of crystallographic and electron diffraction structural determinations on metallocenes have been reported. In the crystalline state ferrocene has the anti-prismatic (staggered) conformation [5.1.6A] whilst ruthenocene and osmocene have the eclipsed conformation [5.1.6B]. An anti-prismatic conformation might be expected to minimize non-bonded

[5.1.6A]   [5.1.6B]

interactions, and the rings are farther apart in ruthenocene and osmocene than in ferrocene (Table 23). The different conformations of ferrocene, on the one hand, and ruthenocene or osmocene, on the other, may not be due to intramolecular forces but to some, at present, unknown factors[528]. Electron diffraction studies on ferrocene suggest that the barrier to rotation of the rings is small. Some crystallographic and electron diffraction data for the metallocenes are given in Table 23.

The anti-prismatic conformation of ferrocene is retained in derivatives, e.g. in ferrocene-disulphonyl chloride which has the transoid structure [5.1.7], in dibenzoylferrocene which has the 1,1'-configuration [5.1.8] and in dibenzoferrocene which has the gauche structure [5.1.9]. Diacetylruthenocene has the cisoid structure [5.1.10].

Some internuclear distances for these substituted ferrocenes and for cyclopentadienyl derivatives of other metals are also given in Table 23.

528 P. J. Wheatley, "π-Complexes of transition metals with aromatic systems", in *Perspectives in Structural Chemistry* (eds. J. D. Dunitz and J. A. Ibers), Wiley, New York (1967).

TABLE 23. SOME DISTANCES (Å) IN π-CYCLOPENTADIENYL–METAL COMPLEXES

	M–C	C–C	Other distances	References
FeCp$_2$ (X-rays)	2.045	1.403	Inter-ring 3.32	J. D. Dunitz, L. E. Orgel and A. Rich, *Acta Cryst.* 9 (1956) 373
FeCp$_2$ (ED)	2.07	1.43	Inter-ring 3.25	S. Shibata, L. S. Bartell and R. M. Gavin, *J. Chem. Phys.* 41 (1964) 717
Ferrocene-1,1'-disulphonyl chloride	2.03	1.38		O. V. Starovskii and Y. T. Struchkov, *Izvest. Akad. Nauk SSSR, Otdel Khim. Nauk* 6 (1960) 1001
1,1-Dibenzoylferrocene	2.05	1.41		Y. T. Struchkov and T. L. Khotsianova, *Kristallografiya* 2 (1957) 376
1,1'-(Tetramethylethylene)ferrocene	1.97–2.11	1.45		M. Rosenblum, *Chemistry of the Iron Group Metallocenes*, Part I, John Wiley, New York, 1965
RuCp$_2$	2.21	1.43	Inter-ring 3.68	G. L. Hardgrove and D. H. Templeton, *Acta Cryst.* 12 (1959) 28
OsCp$_2$	2.22		Inter-ring 3.71	F. Jellinek, *Z. Naturforsch.* 14b (1959) 737
NiCp$_2$	2.18			I. A. Ronova and N. V. Alekseev, *Zhur. Strukt. Khim.* 7 (1966) 886
Fe$_2$Cp$_2$(CO)$_4$	2.11	1.41		O. S. Mills, *Acta Cryst.* 11 (1958) 620
TiCl$_3$Cp	2.38			G. Allegra and P. Ganis, *Atti Accad. Nazl. Lincei Rend. Classe Sci. Fis. Mat. e Nat.* 33 (1962) 303, 438
Mo$_2$Cp$_2$(CO)$_6$	2.30–2.38	1.37–1.44	Mo–Mo 3.22	F. C. Wilson and D. P. Shoemaker, *J. Chem. Phys.* 27 (1958) 809
NiCp(NO)	2.11			A. P. Cox, L. F. Thomas and J. Sheridan, *Nature* 181 (1958) 1157
CrClCp(NO)$_2$	2.20			O. L. Carter, A. T. McPhail and G. A. Sim, *Chem. Commun.* (1966) 49

SO₂Cl

SO₂Cl

[5.1.7]

COPh
COPh

[5.1.8]

Fe

[5.1.9]

MeOC COMe

[5.1.10]

5.1.3b. *Bonding*

There have been many theoretical treatments of the bonding in ferrocene and in other cyclopentadienyl–metal complexes[513]. Although the various treatments differ in some important details the basic approach is similar. The five π-molecular orbitals of the cyclopentadienyl ring are formed from the p_z orbitals, as shown schematically in Fig. 39.

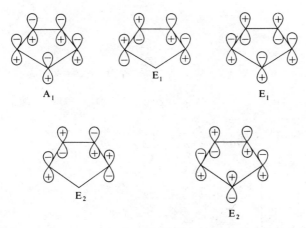

FIG. 39. Showing how the five π-molecular orbitals of the cyclopentadienyl ring are generated from the p_z orbitals.

In ferrocene the linear combination of these orbitals for the two rings gives ten molecular orbitals A_{1g}, A_{2u}, E_{1g}, E_{1u}, E_{2g} and E_{2u} (the E-orbitals being doubly degenerate). Pictorial representations of the A_{1g} and A_{2u} orbitals are given in [5.1.11] and [5.1.12] respectively.

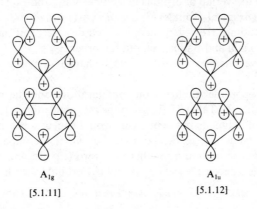

A_{1g}

[5.1.11]

A_{1u}

[5.1.12]

The other bis(cyclopentadienyl) orbitals (E_{1g}, E_{1u}, E_{2g} and E_{2u}) are generated similarly. The metal orbitals can be classified as $A_{1g}(4s, 3d_{z^2})$, $E_{1g}(3d_{xz}, 3d_{yz})$, $E_{2g}(3d_{x^2-y^2})$, $A_{2u}(4p_z)$ and $E_{1u}(4p_x, 4p_y)$. Metal and ring orbitals of the same symmetry class (A_{1g}, E_{1g}, etc.) can in principle combine to give bonding and anti-bonding orbitals, but the contribution towards the metal–ligand binding energy is governed by the relative energies of the interacting metal and ligand orbitals and by the overlap integrals. The many theoretical treatments of the bonding in metallocenes have been discussed critically by Rosenblum[513]. There is not the space to discuss all of these treatments here, but by way of example the molecular orbital diagram for ferrocene, calculated by Shustorovich and Dyatkina[529], is shown in Fig. 40.

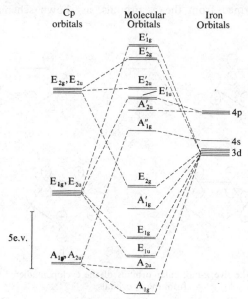

FIG. 40. Molecular orbital diagram for ferrocene.

Although there are some fundamental differences in the several molecular orbital calculations on ferrocene, there seems to be agreement on the following[1,513]: (1) most of the bonding energy is due to overlap of orbitals of symmetry E_{1g} and E_{1u}; (2) there is little if any mixing of the $3d_{z^2}$ and $4s$ metal orbitals; (3) the $3d_{z^2}$ (A_{1g}) and d_{xy}, $d_{x^2-y^2}$ (E_{2g}) metal orbitals are essentially non-bonding. However, there is much disagreement on the relative energies of the orbitals[513], e.g. Dahl and Ballhausen put the order $E_{2g} > E_{1g} > A_{1g}$ (most stable) for ferrocene. The relative energies of the orbitals change in going from nickelocene to titanocene.

It is convenient to consider the electron configurations and the magnetic moments of the metallocenes together. In Table 24 are given the magnetic moments, number of unpaired electrons and expected magnetic moments. One would expect some moments to be greater than the "spin only" value due to orbital contribution. If the unpaired electron is in an orbital where the magnetic quantum number m is zero (e.g. $4s$, $4p_z$, $4d_{z^2}$), then no orbital contribution would be expected; if the $d_{x^2-y^2}$ and d_{xy} orbitals each have one electron, then

529 E. M. Shustorovich and M. E. Dyatkina, *Zhur. Strukt. Khim.* 7 (1966) 139.

TABLE 24. MAGNETIC MOMENTS AND POSSIBLE ELECTRON CONFIGURATIONS FOR SOME METALLOCENES
AND THEIR CORRESPONDING IONS[1, 510, 511, 513]

	Electron configuration	Number unpaired spins	μ (BM) spin only value	μ (BM) expected	μ (BM) found
$TiCp_2^+$	$(E_{2g})^1$	1	1.73	>1.73	2.3
$TiCp_2$	?	0	0	0	0
VCp_2^+	$(E_{2g})^2$	2	2.83	2.83	2.86
VCp_2	$(A_{1g})^1(E_{2g})^2$	3	3.87	3.87	3.78 (3.84)
$CrCp_2^+$	$(A_{1g})^1(E_{2g})^2$	3	3.87	3.87	3.73 (3.81)
$CrCp_2$	$(A_{1g})^2(E_{2g})^2$ or $(A_{1g})^1(E_{2g})^3$	2	2.83	>2.83	3.02
$MnCp_2$	ionic	5	5.9	5.9	5.8–5.9 [a]
$FeCp_2^+$	$(A_{1g}')^2(E_{2g})^3$	1	1.73	>1.73	2.34 (2.26)
$FeCp_2$	$(A_{1g}')^2(E_{2g})^4$	0	0	0	0
$CoCp_2^+$	$(A_{1g}')^2(E_{2g})^4$	0	0	0	0
$CoCp_2$	$(A_{1g}')^2(E_{2g})^4(A_{1g}^*)^1$	1	1.73	1.73	1.76
$NiCp_2$	$(A_{1g}')^2(E_{2g})^4(E_{1g}')^2$ or $(A_{1g}')^2(E_{2g})^4(E_{2u})^2$	2	2.83	2.83	2.86
$TiCp_2$picrate	$(E_{2g})^1$	1	1.73	1.73	2.3
VCp_2^+	$(E_{2g})^2$	2	2.83	ca. 2.83	2.86±0.06

[a] Anti-ferromagnetic with a Néel temperature of 134°, but some magnetic interaction occurs above the Néel temperature. Above the melting point the Curie law is obeyed when $\mu = 5.86$ BM. Also in solution $\mu = ca.$ 5.8 BM.

again no orbital contribution would be expected since the effect of one electron will be cancelled by the effect of the other. With unpaired electrons in other situations, however, some orbital contribution would be expected. The presence or absence of "orbital contribution" has been used to decide between possible electronic ground states. Thus the ferricineum ion has a magnetic moment greater than 2.25 BM, showing some orbital contribution and favouring a configuration of $(A_{1g})^2(E_{2g})^3$ rather than $(E_{2g})^4(A_{1g})^1$ [530].

The e.s.r. spectrum of vanadocene confirms the presence of three unpaired electrons and shows that these electrons spend less than 1% of their time in the 4s orbital.

5.1.3c. Stability

The enthalpy of formation of gaseous ferrocene has been determined as $\Delta H_{298} = +50.61$ kcal mole⁻¹ and $+51.3$ kcal mole⁻¹ [531]. For gaseous nickelocene the enthalpy of formation is $\Delta H_{298} = 85.9$ kcal mole⁻¹ (values from spectroscopic data). These results show that the metal–ring bonding is less strong in nickelocene than in ferrocene.

5.1.4. Physical Measurements

The magnetic moments of dicyclopentadienyl–metal complexes have already been discussed. ESR and thermochemical studies have also been mentioned briefly above; for further details the reader is recommended to consult one of the reviews[1, 510–513].

[530] D. A. Levy and L. E. Orgel, *Mol. Phys.* 4 (1961) 93.
[531] J. W. Edwards and G. L. Kington, *Trans. Faraday Soc.* 58 (1962) 1323

5.1.4a. *Infrared and Raman Spectra*

For a thorough discussion of the applications of infrared and Raman spectroscopy to cyclopentadienyl–metal chemistry see the review by Fritz[532]. The infrared and Raman spectra of ferrocene and tetradeuterioferrocene have been studied in detail[97] and assignments made[532, 533]. Similarly, assignments have been made for the infrared, and sometimes the Raman, spectra of other cyclopentadienyl–metal complexes[532].

Table 25 gives the infrared absorption frequencies and assignments for ferrocene, ruthenocene and nickelocene. From these frequencies it has been estimated that the metal–ring bonding is stronger in ruthenocene than in ferrocene, but weaker in nickelocene than in ferrocene[513, 532, 533].

TABLE 25. SOME INFRARED ACTIVE FREQUENCIES (cm^{-1}) OF FERROCENE, RUTHENOCENE AND NICKELOCENE [513, 532–534]

	$FeCp_2$	$RuCp_2$	$NiCp_2$
C–H stretch	3086	3097	3052
C–C stretch	{ 1408 1351	1410	1421
Asymm. ring breath	1104	1101	1109
C–H bond ($\parallel$) [a]	1001	1002	1002
C–H bond ($\perp$) [a]	814 (811)	821	772
Antisymm. ring tilt	490	446 (528)	?
Antisymm. metal–ring stretch	478	379 (446)	355

[a] $\parallel$ and $\perp$ refer to vibrations parallel or perpendicular to the five-fold axis of symmetry.

Infrared absorption spectroscopy is, of course, very useful in the study of substituted cyclopentadienyl–metal complexes. One important rule for substituted ferrocenes is the so-called "9–10 μ rule". Two absorption peaks near 9 and 10 μ (1100 and 1000 cm^{-1}) are found for the unsubstituted cyclopentadienyl ring (see Table 25), but are absent when the ring is substituted[513]. This rule can be used to distinguish between the two possibilities [5.1.13] and [5.1.14] for disubstituted ferrocene derivatives; only [5.1.13] would show bands at 9 and 10 μ.

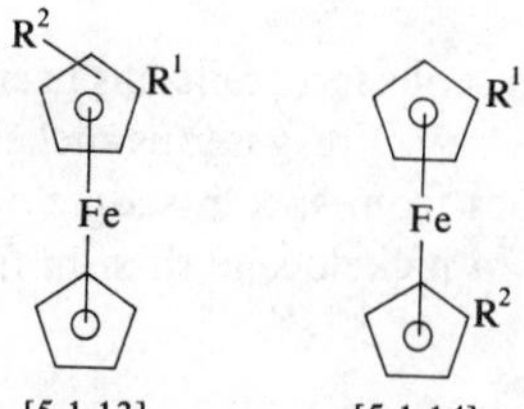

[5.1.13]   [5.1.14]

There is little coupling between the vibrational modes of substituents on one ring with those of substituents on the second ring.

532 H. P. Fritz, *Advances Organomet. Chem.* **1** (1964) 240.
533 E. R. Lippincott and R. D. Nelson, *Spectrochim. Acta* **10** (1958) 307.
534 T. V. Long and F. R. Huege, *Chem. Commun.* (1968) 1239.

For a discussion of the infrared absorption bands v(M–H) of cyclopentadienylmetal hydrides, v(CO) of cyclopentadienylmetal carbonyls or v(NO) of cyclopentadienylmetal nitrosyls, etc., see the appropriate chapter.

5.1.4b. *Nuclear Magnetic Resonance Spectroscopy*

The chemical shift (τ) values for the iron group metallocenes in carbon tetrachloride solution are iron (5.96), ruthenium (5.58) and osmium (5.29). The five protons of a π-cyclopentadienyl ring in a metal complex remain equivalent. As discussed in section 1, σ-cyclopentadienyl–metal complexes (h'-C_5H_5) show fluxional behaviour and variable temperature n.m.r. spectra.

In a monosubstituted ferrocene the 2- and 3-protons may have appreciable chemical shift differences. If the substituent is electron withdrawing, the proton in the 2-position will absorb at lower field and the four-ring protons will give an AA'BB' pattern. This pattern frequently consists of a pair of triplets, since the chemical shift (Δv) difference between A and B protons is much larger than the coupling constants and $J_{adj} \approx J_{cross}$.

The proton magnetic resonance spectrum of the ferricenium ion is too broad to be observed, but by measuring the chemical shift of the Cp protons for a $FeCp_2/FeCp_2^+$ mixture and extrapolating to zero concentration of $FeCp_2$ the chemical shift for $FeCp_2^+$ has been estimated at $-10\ \tau$ [535].

Wide-line n.m.r. measurements on ferrocene are consistent with free rotation of the rings[536].

Some n.m.r. contact shifts in the bis(methylcyclopentadienyl)–metal complexes $M(C_6H_7)_2$ (M = V, Cr, Co or Ni) have also been measured. Both the methyl and ring proton resonances are extremely broad. The methyl resonances are shifted downfield with respect to the aromatic protons of toluene as internal standard. For vanadium and chromium the methylcyclopentadienyl ring protons are shifted downfield, but for cobalt or nickel they are shifted upfield. The shifts have not been satisfactorily explained.

The cyclopentadienyl resonance can be split by coupling with a metal nucleus of spin $I = \frac{1}{2}$. For $RhCp(C_5H_6)$ the coupling $J(^{103}Rh–H)$ is less than 1 Hz, but with $PtICp(CO)$ $J(^{195}Pt–H)$ is 18.5 Hz [537]. With $ReHCp_2$ the cyclopentadienyl resonance occurs at τ 6.0 and is split by the single (hydridic) hydrogen.

5.1.4c. *Mass Spectrometry*

The mass spectra of MCp_2 (M = Mg, V, Cr, Mn, Fe, Co Ni or Ru) show the molecular ion and fragments due to the loss of the cyclopentadienyl group[538]. The low relative abundance of the molecular ion from $MgCp_2$ and $MnCp_2$, compared with the other compounds MCp_2, is in accordance with the relatively weak and ionic metal–ring bonding. The appearance potential of MCp^+ or M^+ from MCp_2 is less for M = Ni than for M = Fe, in agreement with the weaker metal–ligand bonding. The fragmentation patterns of $CoCp(CO)_2$, $MnCp(CO)_3$ and $VCp(CO)_4$ show that the carbonyls are more readily lost than the ring[539].

[535] H. M. Rosenberg and C. Riber, *Michrochem. J.* **6** (1962) 103.
[536] L. N. Mulay, E. G. Rochow and E. O. Fischer, *J. Inorg. Nucl. Chem.* **4** (1957) 231.
[536a] M. F. Rettig and R. S. Drago, *Chem. Commun.* (1966) 891.
[537] H. P. Fritz and C. G. Kreiter, *Chem. Ber.* **96** (1963) 2008.
[538] M. I. Bruce, "Mass spectra of organometallic compounds", in *Advances Organomet. Chem.* **6** (1968) 273.
[539] R. E. Winters and R. W. Kiser, *J. Organomet. Chem.* **4** (1965) 190.

Similar behaviour is shown by $Fe_2Cp_2(CO)_4$ and $Mo_2Cp_2(CO)_6$. The absence of any dimetallic fragments in the mass spectrum of $Cr_2Cp_2(CO)_6$ suggests that the complex is monomeric in the vapour state and that the Cr–Cr bond is weak[540]. The fragmentation patterns of some organometallic complexes sometimes show ions of greater mass than the parent complex. It has been suggested that these ions of high mass are formed by ion association. A more likely explanation is that the compound being investigated decomposes to something of greater mass in the spectrometer inlet system, e.g. $Ni_2Cp_2(CO)_4$ shows peaks in its mass spectrum corresponding to greater mass than $Ni_2Cp_2(CO)_4$ and probably arises by fragmentation of $Ni_3Cp_3(CO)_2$, formed in the inlet system.

The mass spectra of a large number of substituted ferrocenes have been studied. It has been suggested that the presence or absence of the ion $FeCp^+$ is a more sensitive method of distinguishing between homo- and hetero-annular substituted ferrocene than the "9–10 μ rule"[538].

5.1.5. Reactions

5.1.5a. *Electron Transfer Reactions*

All the transition metal metallocenes, MCp_2, can be oxidized to cations and frequently the oxidations are reversible. Ferrocene is easily oxidized to the ferrocenium ion ($FeCp_2^+$), which is blue or green in dilute solution but blood red in more concentrated solutions[513]. The kinetics of the oxidation of $FeCp_2 \rightarrow FeCp_2^+$ by iodine have been studied. The electron exchange between ferrocene and the ferrocenium ion is second order and very fast ($k > 7 \times 10^6$ at 25°)[541]. Even at $-75°$ the half-life of the exchange is only a few milliseconds. Similarly, the electron exchange between $CoCp_2$ and $CoCp_2^+$ is very fast.

The redox potentials for the process

$$MCp_2 \rightarrow MCp_2 + e$$

have been determined polarographically for a number of metallocenes[510, 511]. Some values are given in Table 26.

Chronopotentiometric studies of the iron group metallocenes in acetonitrile suggest an increase in oxidation potential from ferrocene to osmocene. There have been a large number of studies on the redox potentials of substituted ferrocene/substituted ferrocenium ion systems. Frequently the potentials correlate well with Hammett and Taft constants for the substituents, but not always[513, 542, 543].

Chronopotentiometric studies on ruthenocene show that a dipositive ion $[RuCp_2]^{2+}$ is formed. Osmocene similarly gives $[OsCp_2]^{2+}$ but also $[OsICp_2]^+$ or $[Os(OH)Cp_2]^+$ [513].

5.1.5b. *Charge–Transfer Complexes*

Ferrocene forms charge–transfer complexes with electron acceptors such as tetracyanoethylene (TCNE), 2,3-dichloro-5,6-dicyanoquinone or nitrobenzenes. Mössbauer spectroscopy suggests that the ferrocenium ion is not present in the 1:1 ferrocene:TCNE complex[544]. The X-ray structure shows the TCNE to be bonded to one of the rings and not to the iron. The TCNE is planar and 3.1 Å from the Cp ring[545].

[540] R. B. King, *J. Am. Chem. Soc.* **88** (1966) 2075.
[541] D. R. Stranks, *Discussions Faraday Soc.* **29** (1960) 73.
[542] T. Kuwana, D. E. Bublitz and G. Hoh, *J. Am. Chem. Soc.* **82** (1960) 5811.
[543] W. F. Little, C. N. Reilley, J. D. Johnson and A. P. Sanders, *J. Am. Chem. Soc.* **86** (1964) 1382.
[544] R. L. Collins and R. Pettit, *J. Inorg. Nucl. Chem.* **29** (1967) 503.
[545] E. T. Adman, *Dissertation Abstracts* B (1968) 2784.

TABLE 26. POLAROGRAPHIC HALF-WAVE POTENTIALS (IN VOLTS) FOR THE PROCESSES $MCp_2^+ + e = MCp_2$ [a]

Measured in water unless stated otherwise

	Reduction	Oxidation	Reference
Ti	-0.44	-0.44	G. Wilkinson and J. M. Birmingham, J. Am. Chem. Soc. 76 (1954) 4281
V	-0.32		G. Wilkinson and J. M. Birmingham, J. Am. Chem. Soc. 76 (1954) 4281
Fe [b]	+0.30	+0.31	J. A. Page and G. Wilkinson, J. Am. Chem. Soc. 74 (1952) 6149
Co	-1.16		J. A. Page and G. Wilkinson, J. Am. Chem. Soc. 74 (1952) 6149
Ni [b]	-0.21	-0.08	G. Wilkinson, P. L. Pauson and F. A. Cotton, J. Am. Chem. Soc. 76 (1954) 1970
Ru [b]	+0.22	+0.26	J. A. Page and G. Wilkinson, J. Am. Chem. Soc. 74 (1952) 6149
Rh	-1.53	—	F. A. Cotton, R. O. Whipple and G. Wilkinson, J. Am. Chem. Soc. 75 (1953) 3586

[a] Measured against the standard calomel electrode.

[b] In 90% ethanol containing a sodium perchlorate/perchloric acid electrolyte.

5.1.5c. *Protonation*

In a strongly acidic boron trifluoride hydrate solution ferrocene and ruthenocene are protonated to give the metal hydride species $[MHCp_2]^+$ [546]. The evidence for the formation of these hydride species comes principally from the n.m.r. spectra of these acidic solutions. The proposed structure for these hydrides is shown in [5.1.15], i.e. with the cyclopentadienyl

[5.1.15]   [5.1.16]

rings pushed over to one side and no longer parallel and the two filled non-bonding valence orbitals concentrated on the other side. Support for this structure comes from the structure of MoH_2Cp_2 [5.1.16], which shows a similar arrangement of the cyclopentadienyl rings to [5.1.15]. Hydrodi(cyclopentadienyl)rhenium is more easily protonated than ferrocene or ruthenocene, being about as strong a base as ammonia. The protonated species $[ReH_2Cp_2]^+$ presumably has an analogous structure to [5.1.16]. MoH_2Cp_2 and WH_2Cp_2 are similarly protonated to give $[MH_3Cp_2]^+$ (M = Mo or W), but TaH_3Cp_2 (d°-electron configuration) is not. For a more complete discussion of the formation of metal hydrides by protonation of cyclopentadienylmetal or cyclopentadienylmetal carbonyl complexes see the appropriate chapter.

5.1.5d. *Reduction*

Ferrocene is remarkably resistant to catalytic hydrogenation. Thus a rhodium catalyst/ hydrogen system which will hydrogenate benzene in 5 min will not hydrogenate ferrocene after several weeks. However, alkali metals in amine solutions, such as lithium in ethyl-amine, will reduce ferrocene rapidly to iron and the cyclopentadienide anion[547].

5.1.5e. *Cleavage*

The cyclopentadienyl ring is difficult to cleave from the metal in the iron group metallo-cenes and in the cobalticineum ion. In some cyclopentadienylmetal complexes, however, the ring is labile. For example, manganocene, in which the metal–ring bonding is mainly ionic, is easily hydrolysed by water and reacts very rapidly with ferrous chloride to give ferrocene[548]. Vanadocene and chromocene also have labile cyclopentadienyl rings; chromo-cene reacts with ferrous chloride to give ferrocene[510, 511]. Hydrogen chloride removes the cyclopentadienyl ring from $Pd(C_3H_5)Cp$ to give $Pd_2Cl_2(C_3H_5)_2$ [427]. $Pd(C_3H_5)Cp$ also reacts with ferrous chloride to give ferrocene.

A cyclopentadienyl ring can be cleaved from $TiCl_2Cp_2$ by a variety of reagents, e.g. EtOH, $TiCl_4$, Br_2 or $Al_2Cl_2Bu^i_4$ [549], as shown.

[546] T. J. Curphey, J. O. Santer, M. Rosenblum and J. H. Richards, *J. Am. Chem. Soc.* **82** (1960) 5249.
[547] D. S. Trifan and L. Nicholas, *J. Am. Chem. Soc.* **79** (1957) 2746.
[548] G. Wilkinson, F. A. Cotton and J. M. Birmingham, *J. Inorg. Nucl. Chem.* **2** (1956) 95.
[549] R. Feld and P. L. Cowe, *The Organic Chemistry of Titanium*, Butterworths, London (1965).

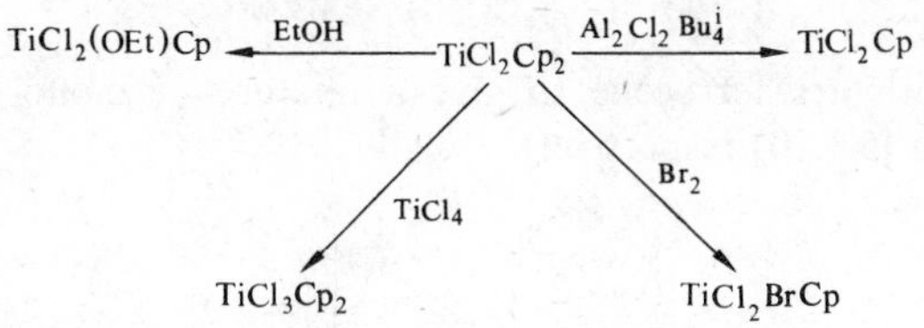

$$TiCl_2(OEt)Cp \xleftarrow{\text{EtOH}} TiCl_2Cp_2 \xrightarrow{Al_2Cl_2Bu_4^i} TiCl_2Cp$$

with $TiCl_4$ branching down to $TiCl_3Cp_2$ and Br_2 branching down to $TiCl_2BrCp$

There are several reactions of chromocene in which a cyclopentadienyl–chromium ring is broken, e.g.

$$CrCl_2Cp(amine) \xleftarrow{\text{CCl}_4/\text{amine}} CrCp_2 \xrightarrow{\text{CCl}_4} [CrCl_3Cp]^-$$

Nickelocene also undergoes reactions involving cleavage of one of the cyclopentadienyl rings.

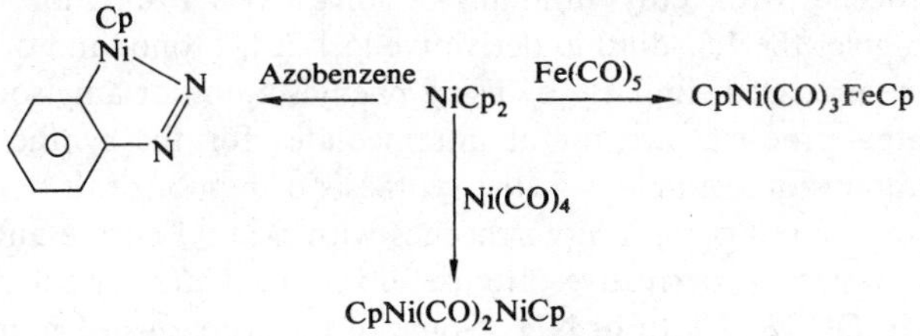

$$\xleftarrow{\text{Azobenzene}} NiCp_2 \xrightarrow{\text{Fe(CO)}_5} CpNi(CO)_3FeCp$$

$$\downarrow Ni(CO)_4$$

$$CpNi(CO)_2NiCp$$

5.1.5f. *Conversion to Cyclopentadiene–Metal Complexes*

Attack on cobaltocene to give substituted cyclopentadiene–metal compounds has been discussed in section 4, e.g. the borohydride reduction of $CoCp_2$ to give $CoCp(C_5H_6)$.

Rhodocene is an extremely reactive "radical" which can be isolated at liquid nitrogen temperatures but dimerizes to give [5.1.17] at room temperature[550].

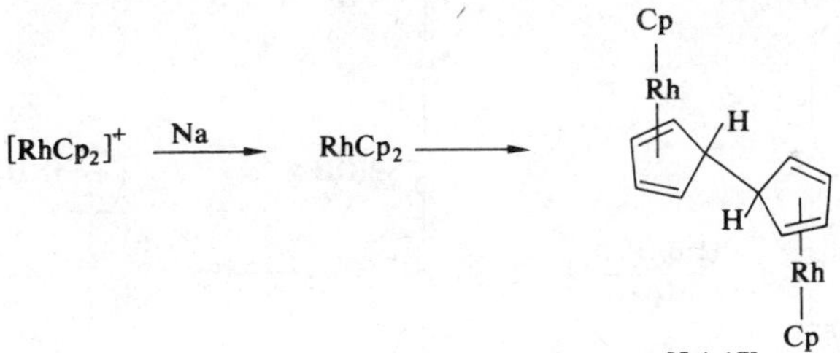

$$[RhCp_2]^+ \xrightarrow{\text{Na}} RhCp_2 \longrightarrow$$

[5.1.17]

Iridocene behaves similarly.

Sodium borohydride reduction of $[FeCp(CO)_2(PPh_3)]^+$ gives the cyclopentadiene complex [5.1.18] [551].

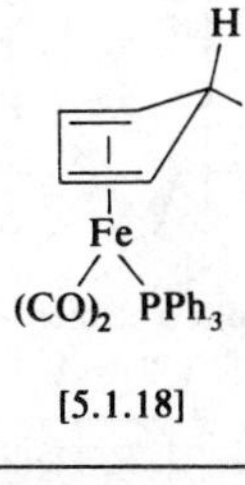

[5.1.18]

550 E. O. Fischer and H. Wawersik, *J. Organomet. Chem.* 5 (1966) 559.
551 A. Davison, M. L. H. Green and G. Wilkinson, *J. Chem. Soc.* (1961) 3172.

5.1.5g. *Metallation*

n-Butyl-lithium lithiates ferrocene to give a mixture of mono- and 1,1'-di-lithiated products [5.1.19] and [5.1.20] respectively.

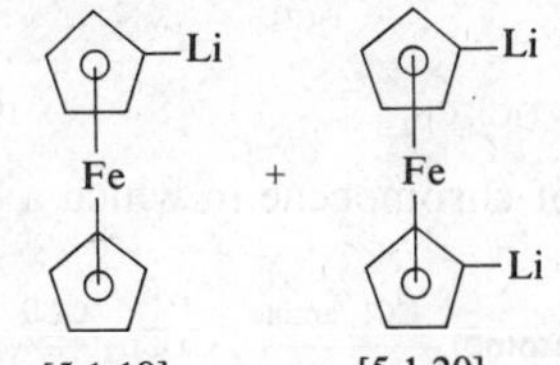

[5.1.19]   [5.1.20]

Mono-lithioferrocene [5.1.19], free from [5.1.20], can be prepared by treating monochloromercuriferrocene with ethyl-lithium[552]. Similarly 1,1'-di(chloromercuri)ferrocene with ethyl-lithium gives the 1,1'-dilithio derivative [5.1.20]. Mono- and di-sodio derivatives of ferrocene can be prepared similarly by using phenylsodium or amylsodium.

These metallated products are useful intermediates for the synthesis of substituted ferrocenes. One important example is in the synthesis of mono- or di-boronic acid derivatives by treating the mono- or di-lithioferrocenes with n-butyl borate and hydrolysing the product. These boronic acid derivatives are useful intermediates for substituting the ferrocene rings with OH, OCOR, Cl, Br or NH_2. Some of the reactions of mono-lithioferrocene and ferroceneboronic acid are summarized in Fig. 41 [513, 514, 553].

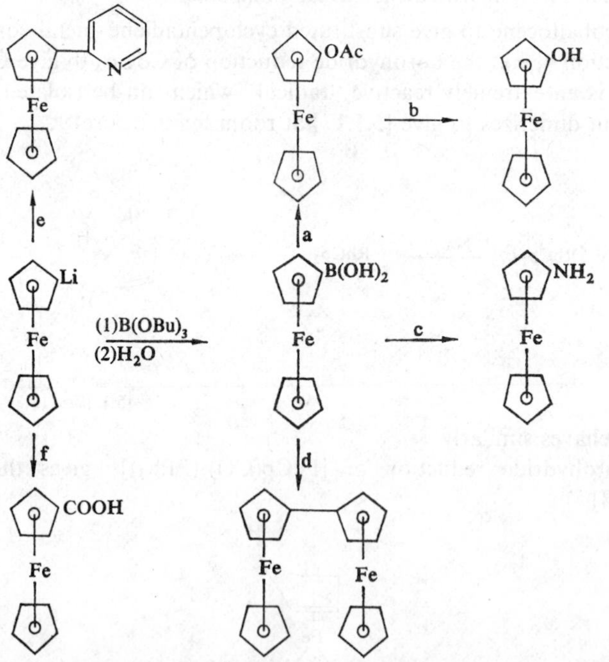

FIG. 41. Some reactions of lithioferrocene and ferroceneboronic acid. (a) $Cu(OAc)_2$; (b) hydrolysis; (c) copper phthalimide followed by hydrolysis; (d) Ag_2O/NH_3; (e) pyridine; (f) CO_2.

552 D. Seyferth, H. D. Hoffman, R. Burton and J. F. Helling, *Inorg. Chem.* **1** (1962) 227.
553 A. N. Nesmeyanov, V. A. Sazonova and A. V. Gerasimenko, *Dokl. Akad. Nauk SSSR Otdel Khim. Nauk* **147** (1962) 634.

Hydrodi(cyclopentadienyl)rhenium is also lithiated by an excess of n-butyl-lithium to give the 1,1'-dilithiated derivative [5.1.21].

[5.1.21]

When ferrocene is treated with mercuric acetate a mixture of acetoxy- and 1,1'-di-(acetoxymercuri)-ferrocenes is formed. This mixture when treated with potassium or lithium chloride is converted into a mixture of the corresponding chloromercuri derivatives[554]. Chloromercuriferrocene has been converted into many other substituted ferrocenes; some examples are given in Fig. 42.

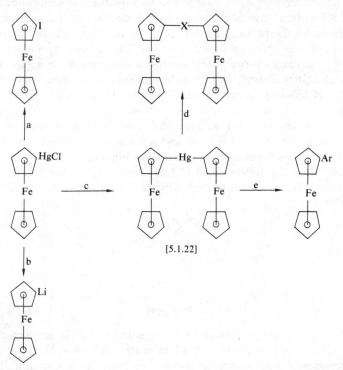

[5.1.22]

FIG. 42. Some reactions of chloromercuriferrocene and diferrocenylmercury [5.1.22]. (a) I_2; (b) LiEt; (c) reduction by $Na_2S_2O_3$; (d) sulphur gives X = S; $SeCl_4$ gives X = Se; (CNS)$_2$ gives X = S–S, etc.; (e) $HgAr_2/Ag$.

5.1.5h. Aromatic Substitution Reactions

Shortly after its discovery dicyclopentadienyliron was found to undergo Friedel–Crafts acylation very readily. The name ferrocene was coined because of this aromatic behaviour.

[554] M. D. Rausch, M. Vogel and H. Rosenberg, J. Org. Chem. 22 (1957) 900.

Since then a very extensive organic (aromatic) chemistry of ferrocene has been developed. Some of this chemistry is discussed below. To discuss it in detail would be out of place here, but several reviews are available[513–516, 555, 556]. Although most of the work has been with ferrocene or its derivatives, cyclopentadienyl rings bonded to other metals also have aromatic character, e.g. in

$VCp(CO)_4$, $CrCp(CO)_2NO$, WH_2Cp_2, $MnCp(CO)_3$, $TiCp(CO)_3$, $ReCp(CO)_3$ and $ReHCp_2$ [513, 514, 557].

Some aromatic substitution reactions of ferrocene. Ferrocene is very readily acylated by acid anhydrides or acid halides in the presence of the usual Friedel–Craft catalysts; boron trifluoride, aluminium chloride, etc. Competitive studies show that ferrocene acylates 3.3×10^6 times faster than benzene[558]. There is a tendency to produce 1,1'-disubstituted products, but several procedures for getting monoacyl derivatives have been devised; one of these is to treat ferrocene with the acid anhydride and phosphoric acid[559].

1,1'-Diacylferrocenes are formed by treating ferrocene with the acid anhydride and aluminium chloride. A very small amount (1–2%) of 1,2-diacetylferrocene is formed, in addition to 1,1'-diacetylferrocene, when ferrocene is treated with acetic anhydride. Acylation of substituted ferrocenes gives the heteroannular 1'-acyl isomer if the substituent is electron withdrawing. An alkyl substituent tends to promote substitution in the 3-position, although substitution in all the other positions also occurs. The carbonyl groups of acyl-substituted ferrocenes undergo reactions completely analogous to those given by acyl-substituted arenes. The organic chemistry of acylated ferrocenes and their derivatives is very extensive[513, 514].

Alkylation of ferrocene under Friedel–Craft conditions tends to give mixtures of poly-alkylated ferrocenes. These can be separated, but many other, and better, routes to alkylated ferrocenes are available, e.g. via acylated ferrocenes or by using alkyl-substituted cyclo-pentadienes. There are several routes to aryl-substituted ferrocenes. The most commonly used method is to treat ferrocene with an aryldiazonium salt.

Biferrocenyl [5.1.23], terferrocenyl [5.1.24] and polyferrocenyl can be formed in a number of ways. One method is to treat a mixture of bromo- and 1,1'-dibromo-ferrocene with copper[560].

[5.1.23]   [5.1.24]

Haloferrocenes cannot be formed by direct halogenation of ferrocene because of the ease of oxidation to the ferrocenium ion. Many routes are known, however; one is to treat a ferroceneboronic acid with a cuprous halide. Similarly, direct nitration of ferrocene has

555 K. Plesske, *Angew. Chem. Int. Edn.* 1 (1962) 312, 394.

556 M. D. Rausch, *Reactions of Coordinated Ligands*, Advances in Chemistry Series, No. 37, Am. Chem. Soc. (1963).

557 R. L. Cooper, M. L. H. Green and J. T. Moelwyn-Hughes, *J. Organomet. Chem.* 3 (1965) 261.

558 M. Rosenblum, J. O. Santer and W. G. Howells, *J. Am. Chem. Soc.* 85 (1963) 1450.

559 P. J. Graham, R. V. Lindsey, G. W. Parshall, M. L. Peterson and G. M. Whitman, *J. Am. Chem. Soc.* 79 (1957) 3416.

560 A. N. Nesmeyanov, V. N. Drozd, V. A. Sazonova, V. I. Romanenko, A. K. Prokofev and L. A. Nikonova, *Chem. Abs.* 59 (1963) 7556.

not yet proved possible because the ferrocene is oxidized to the ferrocenium ion. However, nitroferrocene is formed by treating lithioferrocene with n-propyl nitrate or with dinitrogen tetroxide at $-70°$. Direct sulphonation of ferrocene is possible, however, either by treatment with chlorosulphonic acid or with a dioxan-sulphur trioxide complex in ethylene dichloride. Mono- and the 1,1'-disulphonic acids are known.

Amino-, hydroxy- and carboxylato-ferrocene derivatives have been mentioned briefly in Fig. 41. Ferrocenes substituted by silicon-, germanium-, phosphorus-, arsenic-, sulphur- or selenium-containing groups are known. Many compounds in which one of the cyclopentadienyl rings is fused or bonded to a heterocyclic ring are also known[513].

5.1.5i. *Aromatic Substitution of Cyclopentadienyl–Metal Complexes other than Ferrocene*

Ruthenocene undergoes electrophilic substitution less readily than ferrocene, although the ring (or rings) may be acylated, lithiated, mercuriated or carboxylated, etc. Osmocene on acylation using Friedel–Craft catalysts gives only the monoacyl derivatives, even under forcing conditions. Osmocene can be mono- or di-lithiated by n-butyllithium[513].

The cyclopentadienyl ring of cyclopentadienyl(tricarbonyl)manganese also shows aromatic reactivity. Thus when treated with mercuric acetate in the presence of calcium chloride the monochloromercuriated derivative $Mn(C_5H_4HgCl)(CO)_3$ is formed[514]. The ring is readily acylated by an acid chloride or an acid anhydride in the presence of a Friedel–Craft catalyst[561]. The ring can also be alkylated, e.g. with ethyl bromide and aluminium chloride ethylcyclopentadienyl(tricarbonyl)manganese is formed[562]. Cyclopentadienyl-(tricarbonyl)manganese can also be sulphonated.

The cyclopentadienyl ring of cyclopentadienyl(tricarbonyl)rhenium can be acylated, sulphonated or metallated. Cyclopentadienyl(tricarbonyl)technetium and cyclopentadienyl-(tetracarbonyl)vanadium can be acylated under Friedel–Craft conditions.

Competitive acylation studies show the following order of increasing reactivity:

$$ReCp(CO)_3 < C_6H_6 \sim VCp(CO)_4 < OsCp_2 < MnCp(CO)_3 < RuCp_2 < anisole < FeCp_2 \text{ }[563].$$

5.1.5j. *α-Carbonium Ion Stabilization in Substituted Ferrocenes*

An important feature of the chemistry of substituted ferrocenes is the great stability of α-ferrocenyl carbonium ions. Thus the α-metallocenylethyl acetates [5.1.25] (M = Fe, Ru or Os) are solvolysed in 80% acetone/water by an S_N1 mechanism at faster rates than triphenylmethyl acetate.

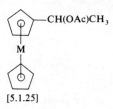

[5.1.25]

[561] E. O. Fischer and K. Pleszke, *Chem. Ber.* **91** (1958) 2719.

[562] A. N. Nesmeyanov, K. N. Anesimov and Z. P. Valueva, *Izv. Akad. Nauk SSSR Otd Khim. Nauk* (1961) 1780; *Chem. Abs.* **56** (1962) 8733.

[563] E. O. Fischer, M. von Foerster, C. G. Kreiter and K. E. Schwarzhans, *J. Organomet. Chem.* **7** (1967) 113.

The rates, relative to the rate for triphenylmethyl acetate, are Fe (6.6), Ru (9.0) and Os (35) [564]. There is a very interesting difference in solvolysis rate between the *exo*- and the *endo*-acetates, [5.1.26] and [5.1.27] respectively. The *exo*-acetate solvolyses 2500 times faster than the *endo*-acetate and both give the *exo*-carbinol [5.1.28], formed by an α-carbonium ion intermediate.

[5.1.26]   [5.1.27]   [5.1.28]

The α-carbonium ion is probably stabilized by delocalization of ring electrons and/or metal electrons into the vacant *p*-orbital of the α-carbon atom.

5.1.5k. *Some Reactions of the Other Ligands*

Cyclopentadienyl–metal complexes undergo many reactions in which other ligands on the metal are displaced. These reactions are more appropriately discussed in other chapters and will only be mentioned here. A carbonyl group of a cyclopentadienylmetal carbonyl can often be displaced by another neutral ligand such as a tertiary phosphine, pyridine, an olefin, a diolefin, an acetylene, etc. An example is

$$CoCp(CO)_2 \xrightarrow{\text{diolefin}} CoCp(\text{diolefin})$$

Such displacements of carbon monoxide are frequently promoted by ultraviolet light.

Binuclear cyclopentadienylmetal carbonyl complexes with metal–metal bonds can be reduced to cyclopentadienylmetal carbonylate ions. These carbonylate ions are useful reactive intermediates[7a]. An example is

$$Mo_2Cp_2(CO)_6 \xrightarrow{\text{Na/Hg}} [MoCp(CO)_3]^- \xrightarrow[\text{halide}]{\text{Alkyl}} MoRCp(CO)_3$$

A halide ligand of a cyclopentadienylmetal halide or of a cyclopentadienylmetal carbonyl halide can often be readily displaced by nucleophiles, e.g. by other halides, or by carboxylate, alkyl, aryl, ethynyl, alkoxy, hydro, hydroxy, an olefin, etc.

5.1.5l. *Some Addition Reactions of Nickelocene*

Nickelocene is a very reactive molecule and a number of reactions involving addition to one of its cyclopentadienyl rings are known[566]. Three of these addition reactions are shown in Fig. 43.

5.2. PENTADIENYL- AND SUBSTITUTED PENTADIENYL–METAL COMPLEXES, EXCLUDING THE CYCLOPENTADIENYLS

The pentadienyliron tricarbonyl cation $[Fe(CO)_3(C_5H_7)]^+$ and its analogues have been mentioned in section 4.1.5. They can be formed either by hydride ion abstraction from a conjugated diolefin–metal complex, $Fe(CO)_3(\text{diolefin})$, or by treating a diolefin carbinol

[564] W. M. Horspool and R. G. Sutherland, *Chem. Commun.* (1967) 786 and references therein.
[565] E. A. Hill and J. H. Richards, *J. Am. Chem. Soc.* **83** (1961) 4216.
[566] D. W. McBride, E. Dudek and F. G. A. Stone, *J. Chem. Soc.* (1964) 1752 and references therein.

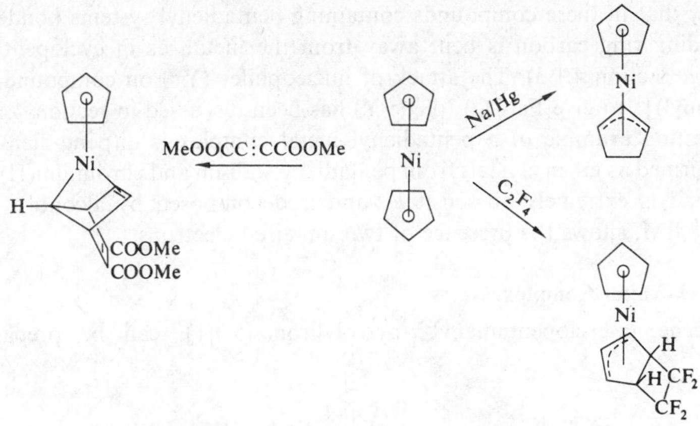

FIG. 43. Some addition reactions of nickelocene.

complex, $Fe(CO)_3$(diolefin carbinol), with acid. The cyclohexadienyl– and cycloheptadienyl–iron tricarbonyl cations, [5.2.1] and [5.2.2] respectively, can be made similarly[567, 568].

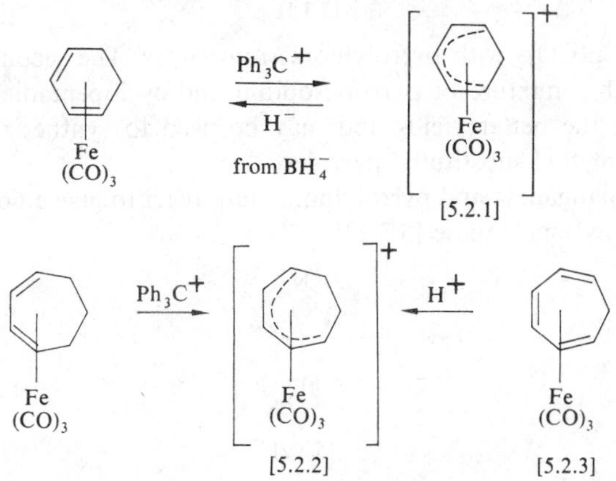

The cycloheptadienyl complex [5.2.2] can also be made by protonation of $Fe(CO)_3$-(cyclohepta-1,3,5-triene), [5.2.3], with fluoroboric acid.

Other cyclohexadienyl–metal complexes are known, e.g.

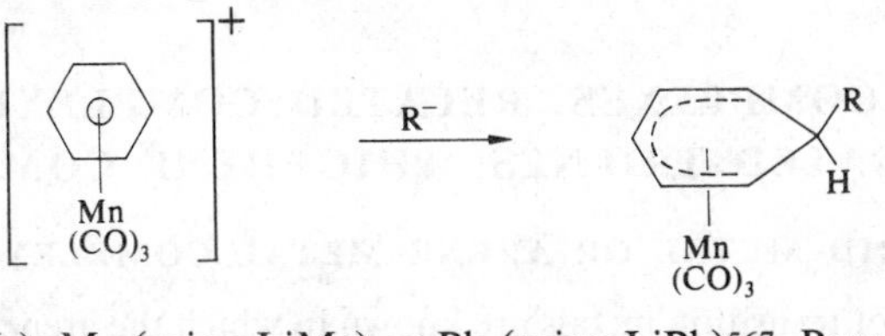

$R = H$ (using $NaBH_4$), Me (using LiMe) or Ph (using LiPh)[567]. Borohydride reduction of $[Re(C_6H_6)_2]^+$ gives $Re(C_6H_6)$(cyclohexadienyl), whilst reduction of $[Ru(C_6H_6)_2]^{2+}$ gives a mixture of Ru(cyclohexadienyl)$_2$ (15%) and Ru(benzene)(cyclohexadiene) (75%)[569].

[567] E. O. Fischer and R. D. Fischer, Angew. Chem. 72 (1960) 919.
[568] H. J. Dauben and D. J. Bertelli, J. Am. Chem. Soc. 83 (1961) 497.
[569] D. Jones, L. Pratt and G. Wilkinson, J. Chem. Soc. (1962) 4458.

It is likely that in these compounds containing pentadienyl systems bonded to a metal the non-bonding ring carbon is bent away from the metal, as in cyclopentadiene–metal complexes (see section 4.1.3). The attack of nucleophiles (Y^-) on compounds of the type $[Fe(CO)_3(dienyl)]^+$ to give $Fe(CO)_3(diene-Y)$ has been discussed in section 4.1.

An interesting example of a pentadienyl–metal complex is di(pentadienyl)chromium, $Cr(C_5H_7)_2$, formed as green crystals from pentadienylsodium and chromium(II) chloride[569a]. The compound is extremely air-sensitive, and is decomposed by alcohols. Its magnetic moment, 2.78 BM, shows the presence of two unpaired electrons.

5.2.1. Pyrrolyl–Metal Complexes

Azaferrocene, π-cyclopentadienyl-π-pyrrolyliron [5.3.1], can be prepared by two methods.

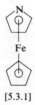

[5.3.1]

One is to treat $FeICp(CO)_2$ with pyrrolylpotassium[514, 570]. The second method is to treat ferrous chloride with a mixture of pyrrolylsodium and cyclopentadienylsodium. The first method gives much the better yields and may be used to synthesize methyl-substituted azaferrocenes from methyl-substituted pyrroles.

Decacarbonyldimanganese and pyrrolylpotassium react to give a 66% yield of azacyclo-pentadienyl(tricarbonyl)manganese [5.3.2][514, 571].

[5.3.2]

Few reactions of these azacyclopentadienyl–metal complexes have been reported. However, azaferrocene is oxidized by dilute nitric acid to an unstable cation. It may also be protonated to give $[FeCp(C_4H_4HN)]^+$.

6. ARENE COMPLEXES, RELATED COMPLEXES WITH CONJUGATED TRIENES, THIOPHENE COMPLEXES

6.1. BENZENOID–METAL OR ARENE–METAL COMPLEXES[510, 511, 527, 572]

Arene complexes of transition metals are known in which the arene occupies three of the normal valence positions of the metal atom, e.g. $Cr(C_6H_6)_2$. In addition, complexes are

569a U. Giannini, E. Pellino and M. P. Lachi, *J. Organomet. Chem.* **12** (1968) 551.
570 K. K. Joshi, P. L. Pauson, A. R. Qazi and W. A. Stubbs, *J. Organomet. Chem.* **1** (1964) 471.
571 K. K. Joshi and P. L. Pauson, *Proc. Chem. Soc.* (1962) 326.
572 H. Zeiss, P. J. Wheatley and H. J. S. Winkler, *Benzenoid–Metal Complexes*, Ronald Press, New York (1966).

known in which there is a weaker interaction or polarization between the metal and the arene ring, e.g. $Ag(C_6H_6)ClO_4$. Compounds of this second type, which exist as stable entities only in the solid state, are discussed briefly in section 6.1.3. In section 6 we are mainly concerned with arene complexes of the first type.

6.1.1. Methods of Preparation

6.1.1a. *The Aluminium Halide/Aluminium Method*

This, the most important method of synthesizing bis-π-arene–metal complexes, is due to E. O. Fischer and W. Hafner[573], who used it to make di-benzenechromium. Chromium trichloride, aluminium chloride, aluminium and benzene when heated together in a bomb at 140° give the di-benzenechromium(I) cation.

$$3CrCl_3 + AlCl_3 + 2Al + 6C_6H_6 \rightarrow 3[Cr(C_6H_6)_2][AlCl_4]$$

The cation can then be reduced to dibenzenechromium with dithionite ion.

$$[Cr(C_6H_6)_2]^+ + S_2O_4^{2-} + 4OH^- \rightarrow Cr(C_6H_6)_2 + SO_3^{2-} + H_2O$$

The yield is improved by using a large excess of aluminium. Alkaline hydrolysis of the reaction mixture gives di-benzenechromium without the need for reduction by dithionite ion.

The aluminium halide/aluminium method has been used to synthesize other compounds of the type $Cr(arene)_2$, with arene = toluene, mesitylene or hexamethylbenzene, etc. The arene must be inert towards aluminium chloride, e.g. when chlorobenzene is used the chlorine is lost, giving di-benzenechromium[574]. The arene ligands of compounds of the type $Cr(arene)_2$ exchange on heating with another arene in the presence of aluminium chloride as catalyst.

$$Cr(arene)_2 + 2arene^* \underset{}{\overset{AlCl_3}{\rightleftharpoons}} Cr(arene^*)_2 + 2arene$$

Hence, since methyl-substituted benzenes form bis-arene–chromium complexes more rapidly than benzene itself, under the $Al/AlCl_3/arene/CrCl_3$ conditions, a trace of mesitylene catalyses the formation of dibenzenechromium and reduces the required temperature markedly[575].

The Fischer–Hafner method has been used for the synthesis of arene derivatives of many metals. Some examples are now given. When titanium tetrachloride is treated with $AlCl_3/Al/C_6H_6$ a violet complex is formed. This has been formulated as a benzene–titanium complex, $TiCl_2(C_6H_6)Al_2Cl_6$ [576], but has also been formulated as a σ-phenylaluminium complex, $TiCl_3 \cdot AlCl_3 \cdot AlCl_2Ph$. Vanadium tetrachloride with $AlCl_3/Al/C_6H_6$ gives $[V^I(C_6H_6)_2][AlCl_4]$. On hydrolysis the di-benzenevanadium(I) cation disproportionates to give di-benzenevanadium(0)[577].

$$5V(C_6H_6)_2^+ \rightarrow 4V(C_6H_6)_2^0 + V^{5+}$$

Di-benzenemolybdenum is prepared in 70% yield by the aluminium chloride/aluminium method. The di-benzenemolybdenum(I) cation is first formed but on alkaline hydrolysis it disproportionates[578].

$$6[Mo(C_6H_6)_2]^+ + 8OH^- \rightarrow 5Mo(C_6H_6)_2 + MoO_4^{2-} + 4H_2O + 2C_6H_6$$

573 E. O. Fischer and W. Hafner, *Z. Naturforsch.* **10b** (1955) 665.
574 R. W. Bush and H. R. Snyder, *J. Org. Chem.* **25** (1960) 1240.
575 E. O. Fischer and J. Seeholzer, *Z. anorg. allgem. Chem.* **312** (1961) 244.
576 H. Martin and F. Vohwinkel, *Chem. Ber.* **94** (1961) 2416.
577 E. O. Fischer and H. P. Kögler, *Chem. Ber.* **90** (1957) 250.
578 E. O. Fischer, F. Scherer and H. O. Stahl, *Chem. Ber.* **93** (1960) 2065.

Di-benzenetungsten or the corresponding cation are formed in only 2% yield from WCl_6/$AlCl_3$/Al/C_6H_6.

Di(arene) complexes of rhenium and technetium have been made by the Fischer–Hafner method. Di(arene)–iron(II) complexes are formed from ferrous halides and the arene in the presence of aluminium chloride[579].

$$FeBr_2 + arene \xrightarrow{AlCl_3} [Fe(arene)_2]^{2+}$$

The di-benzeneiron(II) cation is readily hydrolysed. The bis-hexamethylbenzeneiron(II) cation, however, is more stable to hydrolysis and can be reduced successively by dithionite to the deep violet $Fe(C_6Me_6)_2^{+}$ and the black $Fe(C_6Me_6)_2$ [580]. Di-arene–ruthenium or –osmium complexes are also known.

Cobalt chloride reacts with hexamethylbenzene in the presence of aluminium chloride to give the bis-hexamethylbenzenecobalt(II) complex, $[Co(C_6Me_6)_2]^{2+}[AlCl_4]_2$ [581]. The cation disproportionates in neutral or acid solution to the corresponding cobalt(I) and cobalt(III) cations.

$$[Co(C_6Me_6)_2]^{2+} \rightarrow [Co(C_6Me_6)_2]^{+} + [Co(C_6Me_6)_2]^{3+}$$

A better route to the cobalt(I) cation is to treat cobalt chloride with hexamethylbenzene in the presence of aluminium chloride and aluminium. The cobalt(I) cation can be reduced by sodium in liquid ammonia to bis-hexamethylbenzenecobalt(0). Some of the corresponding rhodium or iridium complexes are also known[581].

6.1.1b. *The Grignard Synthesis*

This is the method employed by Hein in his synthesis of organo-chromium complexes (1919). The final products, for many years thought to be σ-aryl–chromium complexes, were reformulated by Zeiss and Onsager[572] as π-arene–chromium complexes. A careful study of the synthesis of di-arene–chromium complexes by the Grignard method has shown that three steps are involved. Thus when chromium trichloride tetrahydrofuranate is treated with phenylmagnesium bromide (3 mole) at $-25°$, the first product is the red triphenylchromium tetrahydrofuranate[44].

$$CrCl_3 \cdot 3THF + PhMgBr \rightarrow CrPh_3 \cdot 3THF$$

This decomposes on warming or on treatment with ether to a black pyrophoric material of unknown structure. This black material when hydrolysed gives hydrogen and various benzene– or biphenyl–chromium(0) complexes [6.1.1], [6.1.2] and [6.1.3] [123, 582].

579 E. O. Fischer and R. Böttcher, *Z. anorg. allgem. Chem.* **291** (1957) 305.
580 E. O. Fischer and F. Röhrscheid, *Z. Naturforsch.* **17b** (1962) 483.
581 E. O. Fischer and H. H. Lindner, *J. Organomet. Chem.* **1** (1964) 307.
582 J. Hähle and G. Stolze, *Z. Naturforsch.* **19b** (1964) 1081.

Hydrolysis with exposure to air gives some di-arenechromium(I) cations. Bis-biphenyl-chromium [6.1.3] can be extracted from the black intermediate, but the benzene complexes [6.1.1] and [6.1.2] are formed at the hydrolysis step. On deuterolysis only the benzene ligands contain deuterium [123, 582].

Tribenzylchromium tetrahydrofuranate is stable in tetrahydrofuran but rearranges in diethyl ether to give π-arene–chromium complexes in which the arene ligand can be toluene or 2-benzyltoluene.

The Grignard method has been little used for metals other than chromium. However, vanadium tetrachloride reacts with phenylmagnesium bromide to give bis-benzene-vanadium(0) and bis-biphenylvanadium(0)[583].

6.1.1c. *From Acetylenes*

The cyclotrimerization of acetylenes by aryl–chromium complexes to give arenes and arene–chromium complexes has been discussed in section 2.2.5. Organo-manganese, -cobalt and -nickel complexes have also been used to promote the cyclic condensation of acetylenes to arenes and arene–metal complexes[315, 316].

6.1.1d. *Arene–Metal Carbonyl Complexes*

Group VI metal carbonyls react with arenes (Ar) in a high boiling solvent (diglyme, di-n-butyl ether, etc.) or under the influence of ultraviolet light to give arene–metal tricarbonyl complexes.

$$Ar + M(CO)_6 \rightarrow M(CO)_3Ar + 3CO$$

Substituted arenes may be used, e.g. substituted benzenes C_6H_5X with X = F, OH, NH$_2$, COOH, CH$_2$OH, COOMe, NMe$_2$, etc. [572, 584, 585]. Fused aromatic hydrocarbons have been complexed to chromium tricarbonyl. The compounds are of the type $Cr(CO)_3Ar$ with Ar = naphthalene, phenanthrene, anthracene, chrysene or pyrene, but two chromium tricarbonyl moieties may be complexed in complexes of type $Cr_2(CO)_6Ar$ with Ar = bi-phenyl, diphenylmethane or *trans*-stilbene[572]. para-Cyclophanes [6.1.4] react to give either a mono-tricarbonylchromium(0) complex [6.1.5] or a bis-tricarbonylchromium(0) complex [6.1.6].

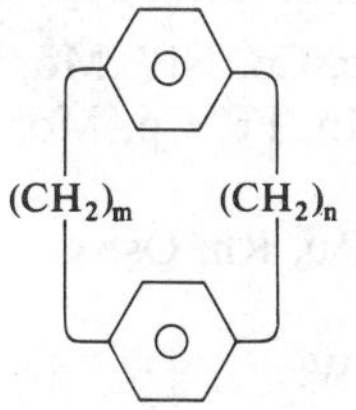

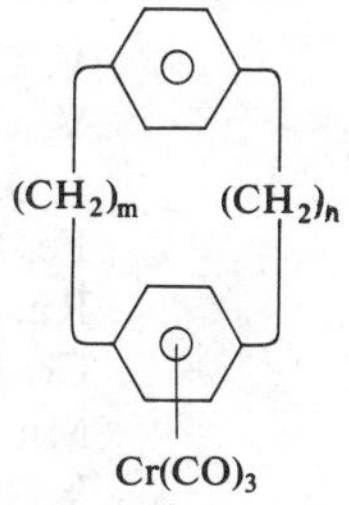

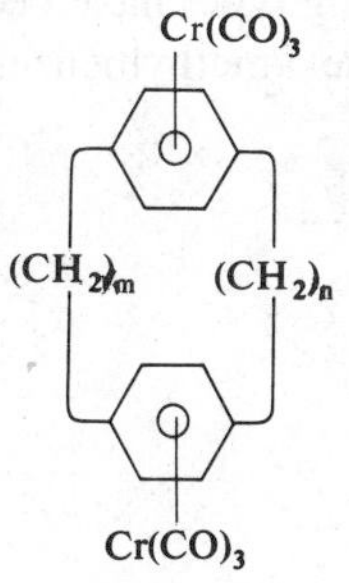

[583] E. Kurras, *Z. anorg. allgem. Chem.* **351** (1967) 268.
[584] B. Nicholls and M. C. Whiting, *J. Chem. Soc.* (1959) 551.
[585] E. O. Fischer, K. Öfele, H. Essler, W. Fröhlich, J. P. Mortensen and W. Semmlinger, *Chem. Ber.* **91** (1958) 2763.

Several compounds of the type tricarbonyl(arene)molybdenum or tricarbonyl(arene)-tungsten are known, although the range of arenes used so far is smaller than with chromium. Hexacarbonylvanadium reacts with arenes to give salts of the type

$$[V(CO)_4(arene)][V(CO)_6] \text{ } [586].$$

Tricarbonyl(arene)manganese(I) cations are formed by treating chloro(pentacarbonyl)-manganese(I) with the arene and aluminium chloride, e.g.

$$MnCl(CO)_5 + C_6H_6 \xrightarrow{AlCl_3} [Mn(CO)_3(C_6H_6)]^+$$

The cation may be isolated as the perchlorate or tetraphenylborate salt[587].

Mercury bis(tetracarbonyl)cobaltate when heated with benzene and aluminium chloride gives a trinuclear cation $[Co_3(CO)_2(C_6H_6)_3]^+$, which possibly has the structure [6.1.7].

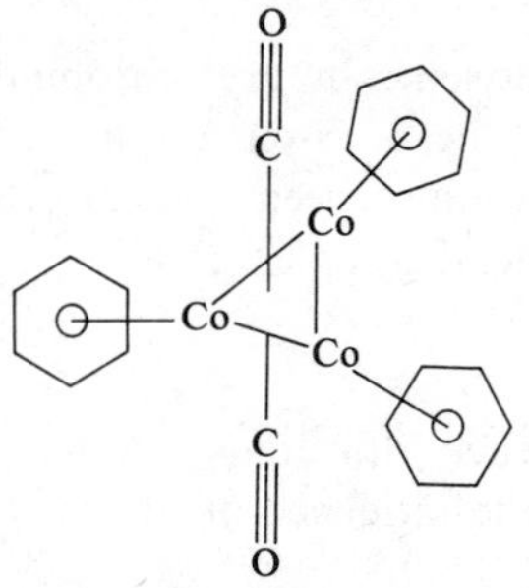

[6.1.7]

6.1.1e. *Miscellaneous*

As described in section 1.1.1, when the ruthenium(II) complex

$$trans\text{-}RuCl_2(Me_2P \cdot CH_2CH_2PMe_2)_2$$

is treated with a sodium–arene, products of composition $Ru(arene)(Me_2PCH_2CH_2PMe_2)_2$ are formed, which seem to be in tautomeric equilibrium with hydro(aryl) complexes of the type $cis\text{-}RuH(aryl)(Me_2PCH_2CH_2PMe_2)_2$ [22].

6.1.2. Types of Arene–Metal Complexes

The following types have been prepared, the arenes most frequently used being benzene, mesitylene or hexamethylbenzene. The references are given in reviews[1, 510, 511, 572].

MAr_2	V, Cr, Fe, Co, Ni?, Mo, W.
MAr_2^+	V, Cr, Mn, Fe, Co, Mo, Ti, Rh, W, Re, Os.
MAr_2^{2+}	Fe, Co, Ru, Rh, Os.
MAr_2^{3+}	Rh, Ir.
$M(CO)_3Ar$	Cr, Mo, W.
$M(CO)_3Ar^+$	Mn.
$M(CO)_4Ar^+$	V [586].
$MCpAr$	Cr, Mo.
$M(CH_3C_5H_4)Ar$	Mn.
$[MCpAr]^+$	Fe.

[586] F. Calderazzo, *Inorg. Chem.* **4** (1965) 223.
[587] G. Winkhaus, L. Pratt and G. Wilkinson, *J. Chem. Soc.* (1961) 3807.

[MCpAr]$^{2+}$ Co, Rh.

[MCpAr(CO)]$^+$ Mo, W.

[Co$_3$(CO)$_2$(C$_6$H$_6$)$_3$]$^+$, [WCl(CO)$_3$(C$_6$Me$_6$)]$^+$

[OsI$_2$(C$_6$H$_6$)]$_x$

6.1.3. Structures, Bonding and Stability

6.1.3a. Structures

The structure of di-benzenechromium(0) has been studied very extensively by X-ray crystallography. There is some disagreement about the structure, but there is no doubt that the rings are mutually parallel and eclipsed. One determination suggests that the C–C bond lengths are alternately long and short, with lengths of 1.436 and 1.366 Å respectively[588]. Another determination suggests that the C–C bonds are essentially equally long at 1.387 Å [589]. The two determinations have been discussed critically and further refinements carried out[527].

A neutron diffraction study of di-benzenechromium shows the C–H bonds are bent towards the metal out of the plane of the ring with a C–H distance of 1.11 Å; no significant differences in the C–C distances round the rings was found[590].

The structures of several complexes of the type Cr(CO)$_3$(arene) have been determined by X-ray diffraction. In tricarbonyl(benzene)chromium the C–C bond lengths are equal (1.41 Å) with the chromium atom situated on the sixfold axis of the benzene molecule at a mean distance of 2.23 Å from each of the six carbon atoms[527, 572]. Other complexes of the type Cr(CO)$_3$(arene) whose structures have been determined include compounds in which the arene is hexamethylbenzene, anisole, o-toluidine, 9,10-dihydrophenanthrene, phenanthrene, anthracene or 1-aminonaphthalene[572]. In the anthracene and phenanthrene complexes the Cr(CO)$_3$ is bonded to a terminal benzene ring. The complexes Cr(CO)$_3$(anisole) and Cr(CO)$_3$(o-toluidine) have the eclipsed conformation [6.1.8].

[6.1.8]   [6.1.9]

It has been suggested that the eclipsed conformation is a result of the electron-releasing substituent (OMe or NH$_2$) directing electron density on to the ortho- and para-carbons; this will favour their being held trans to the electron attracting carbonyl groups. Both tricarbonyl(benzene)chromium and tricarbonyl(hexamethylbenzene)chromium have the staggered conformation [6.1.9].

The structure of the cation Cr(toluene)$_2^+$ shows the C–C (ring) distances to be very nearly equal, at 1.42 Å. The methyl group is bent away slightly (4°) from the metal[592].

[588] F. Jellinek, J. Organomet. Chem. 1 (1963) 43.
[589] F. A. Cotton, W. A. Dollase and J. S. Wood, J. Am. Chem. Soc. 85 (1963) 1543.
[590] G. Albrecht, E. Förster, D. Sippel, F. Eichhorn and E. Kurras, Z. Chem. 8 (1968) 311.
[591] O. L. Carter, A. T. McPhail and G. A. Sim, J. Chem. Soc. (1967) 228.
[592] O. V. Starovskii and Y. T. Struchkov, Dokl. Akad. Nauk SSSR 135 (1960) 620.

A few benzene complexes are known in which the benzene is bonded to the metal atom by weak polarization forces. The first complex of this type to be examined by X-ray diffraction was $Ag(C_6H_6)ClO_4$ [593]. The structure consists of chains of alternating silver ions and benzene molecules with each silver associated with a benzene molecule on either side and vice versa. The two nearest Ag–C distances are 2.50 and 2.63 Å. The structure could not be determined with high accuracy because of disorder in the crystal. Another complex of this type is $CoHg_2(C_6H_6)(SCN)_6$, the X-ray structure of which shows each benzene to be associated with two mercury atoms; the two nearest Hg–C distances are 3.52 and 3.66 Å. In the benzene–copper(I) aluminium chloride complex $CuAlCl_4(C_6H_6)$ the shortest Cu–C distances are 2.15 and 2.30 Å. The C–C distances in the complexed benzene ring are significantly different and suggest that the double bonds have become localized. Some of the distances are shown in [6.1.10]; the three chlorines belong to different $AlCl_4^-$ tetrahedra.

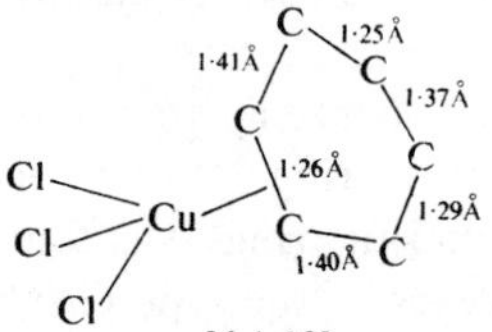

[6.1.10]

Another complex of this type is formed from palladium chloride and benzene, using the reducing Friedel–Craft method. It has the composition $[PdAl_2Cl_7(C_6H_6)]_2$ and the unusual structure [6.1.11].

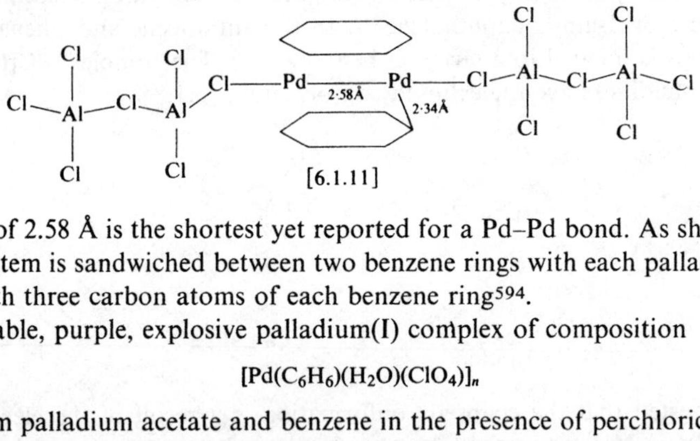

[6.1.11]

The distance of 2.58 Å is the shortest yet reported for a Pd–Pd bond. As shown in [6.1.11], the Pd–Pd system is sandwiched between two benzene rings with each palladium essentially associated with three carbon atoms of each benzene ring[594].

A remarkable, purple, explosive palladium(I) complex of composition

$$[Pd(C_6H_6)(H_2O)(ClO_4)]_n$$

is formed from palladium acetate and benzene in the presence of perchloric acid[595]. In this complex the benzene is possibly associated with the palladium in an analogous fashion to [6.1.11].

6.1.3b. Bonding

The molecular orbital description of the bonding in π-arene–metal complexes is similar to that proposed for cyclopentadienyl–metal complexes. The six p_z-π-orbitals of benzene form six molecular orbitals. In di-benzenechromium there are therefore twelve ring orbitals,

[593] H. G. Smith and R. E. Rundle, *J. Am. Chem. Soc.* **80** (1958) 5075.
[594] G. Allegra, A. Immirzi and L. Porri, *J. Am. Chem. Soc.* **87** (1965) 1394.
[595] J. M. Davidson and C. Triggs, *J. Chem. Soc.* A (1968) 1324.

of symmetry A_{1g}, A_{2u}, E_{1g}, E_{1u}, E_{2g}, E_{2u}, B_{2g} and B_{1u}, with respect to the sixfold axis. These orbitals can interact with the metal $3d$-, $4s$- and $4p$-orbitals[596]. A possible energy level diagram for di-benzenechromium is shown in Fig. 44. The upper, filled, energy levels in di-benzene-chromium are therefore $(E_{2g})^4(A'_{1g})^2$ and it is diamagnetic. Di-benzenevanadium and the di-benzenechromium cation are both paramagnetic with one unpaired electron (μ, ca. 1.7 BM) and they have electron configurations of $(E_{2g})^4(A'_{1g})^1$.

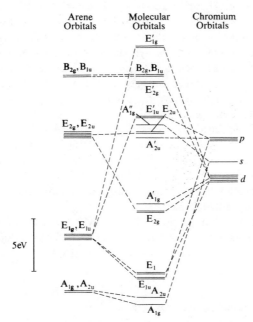

FIG. 44. Energy level diagram for di-benzenechromium.

6.1.3c. *Stability*

The mean dissociation energies ($\bar{D}$) of the metal–ring bonds for a number of di-arene–metal complexes have been derived[597]. The estimated errors are large. Some values of $\bar{D}$ in kcal mole^{-1} are $Cr(C_6H_6)_2$ (40.5 ± 8), $V(C_6H_6)_2$ (70 ± 2) and $Mo(C_6H_6)_2$ (50.7 ± 2) kcal mole^{-1}. The mean metal–ring bonding energy in ferrocene is 69.5 ± 10 kcal mole^{-1}.

6.1.4. **Physical Properties**

6.1.4a. *Nuclear Magnetic Resonance Spectroscopy*

The di-arene–metal complexes are often rather insoluble and many of them are para-magnetic, making their ^1H n.m.r. spectra too broad to be observed. Di-benzenechromium has a narrow ^1H n.m.r. pattern with the resonances shifted ca. 2.9 ppm upfield relative to benzene. The Group VI_A arene–metal tricarbonyl complexes are more soluble and many ^1H n.m.r. spectra have been measured. In general the aromatic hydrogens of the complexed arenes absorb upfield relative to the free arenes and the J-values between the ring protons are

[596] F. A. Cotton, *Chemical Applications of Group Theory*, Interscience, John Wiley, New York (1963).
[597] H. A. Skinner, *Advances Organomet. Chem.* **2** (1964) 49.

reduced by complexation. Nuclear magnetic resonance is useful for determining to which ring the chromium tricarbonyl moiety is attached in complexes of type $Cr(CO)_3(arene)$ where the arene is a condensed arene[598].

6.1.4b. *Infrared and Raman Spectroscopy*

A great deal of effort has been spent on normal coordinate analyses of the infrared and Raman spectra of arene–metal complexes[532, 572]. The dark colour of di-benzenechromium makes its Raman spectrum difficult to measure, but the infrared and Raman spectra of the yellow di-benzenechromium cation (as the iodide salt) have been analysed in detail[532]. The spectra suggest that the ion possesses D_{6h} symmetry. $V(C_6H_6)_2$, $Mo(C_6H_6)_2$ and $W(C_6H_6)_2$ from their infrared and Raman spectra are also thought to have D_{6h} symmetry. However, for $Cr(C_6H_6)_2$ there is the best agreement between the observed and calculated number of bands if the molecule is assumed to have D_{3d} symmetry and D_{3h} for a single ligand, i.e. the ring has undergone a threefold distortion[532].

A detailed study of the infrared spectra of $Cr(CO)_3(C_6H_6)$ and $Cr(CO)_3(C_6D_6)$ shows that the benzene ligand has local C_{3v} symmetry. The "1600" and "1500" bands of benzene are shifted to lower frequencies on complexation to a metal tricarbonyl moiety. Non-complexed arenes show intense bands in the range 850–670 cm^{-1} due to C–H out of plane vibrations. On complexation these bands disappear and are replaced by a very weak absorption within the range 820–795 cm^{-1}.

As might be expected, in complexes of the type $M(CO)_3(arene)$ (M = Cr, Mo or W), the C–O stretching frequencies shift to lower values with electron donating groups on the arene ring. This effect is due to increased back donation from the metal to the π-orbitals of the carbonyl ligands.

6.1.4c. *Electron Spin Resonance Spectroscopy*

The e.s.r. pattern of di-benzenevanadium(0) shows splitting into eight lines by the V-nucleus, $I = 7/2$ [599]. In addition, splitting by the protons can be seen. In the di-benzene-chromium(I) cation the resonance consists of eleven poorly separated lines; thirteen lines would be expected due to the twelve hydrogens. Additional splitting by ^{53}Cr ($I = 3/2$) can also be detected.

6.1.4d. *Electric Dipole Moments*

Dibenzenechromium has a zero dipole moment, as would be expected. Bis(hexamethyl-benzene)cobalt, however, has a dipole moment of *ca.* 1.8 D in benzene or cyclohexane. This suggests some distortion of the molecule from a *trans*-"sandwich" structure[600].

The electric dipole moments of complexes of type $Cr(CO)_3(arene)$ are large, 3.7–6.2 D [572]. Electron-donating groups on the arene ring increase the moment whilst electron-attracting groups decrease it. In benzene solution the dipole moment of $Cr(CO)_3(C_6H_6)$ is 4.92 ± 0.05 D, with the electron drift in the direction shown [6.1.12].

[598] B. Deubzer, E. O. Fischer, H. P. Fritz, C. G. Kreiter, N. Kriebitzsch, H. D. Simmons and B. R. Willeford, *Chem. Ber.* **100** (1967) 3084.

[599] K. H. Hausser, *Z. Naturforsch.* **16a** (1961) 1190.

[600] B. J. Nicholson and H. C. Longuet-Higgins, *Mol. Phys.* **9** (1965) 461.

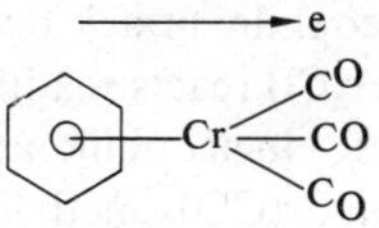

[6.1.12]

6.1.5. Reactions

6.1.5a. *Electron Transfer*

The redox process

$$Cr°(C_6H_6)_2 \underset{+e}{\overset{-e}{\rightleftarrows}} Cr^I(C_6H_6)_2^+$$

is reversible in methanol/benzene solutions; the half-wave reduction potential is 0.8 V, against an 0.2M calomel electrode[601]. Air oxidation will also readily produce the $+I$ state. Increasing methyl-substitution on the benzene ring decreases the ease of reduction whilst phenyl substituents make reduction easier. Arenechromium tricarbonyls do not readily oxidize to $[Cr(CO)_3Ar]^+$.

6.1.5b. *Charge Transfer*

Di-benzenechromium forms a 1:1 complex with tetracyanoethylene. This is probably a salt, i.e. $[Cr(C_6H_6)_2]^+[TCNE]^-$.

6.1.5c. *Electrophilic Substitution*

From dipole moment studies and also from the low pK_a value for $Cr(CO)_3$(benzoic acid) and the high pK_b value for $Cr(CO)_3$(aniline) it is apparent that in arene–metal tricarbonyl complexes (metal = Cr, Mo or W) there is a net drift of electron density from the arene to the metal. Hence complexed arenes are less prone to electrophilic attack than free arenes and no Friedel–Craft acylation products of di-arenemetal(0) complexes have been obtained. However, $Cr(CO)_3$(benzene) can be acetylated under mild conditions to give [6.1.13] [602]. On acetylating $Cr(CO)_3$(toluene) the proportions of *o-*, *m-* and *p*-isomers are

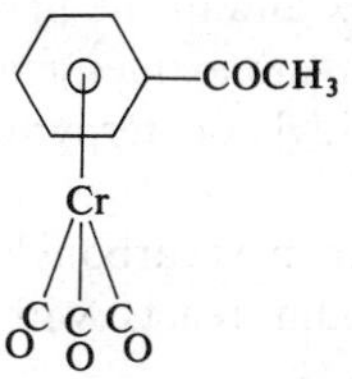

[6.1.13]

39:15:46. Free toluene acetylates faster and the isomers are formed in the proportions of 8:0:92. Arene–metal tricarbonyl complexes are unstable to strong acids and cannot be nitrated or sulphonated[572].

6.1.5d. *Nucleophilic Substitution and Addition Reactions*

Nucleophilic attack on free arenes is a very difficult reaction to bring about. However, as mentioned above, on complexation to a metal there is a net loss of charge to the metal

[601] C. Furlani and E. O. Fischer, *Z. Electro Chem.* **61** (1957) 481.
[602] G. E. Herberich and E. O. Fischer, *Chem. Ber.* **95** (1962) 2803.

and not surprisingly, therefore, nucleophilic attack occurs much more readily[572]. Thus the chlorobenzene complex $Cr(CO)_3(C_6H_5Cl)$ reacts readily with sodium methoxide to give the anisole complex $Cr(CO)_3(C_6H_5OMe)$. Nucleophilic attack on side chain substituents also goes with ease, e.g. the conversion of $Cr(CO)_3$(methyl benzoate) to $Cr(CO)_3$(benzoic acid) by alkaline hydrolysis[603].

Rate studies on the solvolysis of benzylic chloride–chromium tricarbonyl complexes such as [6.1.14] suggest that a carbonium ion, stabilized as shown in [6.1.15], is formed as an intermediate. The final product is the carbinol [6.1.16].

[6.1.14]   [6.1.15]   [6.1.16]

The conversion of $[Mn(CO)_3(C_6H_6)]^+$ to various cycloheptadienyl–manganese complexes $Mn(CO)_3(C_6H_6Y)$ by nucleophilic addition, e.g. using LiPh (Y = Ph) or $LiAlH_4$ (Y = H), has been described in section 5. Similarly, the benzene ring of $[Co(C_4Ph_4)$-(benzene)]^+$ is attacked by nucleophiles (Y$^-$) to give $Co(C_4Ph_4)(C_6H_6Y)$. The di-benzeneruthenium di-cation is reduced by hydride ion to $Ru(C_6H_6)(C_6H_8)$.

6.1.5e. *Displacement of Arene on the Other Ligands*

The arene ligand may be displaced by other ligands from complexes of the type $M(CO)_3$(arene) (M = Cr, Mo or W):

$$Cr(CO)_3(arene) + P(OPh)_3 \rightarrow Cr(CO)_3\{P(OPh)_3\}_3$$

Pyridine, phosphorus trifluoride or tertiary phosphines will similarly displace the arene. The kinetics of such displacement reactions have been studied. They are first order in both the arene–metal tricarbonyl complex and in the ligand which is displacing the arene[604].

Both the arene and the carbon monoxides may be displaced. Thus $Cr(CO)_3(C_6H_6)$ reacts with phenanthroline, bipyridyl or terpyridyl to give $Cr(phen)_3$, $Cr(bipy)_3$ or $Cr(terpy)_2$ respectively.

Di-benzenechromium reacts with hexacarbonylchromium to give tricarbonylbenzene-chromium[605]. Di(mesitylene)vanadium reacts with carbon monoxide to give [V(mesitylene)$_2$][V(CO)$_6$].

A carbon monoxide may be displaced from complexes of the type arene–metal tricarbonyl, e.g.

$$Cr(CO)_3(C_6H_6) \xrightarrow[uv]{L} Cr(CO)_2(L)(C_6H_6)$$

where L = C_2H_4, cyclopentene, phenylacetylene, diphenylacetylene, etc.

$Cr(CO)_3$(hexamethylbenzene) similarly forms complexes of the type $Cr(CO)_2$(olefin)-(hexamethylbenzene) when irradiated with ultraviolet light in the presence of an olefin.

603 D. A. Brown and J. R. Raju, *J. Chem. Soc.* A (1966) 40.
604 F. Zingales, A. Chiesa and F. Basolo, *J. Am. Chem. Soc.* **88** (1966) 2707.
605 E. O. Fischer and K. Öfele, *Chem. Ber.* **90** (1957) 2532.
605ᵃ R. J. Angelici and L. Busetto, *Inorg. Chem.* **7** (1968) 1935.

The tricarbonyl(benzene)manganese(I) cation reacts with cyanide ion as follows:

$$[Mn(CO)_3(C_6H_6)]^+ \xrightarrow{CN^-} MnCN(CO)_2(C_6H_6) + CO$$

6.2.1. Cycloheptatriene Complexes and Complexes with Other Six Electron Donors

Cycloheptatriene reacts with the Group VI metal carbonyls to give compounds of the type $M(CO)_3$(cycloheptatriene) (M = Cr, Mo or W). The structure of the molybdenum complex has been determined by X-rays[606]. The methylene group is folded away from the metal and the C–C distances alternate, as shown in [6.2.1]. The molybdenum is almost

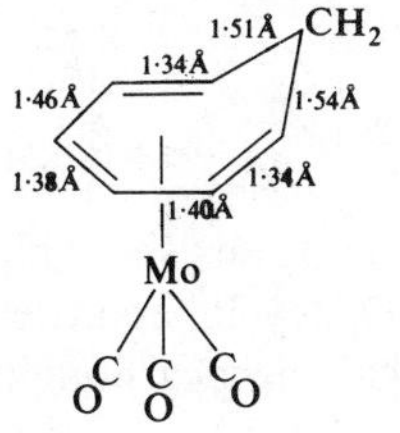

[6.2.1]

equidistant (*ca.* 2.53 Å) from each of the six olefinic carbon atoms and the three carbon monoxides are diametrically opposite the three C=C bonds.

Cycloheptatriene will displace benzene from its complexes

$$C_6H_6Mo(CO)_3 + C_7H_8 \rightarrow C_7H_8Mo(CO)_3 + C_6H_6$$

and 1-phenylcyclohepta-2,4,6-triene reacts with $Mo(CO)_6$ to give a substituted cycloheptatriene rather than an arene complex[170]. Methyl-substituted cycloheptatrienyl complexes are more thermally stable and more resistant to oxidizing agents than the unsubstituted complexes. Presumably the electron-donating methyl groups increase the donor ability of the ring. $Cr(CO)_3(C_7H_8)$ rapidly exchanges with carbon-14-labelled cycloheptatriene either thermally or under ultraviolet irradiation. The corresponding molybdenum complex exchanges much more slowly. The proposed mechanism for the exchange is similar to that postulated for arene exchange[607].

$Mo(CO)_3(C_7H_8)$ is not acylated under Friedel–Craft conditions. Donor ligands (L) such as amines, phosphines, arsines, or stibenes readily displace the ring to yield complexes of the type $Mo(CO)_3L_3$. The carbonyl groups cannot be displaced, however.

The cycloheptatriene complexes react with triphenylmethyl fluorborate to give π-tropylium complexes; see section 7.

Cycloheptatriene reacts with pentacarbonyliron to give $Fe(CO)_3$(cycloheptatriene), in which one double bond remains uncoordinated, as shown in [6.2.2].

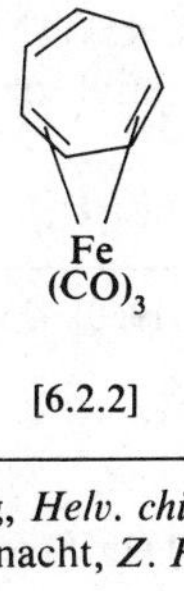

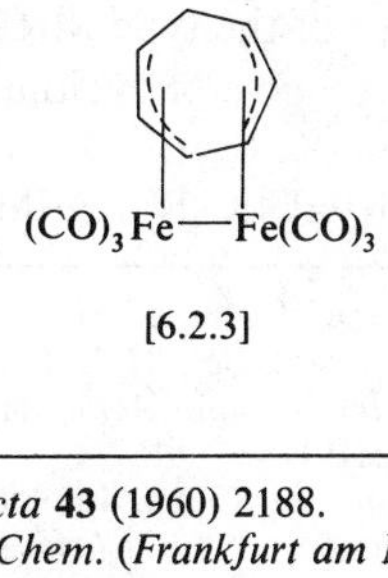

[6.2.2]

$(CO)_3Fe \text{—} Fe(CO)_3$

[6.2.3]

[606] J. D. Dunitz and P. Pauling, *Helv. chim. Acta* **43** (1960) 2188.
[607] W. Strohmeier and H. Mittnacht, *Z. Phys. Chem.* (*Frankfurt am Main*) **34** (1962) 82.

Cycloheptatriene reacts with enneacarbonyldi-iron to give $Fe_2(CO)_6(C_7H_8)$ which possibly has the di(π-allylic) structure [6.2.3].

Hexacarbonylchromium reacts with cyclo-octa-1,3,5-triene to give the red complex $Cr(CO)_3(C_8H_{10})$. This complex has been shown by X-ray diffraction to have the structure [6.2.4].

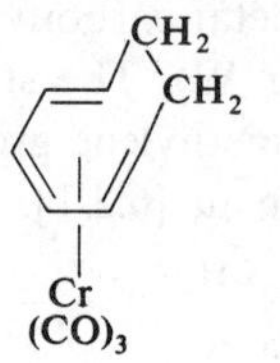

[6.2.4] 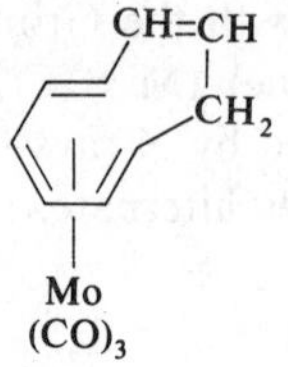   [6.2.5]

The six chromium bonded ring carbon atoms show alternate double and single bond C–C distances, as with the complex $Cr(CO)_3$(cycloheptatriene).

Bicyclo[4,3,0]nonatriene reacts with hexacarbonylmolybdenum to give

$$Mo(CO)_3(C_9H_{10}) \text{ [608]}$$

The structure [6.2.5] has been proposed for this complex.

Hexacarbonylchromium reacts with thiophen to give $Cr(CO)_3(C_4H_4S)$ [609]. This complex is isomorphous with $Cr(CO)_3$(benzene) and has the structure [6.2.6] in which the

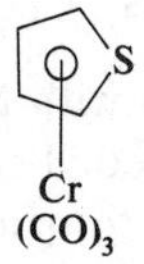

[6.2.6]

thiophen is acting as a six-electron donor. Thiophen, deuteriothiophen, 2-methylthiophen or 3-methylthiophen derivatives of chromium tricarbonyl are formed by displacement of benzene from $Cr(CO)_3(C_6H_6)$ [610].

Tetramethylthiophen reacts with $MnCl(CO)_5$ in the presence of aluminium chloride to give $[Mn(CO)_3(\text{tetramethylthiophen})]^+$ [610a].

7. CYCLOHEPTATRIENYL OR TROPYLIUM COMPLEXES[611]

7.1.1. Preparation

Cycloheptatriene complexes of the type $M(CO)_3(C_7H_8)$ (M = Cr, Mo or W) react with triphenylmethyl fluoroborate to give π-cycloheptatrienyl or tropylium complexes[611, 612], e.g.

$$Mo(CO)_3(C_7H_8) + Ph_3C^+BF_4^- \rightarrow [Mo(CO)_3(C_7H_7)]^+BF_4^- + Ph_3CH$$

[608] R. B. King and F. G. A. Stone, *J. Am. Chem. Soc.* **80** (1960) 4557.
[609] E. O. Fischer and K. Öfele, *Chem. Ber.* **91** (1958) 2395.
[610] A. Mangini and F. Taddei, *Inorg. chim. Acta* **2** (1968) 12.
[610a] H. Singer, *Z. Naturforsch.* **21B** (1966) 810.
[611] M. A. Bennett, *Advan. Organomet. Chem.* **4** (1966) 353.
[612] H. J. Dauben and L. R. Honnen, *J. Am. Chem. Soc.* **80** (1958) 5570.

In contrast, when $Fe(CO)_3$(cycloheptatriene) is treated with triphenylmethyl fluoroborate, the triphenylmethyl cation adds to the triene to give a triphenylmethylcycloheptadienyl complex:

$$Fe(CO)_3(C_7H_8) + Ph_3C^+BF_4^- \rightarrow [Fe(CO)_3(Ph_3C\cdot C_7H_8)]BF_4$$

The complex cation $[Fe(CO)_3(cycloheptatrienyl)]^+$ can be made, however, by treating $Fe(CO)_3$(7-methoxycycloheptatriene) with acid.

$$CH_3O\cdot C_7H_7 + Fe_2(CO)_9 \rightarrow Fe(CO)_3(C_7H_7OMe)$$
(7-methoxycycloheptatriene)

$$\downarrow HBF_4$$
$$[Fe(CO)_3(C_7H_7)]^+BF_4^-$$

The infrared spectrum of the complex ion shows that the tropylium ring has an uncoordinated double bond, and a structure [7.1.1], with the iron bonded to five carbon atoms, has been suggested. However, a rapid fluxional motion of the iron atom around all seven carbon atoms of the ring must be occurring, since a single, sharp resonance is found in the ^1H n.m.r. spectrum.

[7.1.1]

Cycloheptatriene reacts with vanadium hexacarbonyl to give two complexes[613]:

$$V(CO)_6 + C_7H_8 \rightarrow V(CO)_3(C_7H_7) + [V(C_7H_7)(C_7H_8)]^+[V(CO)_6]^-$$

Little is known about the mechanism of this reaction, but it is possible that both products are formed from a common intermediate, $V(CO)_3(C_7H_8)$. The cation $[V(C_7H_7)(C_7H_8)]^+$ is paramagnetic, with one unpaired electron.

No di(π-tropylium) complexes of the type $[M(C_7H_7)_2]^{n+}$ have been prepared. However, some complexes of the type $M(C_7H_7)(\pi\text{-}L)$ are known, where π-L is a polyolefin, a cyclopentadienyl or an arene ring. The complex $V(C_7H_7)(C_7H_8)$ has already been mentioned, and the purple mixed sandwich compound $VCp(C_7H_7)$ is formed by treating $VCp(CO)_4$ with cycloheptatriene[614]. This compound is fairly stable in air and can be sublimed. The yellow cation $[CrCp(C_7H_7)]^+$ and the dark green neutral complex $CrCp(C_7H_7)$ are also known[615, 616]. One method of synthesizing these two complexes is outlined below (C_7H_8 = cyclohepta-1,3,5-triene).

$$CrCl_2Cp(THF) + \textit{iso}\text{-PrMgBr} \xrightarrow{\text{Et}_2O} CrCp(\textit{iso}\text{-Pr})_2Et_2O$$

$$\downarrow C_7H_8$$
$$\downarrow h\nu$$

$$[CrCp(C_7H_7)]^+ \xrightarrow[\text{oxidation}]{\text{aerial}} CrCp(C_7H_8)$$

$$S_2O_4^{2-} \searrow \qquad \swarrow Pt \text{ in } C_6H_6$$

$$CrCp(C_7H_7)$$

613 R. P. M. Werner and S. A. Manastryrskyj, *J. Am. Chem. Soc.* **83** (1961) 2023.
614 R. B. King and F. G. A. Stone, *J. Am. Chem. Soc.* **81** (1959) 5263.
615 E. O. Fischer and S. Breitschaft, *Angew. Chem. Int. Edn.* **2** (1963) 44.
616 R. B. King and M B. Bisnette, *Inorg. Chem.* **3** (1964) 785.

The molybdenum complex $MoCp(C_7H_7)$ is formed, but in only 1.8% yield, from $MoCl_5$, cycloheptatriene, MgBrCp and $MgBrPr^i$. It reacts with iodine to give the paramagnetic ion $[MoCp(C_7H_7)]^+$ [617].

The cation $[CrCp(C_7H_7)]^+$ is also formed by treating $CrCp(C_6H_6)$ with cyclohepta-1,3,5-triene and aluminium chloride[615].

An attempt to prepare dicycloheptatrienechromium from cycloheptatriene, anhydrous chromium trichloride and ethylmagnesium chloride under ultraviolet irradiation gave, instead, Cr(cycloheptatrienyl)(cyclohepta-1,3-diene)[618].

A very interesting ring expansion of a coordinated benzene ring to a coordinated substituted tropylium ring occurs when $CrCp(C_6H_6)$ [7.1.2] is treated with acyl halides under Friedel–Craft conditions. With acetyl chloride the methylheptatrienyl cation [7.1.3] is formed; this may be reduced by dithionite in alkali to the neutral species [7.1.4] [615,619].

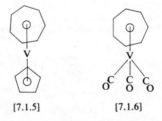

A cycloheptatrienyl–manganese complex $[MnCp(C_7H_6CH_3)]^+$ is similarly formed from $MnCp(C_6H_6)$ and acetyl chloride under Friedel–Craft conditions.

7.1.2. Types of Cycloheptatrienyl–Metal Complexes[611]

$MCp(C_7H_6R)$ R = H usually but can be Me [619]; V [613, 614], Cr [615, 616], Mo [617].
$MCp(C_7H_6R)]^+$ R = H, Me or Ph. Cr [615], Mo [617], Mn [619].
$M(C_7H_7)(CO)_3$ V [613].
$[M(C_7H_7)(CO)_3]^+$ Cr, Mo [611, 612], Fe[611].
$M(C_7H_7)(cyclohepta-1,3-diene)$[618]

7.1.3. Structure and Bonding

The structures of two cycloheptatrienyl complexes have been determined by X-ray diffraction, viz. $VCp(C_7H_7)$ [7.1.5] [620] and $V(C_7H_7)(CO)_3$ [7.1.6] [621].

[7.1.5]   [7.1.6]

[617] E. O. Fischer and H. W. Wehner, *J. Organomet. Chem.* **11** (1968) 29.
[618] E. O. Fischer, A. Reckziegel, J. Müller and P. Goser, *J. Organomet. Chem.* **11** (1968) 13.
[619] E. O. Fischer and S. Breitschaft, *Chem. Ber.* **99** (1966) 2213.
[620] G. Engebretson and R. E. Rundle, *J. Am. Chem. Soc.* **85** (1963) 481.
[621] G. Allegra and G. Perego, *Ric. Sci. Parte II* **1A** (1961) 362.

In both cases the C_7-rings are planar and the C–C distances the same within experimental error. A feature of the mixed sandwich compound [7.1.5] is that the vanadium is closer to the plane of the cycloheptatrienyl ring (1.5 Å) than to the plane of the cyclopentadienyl ring (1.9 Å).

The bonding of the cycloheptatrienyl ring system to a metal is probably similar to the bonding of cyclopentadienyl or benzene rings to metals and will not be discussed in detail here. Cycloheptatrienyl complexes are formed with metals that have few d-electrons and do not seem to be formed by the Group VIII metals.

7.1.4. Infrared and Nuclear Magnetic Resonance Spectra

These will not be discussed in any detail here. Fritz[532] discusses the assignments of infrared absorption bands for some cycloheptatrienyl–metal complexes. The ^1H n.m.r. patterns of most cycloheptatrienyl–metal complexes show that all the hydrogens are equivalent 'and that the cycloheptatrienyl ligand is symmetrically bonded to the metal. The complex $MoCp(C_7H_7)(CO)_2$, however, has a temperature-dependent ^1H n.m.r. pattern. At or about room temperature the pattern consists of a singlet at τ 4.93 due to the cyclopentadienyl ring protons and a singlet at τ 5.21 due to the cycloheptatrienyl ring protons. On cooling to $-40°$ the resonance due to the C_7 ring broadens. This has been interpreted in terms of the h^3-cycloheptatrienyl structure [7.1.7].

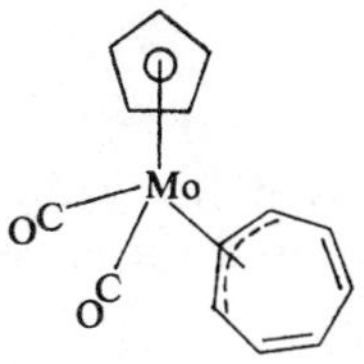

[7.1.7]

7.1.5. Reactions

7.1.5a. Electron Transfer

Examples of the reduction of cycloheptatrienyl cationic complexes to neutral cycloheptatrienyl complexes, and the reverse process, were mentioned in section 7.1.1, e.g. the reduction of $[CrCp(C_7H_7)]^+$ by dithionite ion in alkali to $CrCp(C_7H_7)$ and the oxidation of $MoCp(C_7H_7)$ by iodine to $[MoCp(C_7H_7)]^+$.

7.1.5b. Nucleophilic Attack

The reactions of complexes of the type $[M(CO)_3(C_7H_7)]^+$ (M = Cr or Mo) with nucleophiles have been studied extensively.

With nucleophiles (Y^-) such as H^- or OMe^-, addition occurs to give the *exo*-substituted cycloheptatriene complex [7.1.8].

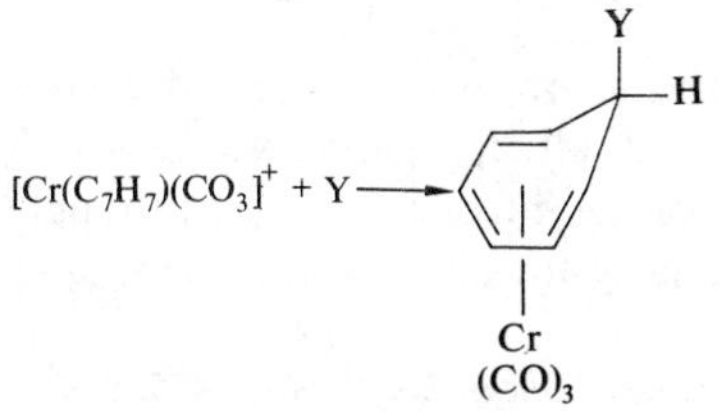

$$[Cr(C_7H_7)(CO_3]^+ + Y \longrightarrow$$

[7.1.8]

A remarkable reaction occurs when cycloheptatrienyl complexes of the type $[M(CO)_3(C_7H_7)]^+$ (M = Cr or Mo) are treated either with sodium cyclopentadienide or with sodio-diethyl malonate. Ring contraction occurs[622], e.g.

$$[Mo(CO)_3(C_7H_7)]^+ + C_5H_5^- \rightarrow Mo(CO)_3(C_6H_6) + C_6H_6$$

The π-arene (benzene) ligand is derived from the cycloheptatrienyl ring and not the cyclopentadienyl ring. This is shown by the following two reactions:

$$[Cr(CO)_3(C_7H_7)]^+ + C_5H_4CH_3^- \rightarrow Cr(CO)_3(C_6H_6)$$
$$[Cr(CO)_3(C_7H_6CH_3)]^+ + C_5H_5^- \rightarrow Cr(CO)_3(C_6H_5CH_3)$$

A mechanism has been proposed for this ring contraction reaction.

It is interesting to note that when a substituted cycloheptatriene C_7H_7Y reacts with $Cr(CO)_6$ or $Cr(CO)_3py_3$ the endo-isomer [7.1.9] is formed, exclusively.

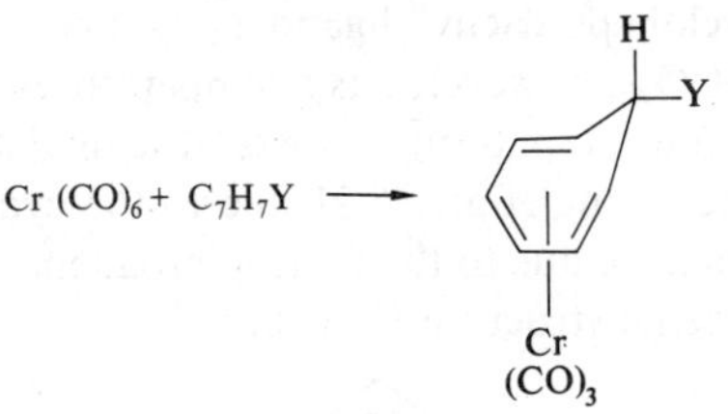

$$Cr(CO)_6 + C_7H_7Y \longrightarrow$$

[7.1.9]

However, treatment of $[Cr(CO)_3(C_7H_7)]^+$ with other nucleophiles can give ditropyl $(C_7H_7-C_7H_7)$ complexes as the main products; although some product of type [7.1.8] is also formed, e.g. with cyanide ion

$$[Cr(CO)_3(C_7H_7)]^+ + CN^- \rightarrow Cr(CO)_3(C_7H_7CN) \quad (18\%)$$
$$Cr(CO)_3(ditropyl) \quad (17\%)$$
$$Cr_2(CO)_6(ditropyl) \quad (46\%)$$

Other reagents which cause coupling to ditropyl complexes include sodamide, phenyl-lithium and sodium acetate.

$[Mo(CO)_3(C_7H_7)]^+$ reacts with halide ions to give $MoX(CO)_2(C_7H_7)$ (X = Cl, Br or I). The iodo-complex reacts with $[Mn(CO)_5]^-$ to give a mixed metal–metal bonded complex $(C_7H_7)(CO)_2Mo-Mn(CO)_5$ or with sodium cyclopentadienide to give $MoCp(C_7H_7)(CO)_2$ (see section 7.1.4).

8. CYCLO-OCTATETRAENE, AZULENE COMPLEXES AND CARBORANE COMPLEXES

Cyclo-octatetraene can bond to transition metals in a variety of ways, e.g. as a 1,5- or a 1,3-diolefin, as a triolefin, as a di-allylic ligand, etc. It is convenient, however, to discuss cyclo-octatetraene–metal complexes in one section, although some have been mentioned briefly in earlier sections. Similarly for azulene and carborane (carbolide) complexes.

622 J. D. Munro and P. L. Pauson, J. Chem. Soc. (1961) 3475.
623 J. D. Munro and P. L. Pauson, J. Chem. Soc. (1961) 3484.

8.1. CYCLO-OCTATETRAENE–METAL COMPLEXES

Free cyclo-octatetraene (COT) exists in the tub conformation with no conjugation between the double bonds. When coordinated to platinum(II), palladium(II), rhodium(I) or cobalt(I) it usually retains the tub conformation and acts as a chelating 1,5-diene system. Thus with $PtCl_4^{2-}$, $PtCl_2(COT)$ is formed and with rhodium trichloride or with $Rh_2Cl_2(cyclo\text{-}octene)_2$, the chloro-bridged complex [8.1.1] is formed[624].

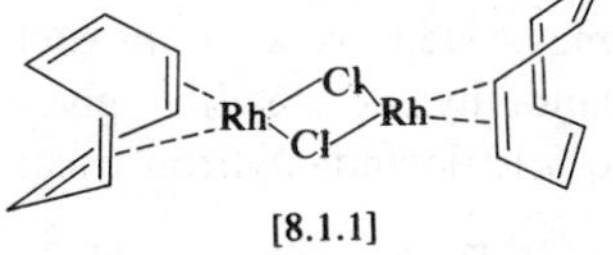

[8.1.1]

The platinum complex $PtI_2(COT)$ reacts with alkyl or aryl Grignard reagents to give mononuclear complexes, $PtR_2(C_8H_8)$ (R = Me, Et or p-tolyl), and also binuclear complexes $Pt_2R_2(C_8H_8)$ (R = Me or Ph). The binuclear complexes probably have the structure [8.1.2] and show no infrared absorption band characteristic of an uncomplexed double bond; the mononuclear complexes show a band at ca. 1635 cm^{-1} due to $\nu(C=C)$ (uncomplexed).

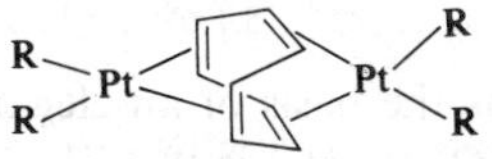

[8.1.2]

Cyclo-octatetraene also acts as a 1,5-chelating diene with cobalt(I), e.g. $CoCp(CO)_2$ reacts with cyclo-octatetraene to give the mononuclear complex [8.1.3] and the binuclear complex [8.1.4] [625].

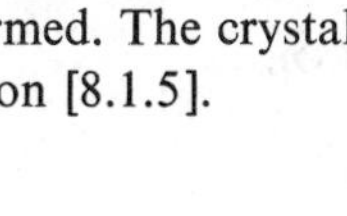

[8.1.3]

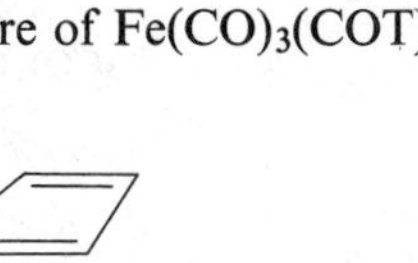

[8.1.4]

Cyclo-octatetraene (or substituted cyclo-octatetraene) give a variety of complexes when treated with iron carbonyls. With $Fe(CO)_5$, two complexes $Fe(CO)_3(COT)$ and $Fe_2(CO)_6(C_8H_8)$ are formed. The crystal structure of $Fe(CO)_3(COT)$ shows that the ring is in the chair conformation [8.1.5].

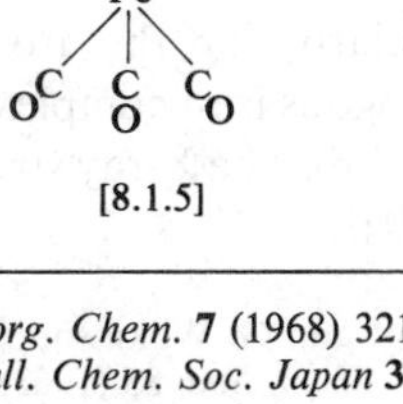

[8.1.5]

624 M. A. Bennett and J. D. Saxby, *Inorg. Chem.* 7 (1968) 321.
625 A. Nakamura and N. Hagihara, *Bull. Chem. Soc. Japan* 33 (1960) 425.

The complex $Fe(CO)_3(COT)$ shows fluxional behaviour in that at room temperature the ^1H n.m.r. spectrum consists of a singlet corresponding to a rapid movement of the iron atom around the ring. On cooling to very low temperatures $(-145°)$ this movement is stopped and the ^1H n.m.r. spectrum can then be interpreted in terms of the structure [8.1.5]. There is disagreement, however, on the nature of the fluxional motion; see Cotton's review[127] for a discussion of the various interpretations. In spite of having two uncomplexed double bonds, $Fe(CO)_3(COT)$ reacts only slowly with bromine and then with evolution of carbon monoxide; the expected addition of bromine to the two uncomplexed double bonds does not occur. The uncoordinated diene system is also unreactive towards most Diels–Alder dienophiles, although tetracyanoethylene readily gives an adduct. Protonation of $Fe(CO)_3(COT)$ gives the bicyclo[5,1,0]octadienyliron tricarbonyl cation [8.1.6] [626].

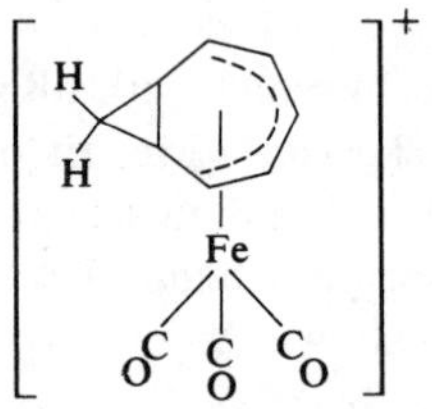

[8.1.6]

This protonation has been used as the basis of an elegant synthesis of homotropone[627].

The crystal structure of $Fe_2(CO)_6(C_8H_8)$ is [8.1.7].

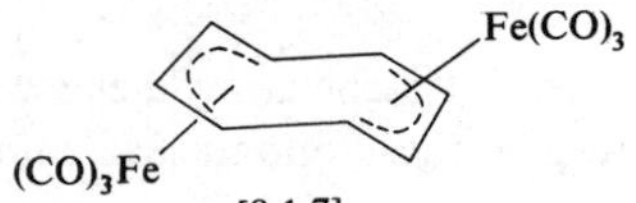

[8.1.7]

Cyclo-octatetraene reacts with $Fe_2(CO)_9$ to give some of the compound [8.1.7] and another compound also of composition $Fe_2(CO)_6(COT)$. The structure of this second iron complex is not known but the structure of what appears to be an analogous ruthenium complex $Ru_2(CO)_6(COT)$ has been determined by X-ray diffraction. In this structure the cyclo-octatetraene ring is non-planar and the two rutheniums are bonded differently. The structure can be represented *approximately* by [8.1.8] with the $Ru_2(CO)_6$ group oblique to a possible symmetry plane of the COT ring.

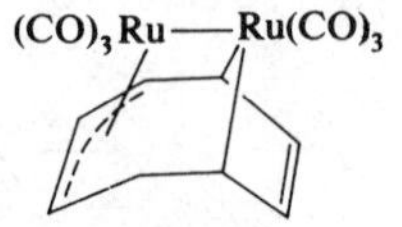

[8.1.8]

The solution ^1H n.m.r. spectrum, however, is of the AA′BB′CC′DD′ type, the molecule showing fluxional behaviour; similarly for the iron complex. On being heated both $Ru_2(CO)_6(COT)$ [8.1.8] and its analogous iron complex lose a molecule of carbon monoxide to give $M_2(CO)_5(COT)$ (M = Ru or Fe). The X-ray structure of the iron complex shows it to have the remarkable structure [8.1.9].

[626] A. Davison, W. McFarlane, L. Pratt and G. Wilkinson, *J. Chem. Soc.* (1962) 4821.
[627] J. D. Holmes and R. Pettit, *J. Am. Chem. Soc.* **85** (1963) 2531.

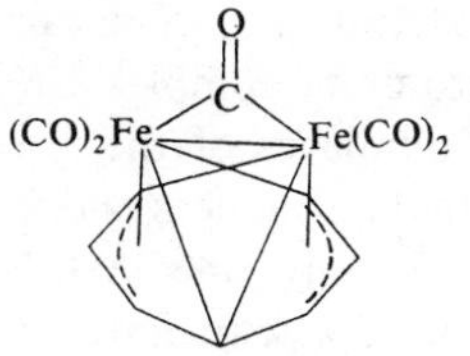

[8.1.9]

Five complexes have been prepared from 1,3,5,7-tetramethylcyclo-octatetraene (TMCOT) and iron carbonyls[628]. The structures of four of these are [8.1.10], [8.1.11], [8.1.12] and [8.1.13]. The structure of the fifth of formula $Fe_2(CO)_6(TMCOT)$ is not known.

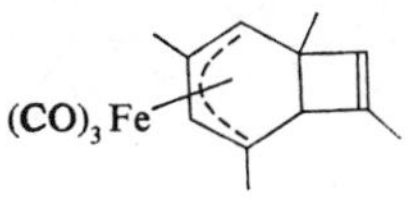

[8.1.10]

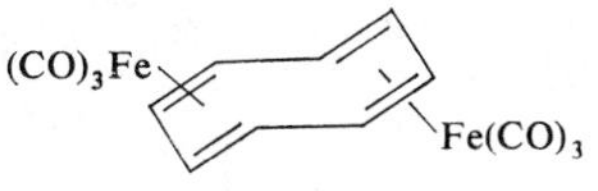

[8.1.11]

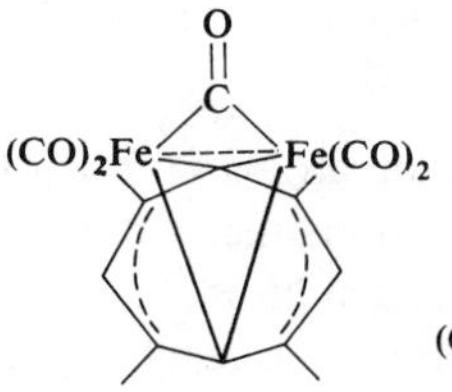

[8.1.12] 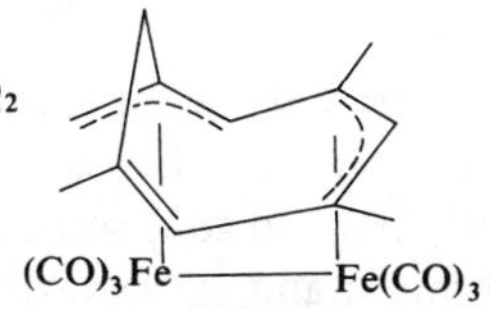   [8.1.13]

Di(cyclo-octatetraene)iron has been prepared by reducing iron halides in the presence of an excess of COT. The structure is [8.1.14], i.e. with one COT ring acting as a triolefin and the other as a diolefin[629].

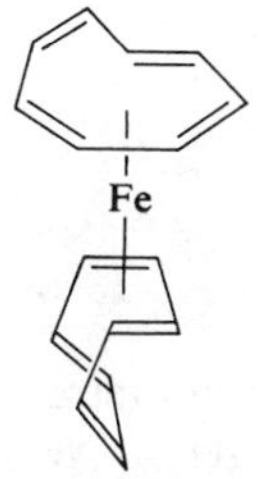

[8.1.14]

The molecule shows fluxional behaviour, i.e. it has a variable temperature n.m.r. spectrum.

Cyclo-octatetraene also acts as a six-electron donor in complexes of the type $M(CO)_3(C_8H_8)$ (M = Cr, Mo or W). These are prepared from COT and $M(CO)_3(MeCN)_3$ and presumably have the structure [8.1.15].

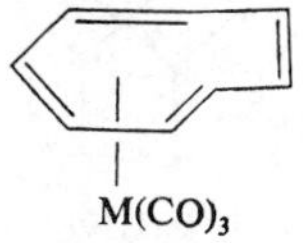

[8.1.15]

[628] F. A. Cotton and M. D. LaPrade, *J. Am. Chem. Soc.* **90** (1968) 2026.
[629] G. Allegra, A. Colombo, A. Immirzi and I. W. Bassi, *J. Am. Chem. Soc.* **90** (1968) 4455.

Analogous complexes are formed from 1,3,5,7-tetramethylcyclo-octatetraene (TMCOT) and the X-ray structure of the chromium complex Cr(CO)$_3$(TMCOT) shows it to have a structure analogous to [8.1.15]. Complexes of the type M(CO)$_3$(TMCOT) are fluxional; at low temperatures four sharp methyl resonances and four sharp ring proton resonances are observed, but at high temperatures only one methyl resonance and one ring proton resonance are observed. At intermediate temperatures the behaviour is not fully understood[127, 630].

Another compound formed from Ru$_3$(CO)$_{12}$ and cyclo-octatetraene is Ru$_3$(CO)$_4$(COT)$_2$. This is a fluxional molecule with a variable temperature n.m.r. spectrum but with all sixteen hydrogens apparently equivalent at room temperature. Its structure has been determined by X-ray crystallography and is shown diagrammatically in [8.1.16].

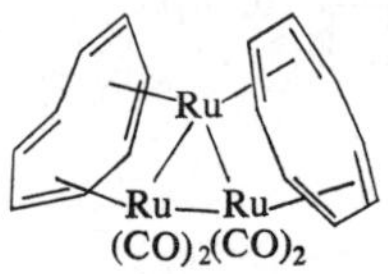

[8.1.16]

The structure consists of a triangle of ruthenium atoms as in Ru$_3$(CO)$_{12}$. Each COT ring is bonded to two ruthenium atoms in a delocalized manner but no two ring carbon atoms of a COT ring are equivalently bonded to a ruthenium atom[631].

Dodecacarbonyltriosmium and cyclo-octatetraene react together under the influence of ultraviolet light to give a complex Os(CO)$_3$(COT); this isomerizes on heating to a second complex. The structure [8.1.17] has been proposed for the first complex and [8.1.18] for the second[416].

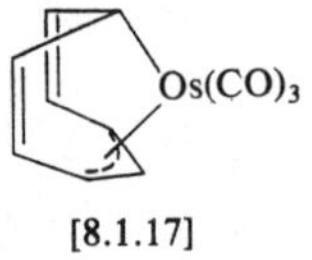

[8.1.17]

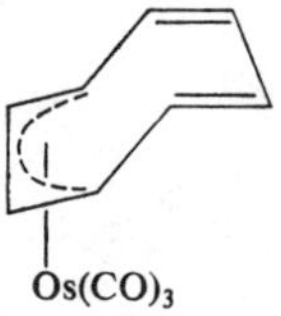

[8.1.18]

Cyclo-octatetraene reacts with titanium tetra-n-butoxide and triethylaluminium to give two compounds, Ti$_2$(C$_8$H$_8$)$_4$ and Ti$_2$(C$_8$H$_8$)$_3$. An X-ray diffraction study shows the second compound to have the remarkable structure [8.1.19] with two planar COT rings and a bridging COT ring.

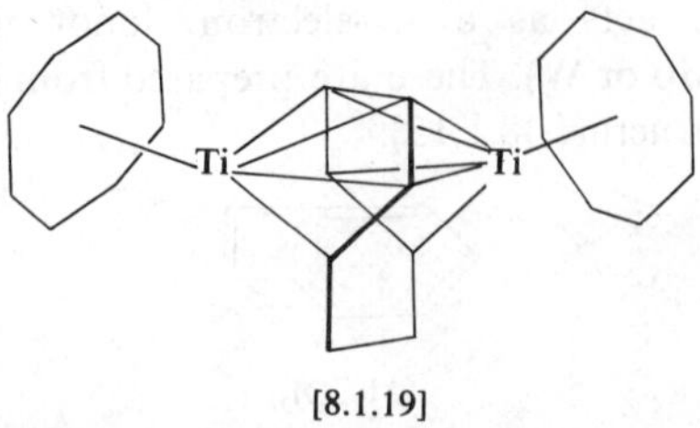

[8.1.19]

630 F. A. Cotton, J. W. Faller and A. Musco, J. Am. Chem. Soc. 88 (1966) 4506.
631 M. J. Bennett, F. A. Cotton and P. Legzdins, J. Am. Chem. Soc. 89 (1967) 6797.

8.2. AZULENE COMPLEXES

Azulenes (az) can bond to transition metals either via the seven-membered ring or via the five-membered ring. They react with hexacarbonylmolybdenum to give complexes of the type $Mo_2(CO)_6(az)$. The complex $Mo_2(CO)_6$(azulene) is diamagnetic, highly polar ($\mu = 7.87 \pm 0.07$ D) and decomposes above 150°, liberating azulene. The structure [8.2.1]

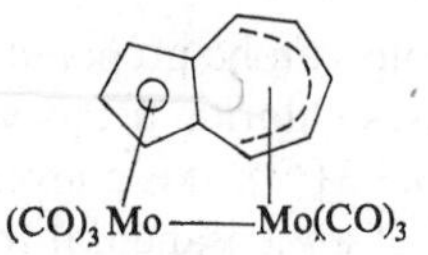

(CO)$_3$Mo——Mo(CO)$_3$

[8.2.1]

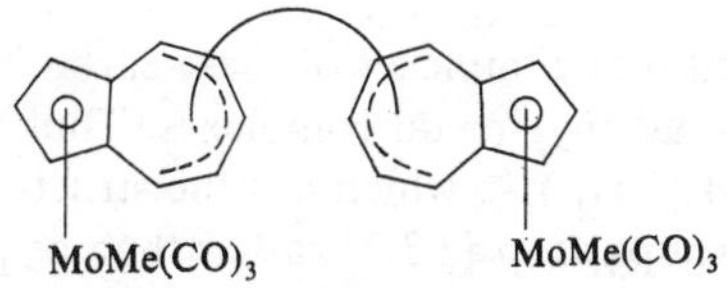

MoMe(CO)$_3$ MoMe(CO)$_3$

[8.2.2]

has been suggested[632]. When $Mo_2(CO)_6$(azulene) is treated successively with sodium and then methyl iodide a complex $Mo_2Me_2(CO)_6(C_{10}H_7)_2$ is formed. This has the structure [8.2.2], as shown by X-ray diffraction[633].

Azulenes react with $Fe(CO)_5$ to give complexes of type $Fe_2(CO)_5(az)$. The complex formed from azulene itself has been shown by X-ray diffraction[432] to have the structure [3.1.15] (see section 3.1.3a).

When ferric chloride is treated with isopropylmagnesium bromide and azulene an orange brown complex $C_{12}H_{16}Fe$ is formed. This complex was originally given the wrong structure but is now known to be a ferrocene derivative [8.2.3] by X-ray crystallography. When chromic chloride is treated similarly it gives dark green crystals of formula $C_{20}H_{18}Cr$. This compound is probably a substituted cyclopentadienylcycloheptatrienyl complex of structure [8.2.4] [634].

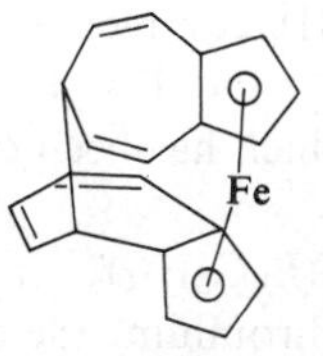

[8.2.3] 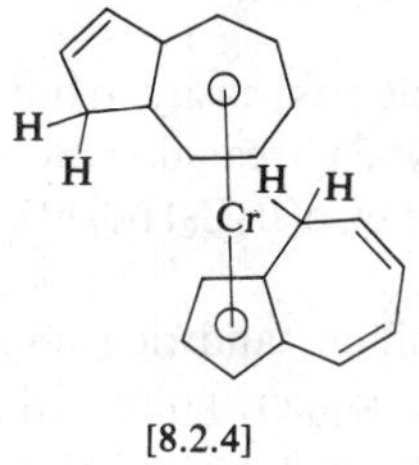   [8.2.4]

A complex $CrCp(C_{10}H_9)$ is also known, where $C_{10}H_9$ = an azulenium ligand, bonded to the chromium via the seven-membered ring.

8.3. CARBORANE–METAL COMPLEXES[635]

An interesting development in transition metal chemistry is the discovery that carborane groupings can "sandwich" bond to transition metals. The isomeric 1,2- and 1,7-dicarbollide ions $B_9C_2H_{11}^{2-}$ are thought to have the eleven-particle icosahedral structures, [8.3.1] and [8.3.2] respectively, each with an open pentagonal face and available orbitals which are similar to those present in the cyclopentadienide ion.

632 R. Burton, L. Pratt and G. Wilkinson, J. Chem. Soc. (1960) 4290.
633 P. H. Bird and M. R. Churchill, Chem. Commun. (1967) 705.
634 E. O. Fischer and J. Müller, J. Organomet. Chem. 1 (1964) 464.
635 M. F. Hawthorne, Accounts Chem. Res. 1 (1968) 281.

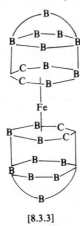

[8.3.1]   [8.3.2]

It has been found that complexes of these carbollide ions can be prepared which are analogous to cyclopentadienyl–metal complexes. Thus ferrous chloride reacts with $1,2\text{-}B_9C_2H_{11}^{2-}$
to give $[Fe(1,2\text{-}B_9C_2H_{11})]^{2-}$ which has the structure [8.3.3] [636]. It is convenient to represent
the carbollide groups as in [8.3.3] rather than as part of an icosahedral framework [8.3.1].

[8.3.3]

The iron(II) complex is easily oxidized to the iron(III) complex $[Fe(1,2\text{-}B_9C_2H_{11})_2]^-$.
Ferrous chloride when treated with a mixture of $C_5H_5^-$ and $B_9C_2H_{11}^{2-}$ gives the mixed
sandwich complex $FeCp(B_9C_2H_{11})$ [635], the structure of which has been confirmed by X-ray
analysis[637].

Other bis-carbollide sandwich complexes of type $M(B_9C_2H_{11})_2^{n-}$ have been prepared
where M = cobalt, copper, gold, nickel, palladium, or chromium; the charge n may have
more than one value for a given metal, i.e. the metal may have more than one valence state.

Other carbollide–metal complexes include Pd(1,2-dimethylcarbollide)(tetraphenylcyclobutadiene), $[Re(1,2\text{-carbollide})(CO)_2]^-$ and $[M(1,2\text{-carbollide})(CO)_3]^{2-}$ (M = Cr,
Mo or W) [635]. The $[B_7C_2H_{11}]^{2-}$ ion has also been shown to bond to transition metals, e.g.
in the complex $CoCp(B_7C_2H_9)$. Another ligand $B_8C_2H_{10}^{4-}$ bridges two cobalt atoms in the
complex ion $[B_9C_2H_{11}CoB_8C_2H_{10}CoB_9C_2H_{11}]^-$ [638]. Several other types of carborane–
metal complexes have been prepared[635].

[636] M. F. Hawthorne, D. C. Young and P. A. Wegner, *J. Am. Chem. Soc.* **87** (1965) 1818.
[637] A. Zalkin, D. H. Templeton and T. E. Hopkins, *J. Am. Chem. Soc.* **87** (1965) 3988.
[638] J. N. Francis and M. F. Hawthorne, *J. Am. Chem. Soc.* **90** (1968) 1663.

INDEX

Contents of Comprehensive Inorganic Chemistry

INDEPENDENT OPINION

"These books are attractively bound and have clear print. Since the length and cost are not prohibitive, this set of books should be well within the budget of most libraries. Not only will the professional chemist find these books useful, but students and other readers will find them a valuable reference source. (Comprehensive Inorganic Chemistry) should be found in every undergraduate and graduate library, as well as industrial libraries. Many professional chemists may even consider them for personal libraries. Highly recommended."

Choice—*A publication of the Association of College and Research Libraries.*

INDEPENDENT OPINION

Volume 1   1467 pp + index

"This covers the chemistry of hydrogen, the noble gases, and of the elements of Groups IA, IIA, IIIB, carbon and silicon. The first three chapters deal with hydrogen, hydrides, deuterium and tritium and the fourth is an interesting discussion of the proton, protonic acids and the hydrogen bond. Two chapters follow on the inert gases, including interesting and extensive recent knowledge about their compounds set out by N. Bartlett and F. O. Sladky. Four chapters on the alkalis and alkaline earths contain a wealth of detail, although perhaps along traditional lines. N. N. Greenwood has written an excellent account on boron chemistry of book length in itself, and another chapter deals at length with much new information about aluminium, gallium, indium and thallium. Chapters 13 and 15 deal with carbon and silicon. Here it might have been expected that more would have been included on the high pressure chemistry of carbon and silicates, and mineral chemistry. Chapter 14 by M. L. H. Green and P. Powell is a useful introduction to the organic chemistry of the metallic elements, along modern lines of ligand field theory and ideas about metal complexes.

Throughout this volume, with its different authors, it is perhaps inevitable that there is some lack of uniformity in the extent of detail given. There are also a few lapses in symbolism, notation, and uniformity of units. Yet the whole must be regarded as a highly commendable collection of material which will be valuable to chemists of all kinds."

Professor Sir Harold Thompson FRS
Oxford

Volume 2   1594 pp + index

"Volume 2 is concerned with the chemistry of the elements of Groups IV, V, VI, VII. The general impression on reading the various chapters of this volume is the great effectiveness in reporting a considerable amount of chemistry in a very digestible form. The systematic presentation applied to each chapter allows a rapid assessment of the appropriate chemical information, and the text is well documented with reference to the original literature plus good review articles for a more detailed coverage. Perhaps a minor criticism of this work lies in the indexing; the subject index is relatively sparse for a text of such magnitude and it would have been of considerable utility to have a formulae index to the text.

The area of chemistry covered by this work is obviously very great, but it does appear to have dealt with it in a very succinct manner for the majority of the text, which extends to approximately 1500 pages. The two major chapters in the book are associated with the chemistry of nitrogen, approximately 240 pages, and the chemistry of the halogens (excepting fluorine), approximately 500 pages. The book thus encompasses in one volume what would normally be a series of books. Both of the above mentioned chapters are admirable and the authors, K. Jones on nitrogen with A. J. Downs and C. J. Adams on the halogens, are to be commended on both the presentation and coverage. The chapter on the halogens illustrates the real strength of the series, in that detailed chemical information is not only presented but discussed in physico-chemical theoretical terms. A scientific compendium of this size often suffers from the "catalogue" approach, but the present text presents the chemistry in critical mode with a realistic assessment of the various physical methods used in property determination. Thus the properties of the halogens are discussed in terms of bond energies, bond lengths, vibrational properties, e.s.r., n.m.r., n.q.r. and Mossbauer spectroscopy, electronic and magnetic properties and dipole moments allowing a detailed appraisal of the use of various modern methods in studying the chemical properties reviewed.

Considering the magnitude of the task undertaken, it is extremely pleasing to note the number of chapters referring to papers in the 1971 period— a truly great commendation on the overall editorship of these volumes. Perhaps a general note in each volume stating the period covered by the references would have been of help. In general this work provides a welcome and unique addition to the inorganic literature."

Professor J. Lewis FRS
Cambridge

Volume 3   1370 pp + index

"This volume covers the chemistry of the elements of the d-block of the Periodic Table (the transition elements), with the exception of the Lanthanide elements (Vol. 4), the Actinide elements (Vol. 5) and some special aspects which are common to many of the transition elements (Vol. 4). The volume is therefore concerned specifically with the three elements which characterize each of the ten transition groups, and the chapters are mostly grouped in this way. However, the six platinum metals are treated in one chapter which is the best way to fit these similar elements into the overall scheme which is standard for all five volumes. There are altogether 17 chapters, written by 14 authors who are internationally recognized 1370 pages of text and a useful 17 page index.

The five volumes are quite remarkable, in that they can justifiably claim to be comprehensive, yet at the same time remain interesting and readable; they are probably unique in this respect. Volume 3 serves as an excellent source-book for the essential physical constants of all important compounds (simple and complex) of the transition metals. These are arranged so that significant comparisons are made wherever possible, and there are extensive references. It says much for the ingenuity of the editors, authors, and particularly the printers that the presentation of such an amount of information has been possible, while still maintaining the readability of the text. Throughout the volume chemical properties and reactions are discussed and interpreted rather than listed. The need for skilled correlation of data is particularly important in Volume 3, since it is in the area of the transition elements that a major part of the research work in inorganic chemistry has been published in recent years, and in this area also there has been a major interaction of inorganic with theoretical chemistry.

This volume must surely become the first point of reference for research workers and teachers alike. The transition elements play an important role in Pure and Applied Chemistry, Physics, Materials Science and Biology, and the authors clearly intend their chapters to be of value to this wide audience. Teachers at any level will also appreciate the very high quality of the general presentation, discussion, formulae and diagrams. Apart from reference to the original literature, few scientists will find it necessary to look outside this volume for their material."

Professor C. C. Addison FRS
Nottingham

Volume 4   994 pp + index

"Volume 4 is concerned with the general chemistry of the lanthanides and some special topics in transition metal chemistry.

Therald Moeller has packed a great amount of the fundamental chemistry of the lanthanides into his 101 pages in an interesting and scholarly manner with tables of essential data. Important recent developments in their organometallic chemistry have come too late to be included, but the chapter provides a useful fairly detailed first reference to their inorganic chemistry.

The subjects of the surveys are topical and obviously bear the mark of the late Sir Ronald Nyholm. They vary considerably in detail of treatment, interest and authority. Generally they emphasize recent work until about 1969–70 but rarely show a sense of history. They vary in length from 60 to 200 pages, mostly around 100 pages. They are authoritative and useful surveys all giving numerous references to recent reviews and

original work. The authors are well known chemists whose style and subject matter are familiar to most inorganic chemists. There are eight surveys as follows:—

Carbonyls, cyanides, isocyanides and nitrosyls by W. Griffith. Compounds of the transition elements involving metal-metal bonds by D. L. Kepert and K. Vrieze. Transition metal hydrogen compounds by J. C. Green and M. L. H. Green. Non-stoichiometric compounds: an introductory essay by D. J. M. Bevan. Tungsten bronzes, vanadium bronzes and related compounds by P. Hagenmuller. Isopolyanions and heteropolyanions by D. L. Kepert. Transition metal chemistry by B. F. G. Johnson. Organo-transition metal compounds and related aspects of homogeneous catalysis by B. L. Shaw and N. I. Tucker.

This volume has its own subject index of sixteen and a half pages, and is well produced with numerous tables of data and references provided at the foot of each page."

Professor J. Chatt FRS
Sussex

Volume 5   635 pp + Master index

"Volume 5 is devoted to the Actinides (635 pp) and the Master Index (78 pp). The latter serves little purpose since it merely indicates the subsections of CIC, and thus repeats the indexes in each individual volume. Indeed, as the treatment of each element or series of elements follows a standard pattern, the volumes are essentially self-indexing anyway. A one-page table of contents at the beginning of Volume 5 would have been more helpful and is a curious omission. The running headings at the top of each double page are also singularly uninformative, only three being used: 'The Elements' for 102 pages, 'Compounds' for 361 pages and 'Solution Chemistry' for the remaining 171 pages.

The treatment of actinium and the actinides (elements 89–103) is both readable and authoritative. Nine of the contributors are from AERE, Harwell, and the other five (with one exception) are from nuclear chemistry institutes in Sweden and Germany. In reviewing these 5f elements it is salutary to recall that the majority have been synthesized for the first time within the last 30 years—yet the number of compounds known and the amount of information on them has already outstripped the more limited chemistry of their 4f congeners, the lanthanides. The authors have done a magnificent job in assembling, collating, assessing, and systematizing a vast amount of data on the physical and chemical properties of these elements and their numerous compounds. The work, which is extensively referenced, will undoubtedly remain the standard first source of information in this area for many years to come."

Professor N. N. Greenwood FRIC
Leeds

Group:		IA	IIA	IIIB	IVB	VB	VIB	VIIB
Principal quantum number (Period)	Valence shell	s^1	s^2	$d^1s^2f^x$	d^2s^2	$(d^3s^2)^†$	$(d^5s^1)^†$	d^5s^2
$n = 1$	$1s$	1 H 1.008						
$n = 2$	$2s2p$	3 Li 6.94	4 Be 9.01					
$n = 3$	$3s3p$	11 Na 22.99	12 Mg 24.31					
$n = 4$	$4s3d4p$	19 K 39.10	20 Ca 40.08	21 Sc 44.96	22 Ti 47.90	23 V 50.94	24 Cr 51.996	25 Mn 54.94
$n = 5$	$5s4d5p$	37 Rb 85.47	38 Sr 87.62	39 Y 88.91	40 Zr 91.22	41 Nb 92.91	42 Mo 95.94	43 Tc 99
$n = 6$	$6s4f5d6p$	55 Cs 132.91	56 Ba 137.34	57* La 138.91	72 Hf 178.49	73 Ta 180.95	74 W 183.85	75 Re 186.2
$n = 7$	$7s5f6d7p$	87 Fr	88 Ra 226	89⊙ Ac				

		58 Ce 140.12	59 Pr 140.91	60 Nd 144.24
*Lanthanide Series				
⊙Actinide Series		90 Th 232.04	91 Pa	92 U 238.03

†Variable valence shells